Genetic Algorithms in Molecular Modeling

Principles of QSAR and Drug Design

GENETIC ALGORITHMS IN MOLECULAR MODELING

Edited by

James Devillers
CTIS, Lyon, France

ACADEMIC PRESS
Harcourt Brace & Company, Publishers
London San Diego New York
Boston Sydney Tokyo Toronto

ACADEMIC PRESS LIMITED
24–28 Oval Road
London NW1 7DX

United States Edition published by
ACADEMIC PRESS INC.
San Diego, CA 92101

This book is printed on acid-free paper

A catalogue record of this book is available from the British Library

ISBN 0–12–213810–4

Typeset by Florencetype Ltd, Stoodleigh, Devon
Printed and bound in Great Britain by Hartnolls Ltd, Bodmin, Cornwall

Contents

A colour plate section appears between pages 212–213.

Contributors

J.M. Caruthers, Laboratory for Intelligent Process Systems, School of Chemical Engineering, Purdue University, West Lafayette, IN 47907, USA.
K. Chan, Laboratory for Intelligent Process Systems, School of Chemical Engineering, Purdue University, West Lafayette, IN 47907, USA.
J. Devillers, CTIS, 21 rue de la Bannière, 69003 Lyon, France.
D. Domine, CTIS, 21 rue de la Bannière, 69003 Lyon, France.
W.J. Dunn, College of Pharmacy, University of Illinois at Chicago, 833 S. Wood Street, Chicago, IL 60612, USA.
R.C. Glen, Tripos Inc., St Louis, MO 63144, USA.
H. Hamersma, Department of Computational Medicinal Chemistry, NV Organon, P.O. Box 20, 5340 BH Oss, The Netherlands.
A.J. Hopfinger, Laboratory of Molecular Modeling and Design, M/C 781, The University of Illinois at Chicago, College of Pharmacy, 833 S. Wood Street, Chicago, IL 60612-7231, USA.
G. Jones, Krebs Institute for Biomolecular Research and Department of Information Studies, University of Sheffield, Western Bank, Sheffield S10 2TN, UK.
R. Leardi, Istituto di Analisi e Tecnologie Farmaceutiche ed Alimentari, Università di Genova, via Brigata Salerno (ponte), I–16147 Genova, Italy.
B.T. Luke, International Business Machines Corporation, 522 South Road, Poughkeepsie, NY 12601, USA.
T.D. Muhammad, Department of Biological Chemistry, Finch University of Health Sciences/The Chicago Medical School, 3333 Green Bay Road, North Chicago, IL 60064, USA.
H.C. Patel, Laboratory of Molecular Modeling and Design, M/C 781, The University of Illinois at Chicago, College of Pharmacy, 833 S. Wood Street, Chicago, IL 60612-7231, USA.
C. Putavy, CTIS, 21 rue de la Bannière, 69003 Lyon, France.
D. Rogers, Molecular Simulations Incorporated, 9685 Scranton Road, San Diego, CA 92121, USA.
A. Sundaram, Laboratory for Intelligent Process Systems, School of Chemical Engineering, Purdue University, West Lafayette, IN 47907, USA.
V.J. van Geerestein, Department of Computational Medicinal Chemistry, NV Organon, P.O. Box 20, 5340 BH Oss, The Netherlands.
S.P. van Helden, Department of Computational Medicinal Chemistry, NV Organon, P.O. Box 20, 5340 BH Oss, The Netherlands.

V. Venkatasubramanian, Laboratory for Intelligent Process Systems, School of Chemical Engineering, Purdue University, West Lafayette, IN 47907, USA.

D.E. Walters, Department of Biological Chemistry, Finch University of Health Sciences/The Chicago Medical School, 3333 Green Bay Road, North Chicago, IL 60064, USA.

P. Willett, Krebs Institute for Biomolecular Research and Department of Information Studies, University of Sheffield, Western Bank, Sheffield S10 2TN, UK.

Preface

Genetic algorithms are rooted in Darwin's theory of natural selection and evolution. They provide an alternative to traditional optimization methods by using powerful search techniques to locate optimal solutions in complex landscapes. The popularity of genetic algorithms is reflected in the ever-increasing mass of literature devoted to theoretical works and real-world applications on various subjects such as financial portfolio management, strategy planning, design of equipment, and so on. Genetic algorithms and related approaches are also beginning to infiltrate the field of QSAR and drug design.

Genetic Algorithms in Molecular Modeling is the first book on the use of genetic algorithms in QSAR and drug design. Comprehensive chapters report the latest advances in the field. The book provides an introduction to the theoretical basis of genetic algorithms and gives examples of applications in medicinal chemistry, agrochemistry, and toxicology. The book is suited for uninitiated readers willing to apply genetic algorithms for modeling the biological activities and properties of chemicals. It also provides trained scientists with the most up-to-date information on the topic. To ensure the scientific quality and clarity of the book, all the contributions have been presented and discussed in the frame of the *Second International Workshop on Neural Networks and Genetic Algorithms Applied to QSAR and Drug Design* held in Lyon, France (June 12–14, 1995). In addition, they have been reviewed by two referees, one involved in molecular modeling and another in chemometrics.

Genetic Algorithms in Molecular Modeling is the first volume in the series *Principles of QSAR and Drug Design*. Although the examples presented in the book are drawn from molecular modeling, it is suitable for a more general audience. The extensive bibliography and information on software availability enhance the usefulness of the book for beginners and experienced scientists.

James Devillers

1 Genetic Algorithms in Computer-Aided Molecular Design

J. DEVILLERS
CTIS, 21 rue de la Bannière, 69003 Lyon, France

Genetic algorithms, which are based on the principles of Darwinian evolution, are widely used for combinatorial optimizations. We introduce the art and science of genetic algorithms and review different applications in computer-aided molecular design. Information on software availability is also given. We conclude by underlining some advantages and drawbacks of genetic algorithms.

KEYWORDS: *computer-aided molecular design; genetic algorithms; QSAR; software.*

INTRODUCTION

The design of molecules with desired properties and activities is an important industrial challenge. The traditional approach to this problem often requires a trial-and-error procedure involving a combinatorially large number of potential candidate molecules. This is a laborious, time-consuming and expensive process. Even if the creation of a new chemical is a difficult task, in many ways it is rule-based and many of the fundamental operations can be embedded in expert system procedures. Therefore, there is considerable incentive to develop computer-aided molecular design (CAMD) methods with a view to the automation of molecular design (Blaney, 1990; Bugg *et al.*, 1993).

In the last few years, genetic algorithms (Holland, 1992) have emerged as robust optimization and search methods (Lucasius and Kateman, 1993, 1994). Diverse areas such as digital image processing (Andrey and Tarroux, 1994), scheduling problems and strategy planning (Cleveland and Smith, 1989; Gabbert *et al.*, 1991; Syswerda, 1991; Syswerda and Palmucci, 1991; Easton

In, *Genetic Algorithms in Molecular Modeling* (J. Devillers, Ed.)
Academic Press, London, 1996, pp. 1–34.
ISBN 0-12-213810-4

and Mansour, 1993; Kidwell, 1993; Kobayashi *et al.*, 1995), engineering (Bramlette and Bouchard, 1991; Davidor, 1991; Karr, 1991; Nordvik and Renders, 1991; Perrin *et al.*, 1993; Fogarty *et al.*, 1995), music composition (Horner and Goldberg, 1991), criminology (Caldwell and Johnston, 1991) and biology (Hightower *et al.*, 1995; Jaeger *et al.*, 1995) have benefited from these methods. Genetic algorithms have also largely infiltrated chemistry, and numerous interesting applications are now being described in the literature (e.g. Lucasius and Kateman, 1991; Leardi *et al.*, 1992; Li *et al.*, 1992; Hartke, 1993; Hibbert, 1993a; Wehrens *et al.*, 1993; Xiao and Williams, 1993, 1994; Chang and Lewis, 1994; Lucasius *et al.*, 1994; Mestres and Scuseria, 1995; Rossi and Truhlar, 1995; Zeiri *et al.*, 1995). Among them, those dedicated to molecular modeling appear promising as a means of solving some CAMD problems (Tuffery *et al.*, 1991; Blommers *et al.*, 1992; Dandekar and Argos, 1992, 1994; Fontain, 1992a,b; Judson, 1992; Judson *et al.*, 1992, 1993; Hibbert, 1993b; Jones *et al.*, 1993; McGarrah and Judson, 1993; Unger and Moult, 1993a,b; Brown *et al.*, 1994; May and Johnson, 1994; Ring and Cohen, 1994; Sheridan and Kearsley, 1995). Under these conditions, this chapter is organized in the following manner. First, a survey of the different classes of search techniques is presented. Secondly, a brief description of how genetic algorithms work is provided. Thirdly, a review of the different applications of genetic algorithms in quantitative structure–activity relationship (QSAR) and drug design is presented. Fourthly, information on software availability for genetic algorithms and related techniques is given. Finally, the chapter concludes by underlining some advantages and drawbacks of genetic algorithms.

CLASSES OF SEARCH TECHNIQUES

Analysis of the literature allows the identification of three main types of search methods (Figure 1). Calculus-based techniques are local in scope and depend upon the existence of derivatives (Ribeiro Filho *et al.*, 1994). According to these authors, such methods can be subdivided into two classes: indirect and direct. The former looks for local extrema by solving the equations resulting from setting the gradient of the objective function equal to zero. The search for possible solutions starts by restricting itself to points with slopes of zero in all directions. The latter seeks local optima by working around the search space and assessing the gradient of the new point, which drives the search. This is simply the notion of 'hill climbing' where the search is started at a random point, at least two points located at a certain distance from the current point are tested, and the search continues from the best of the tested nearby points (Koza, 1992; Ribeiro Filho *et al.*, 1994). Due to their lack of robustness, calculus-based techniques can only be used on well-defined problems (Goldberg, 1989a; Ribeiro Filho *et al.*, 1994).

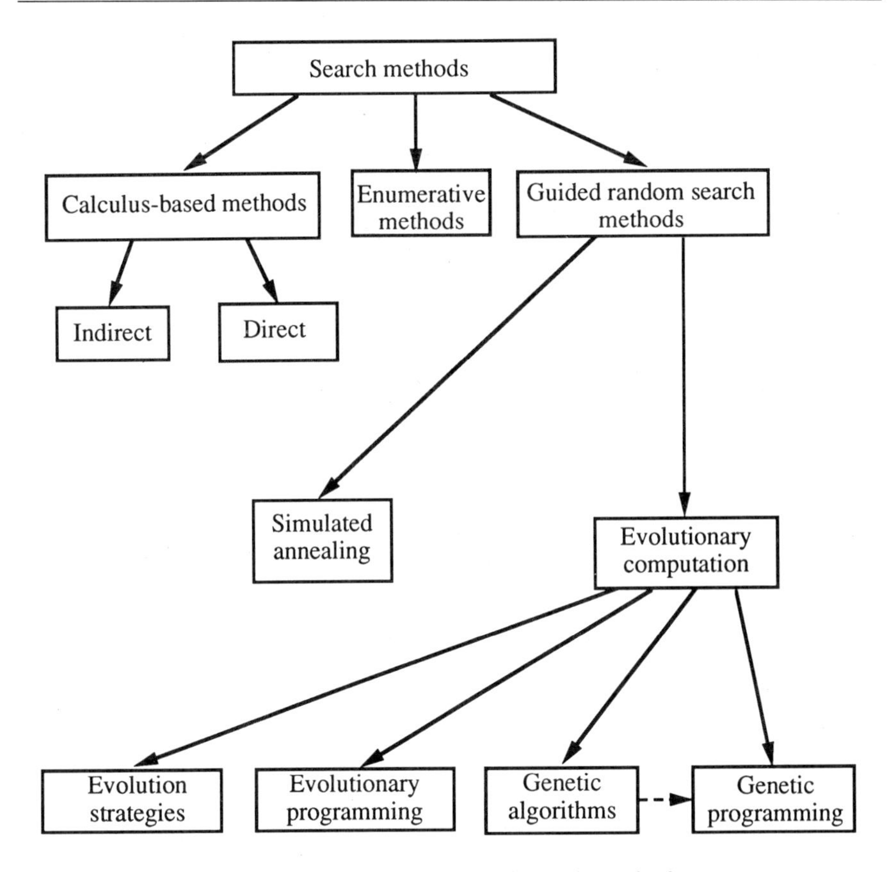

Figure 1 Different classes of search methods.

Enumerative methods (Figure 1) search every point related to an objective function's domain space, one point at a time. They are very simple to implement, but may require significant computation and therefore suffer from a lack of efficiency (Goldberg, 1989a).

Guided random search techniques (Figure 1) are based on enumerative approaches, but use supplementary information to guide the search. Two major subclasses are simulated annealing and evolutionary computation. Simulated annealing is based on thermodynamic considerations, with annealing interpreted as an optimization procedure. The method probabilistically generates a sequence of states based on a cooling schedule to converge ultimately to the global optimum (Metropolis *et al.*, 1953; Kirkpatrick *et al.*, 1983). The main goal of evolutionary computation (de Jong and Spears, 1993) is the application of the concepts of natural selection to a population of structures in the memory of a computer (Kinnear, 1994). Evolutionary computation can be subdivided into evolution strategies, evolutionary programming, genetic algorithms, and genetic programming (Kinnear, 1994; Angeline, 1995).

Evolution strategies were proposed in the early 1970s by Rechenberg (1973). They insist on a real encoding of the problem parameters. Evolution strategies are frequently associated with engineering optimization problems (Kinnear, 1994). They promote mutations rather than recombinations. Basically, evolutionary programming is also sceptical about the usefulness of recombinations but allows any type of encoding (Fogel, 1995). With genetic algorithms, a population of individuals is created and the population is then evolved by means of the principles of variation, selection, and inheritance. Indeed, genetic algorithms differ from evolution strategies and evolutionary programming in that this approach emphasizes the use of specific operators, in particular crossover, that mimic the form of genetic transfer in biota (Porto *et al.*, 1995). Genetic programming (Koza, 1992; Kinnear, 1994) is an extension of genetic algorithms in which members of the population are parse trees of computer programs. Genetic programming is most easily implemented where the computer language is tree structured and therefore LISP is often used (Kinnear, 1994).

MECHANICS OF SIMPLE GENETIC ALGORITHMS

An overview of the natural selection

In nature, the organisms that are best suited to competition for scanty resources (e.g. food, space) survive and mate. They generate offspring, allowing the transmission of their heredity by means of genes contained in their chromosomes. Adaptation to a changing environment is essential for the perenity of individuals of each species. Therefore, natural selection leads to the survival of the fittest individuals, but it also implicitly leads to the survival of the fittest genes. The reproduction process allows diversification of the gene pool of a species. Evolution is initiated when chromosomes from two parents recombine during reproduction. New combinations of genes are generated from previous ones and therefore a new gene pool is created. Segments of two parent chromosomes are exchanged during crossovers, creating the possibility of the 'right' combination of genes for better individuals. Mutations introduce sporadic and random changes in the chromosomes. Repeated selection, crossovers and mutations cause the continuous evolution of the gene pool of a species and the generation of individuals that survive better in a competitive environment. Pioneered by Holland (Holland, 1992), genetic algorithms are based on the above Darwinian principles of natural selection and evolution. They manipulate a population of potential solutions to an optimization (or search) problem (Srinivas and Patnaik, 1994). Specifically, they operate on encoded representations of the solutions, equivalent to the chromosomes of individuals in nature. Each solution is associated with a fitness value which reflects how good it is compared to other solutions in the population. The selection policy is ultimately responsible for ensuring survival of the best fitted individuals. Manipulation of 'genetic material' is performed through crossover and mutation operators. Detailed theoretical discussions of genetic algorithms are beyond the scope of this paper and can be found in numerous books (Goldberg, 1989a; Davis, 1991; Rawlins, 1991; Michalewicz, 1992; Whitley, 1993; Renders, 1995; Whitley and Vose, 1995). In the following paragraph, we only present some basic principles which aid understanding of the functioning of the classical genetic algorithm. However, when necessary, additional bibliographical information is provided in order to give a brief guide into the labyrinth of genetic algorithm research.

How do genetic algorithms work?

A genetic algorithm operates through a simple cycle including the following stages:

- encoding mechanism;
- creation of a population of chromosomes;
- definition of a fitness function;
- genetic manipulation of the chromosomes.

In the design of a genetic algorithm to solve a specific problem, the encoding mechanism is of prime importance. Basically, it depends on the nature of the problem variables. However, traditionally a binary encoding is used. This is particularly suitable when the variables are Boolean (e.g. the presence or absence of an atom in a molecule). Under these conditions, a chromosome consists of a string of binary digits (bits) that are easily interpretable. When continuous variables are used (e.g. physicochemical descriptors), a common method of encoding them uses their integer representation. Each variable is first linearly mapped to an integer defined in a specific range, and the integer is encoded using a fixed number of binary bits. The binary codes of all the variables are then concatenated to obtain the binary string constituting the chromosome. The principal drawback of encoding variables as binary strings is the presence of Hamming cliffs which are large Hamming distances between the binary codes of adjacent integers (Srinivas and Patnaik, 1994). Thus, for example, 011 and 100 are the integer representations of 3 and 4, respectively (Table I), and have a Hamming distance of 3. For the genetic algorithm to improve the code of 3 to that of 4, it must alter all bits simultaneously. Such a situation presents a problem for the functioning of the genetic algorithms. To overcome this problem, a Gray coding can be used (Forrest, 1993). Gray codes have the property whereby incrementing or decrementing any number by 1 is always a one-bit change (Table I). Therefore, adjacent integers always present a Hamming distance of 1.

Table I *Comparison of binary and Gray coded integers.*

Integers	Binary code	Gray code
0	000	000
1	001	001
2	010	011
3	011	010
4	100	110
5	101	111

The initial population of individuals is usually generated randomly. When designing a genetic algorithm model, we have to decide what the population size must be (Goldberg, 1989b). General wisdom dictates that increasing the population size increases its diversity and reduces the probability of a premature convergence to a local optimum. However, this strategy also increases the time required for the population to converge to the optimal regions in the search space. In practice, the above rule of thumb does not always appear to be true, and the most effective population size is dependent on the problem being solved, the representation used, and the choice of the operators (Syswerda, 1991).

The individuals of the population are exposed to an evaluation function that plays the role of the environmental pressure in the Darwinian evolution. This function is called fitness. Based on each individual's fitness, a mechanism of selection determines mates for the genetic manipulation process. Low-fitness individuals are less likely to be selected than high-fitness individuals as parents for the next generation.

A fitness scaling is often used to prevent the early domination of super-individuals in the selection process and to promote healthy competition among near equals when the population has largely converged. Different scaling procedures can be used. Among them we can cite the linear scaling (Goldberg, 1989a), the sigma truncation (Goldberg, 1989a), the power law scaling (Goldberg, 1989a), the sigmoidal scaling (Venkatasubramanian *et al.*, 1994) and the Gaussian scaling (Venkatasubramanian *et al.*, 1995a). Many practical problems contain constraints that must be satisfied in a modeling process. They can be handled directly by the fitness function (Goldberg, 1989a).

The aim of parent selection in a genetic algorithm is to provide more reproductive chances for the most fit individuals. There are many ways to do this. The classical genetic algorithm uses the roulette wheel selection scheme which is exemplified in Table II. It shows a population of six individuals with a set of evaluations totaling 47. The first row of the table allows identification of the individuals, the second contains each individual's fitness, and the third contains the running total of fitness. The second part

Table II *Examples of roulette wheel parent selection.*

Individual	1	2	3	4	5	6
Fitness	7	4	15	8	2	11
Running total	7	11	26	34	36	47
Random number	23	37	11	46		
Individual chosen	3	6	2	6		

of Table II shows four numbers generated randomly from the interval between 0 and 47, together with the index of the individual that would be chosen by roulette wheel parent selection for each of these numbers. In each case, the individual selected is the first one for which the running total of fitness is greater than or equal to the random number (Davis, 1991). With the roulette wheel parent selection, each parent's chance of being selected is directly proportional to its fitness. In the initial generations, the population has a low average fitness value. The presence of a few individuals with relatively high fitness values induces the allocation of a large number of offspring to these individuals, and can cause a premature convergence.

A different problem arises in the later stages of the genetic algorithm, when the population has converged and the variance in individual fitness values becomes small. As stressed in the previous section, scaling transformations of the fitness can be used to overcome these problems.

Another solution consists of using alternate selection schemes. Among these, the most commonly employed is the tournament selection (Angeline, 1995), in which an individual must win a competition with a randomly selected set of individuals. The winner of the tournament is the individual with the highest fitness of the tournament competitors. The winner is then incorporated in the mating pool. The mating pool, being composed of tournament winners, has a higher average fitness than the average population fitness. This fitness difference provides the selection pressure, which drives the genetic algorithm to improve the fitness of each succeeding generation (Blickle and Thiele, 1995; Miller and Goldberg, 1995).

Crossovers and mutations are genetic operators allowing the creation of new chromosomes during the reproduction process. Crossover occurs when two parents exchange parts of their corresponding chromosome. Due to their importance for genetic algorithm functioning, much of the literature has been devoted to different crossover techniques and the analysis of these (Schaffer *et al.*, 1989; Syswerda, 1989, 1993; Schaffer and Eshelman, 1991; Spears and de Jong, 1991; Qi and Palmieri, 1993; Spears, 1993; Jones, 1995; Robbins, 1995). The one-point crossover is the simplest form. It occurs when parts of two parent chromosomes are swapped after a randomly selected point, creating two children (Figure 2). A side-effect of the one-point crossover is that interacting bits that are relatively far apart on the chromosome are more

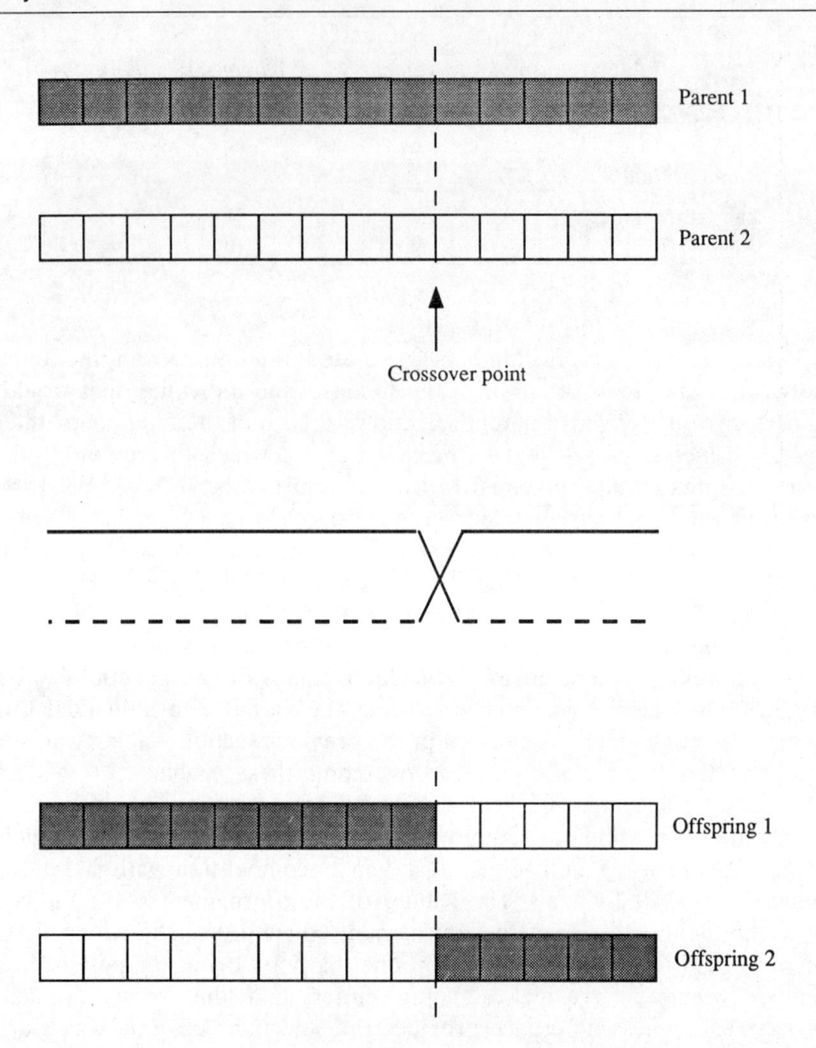

Figure 2 One-point crossover.

likely to be separated than bits that are close together (the converse is true). Because the one-point crossover determines its crossover point randomly along the length of the bit string with a uniform distribution, this operator is unbiased with respect to the distribution of material exchanged. Therefore, it is classically indicated in the genetic algorithm literature that the one-point crossover is characterized by high positional bias and low distributional bias.

In a two-point crossover scheme, two crossover points are randomly chosen and segments of the strings between them are exchanged (Figure 3). The two-point crossover reduces positional bias without introducing any

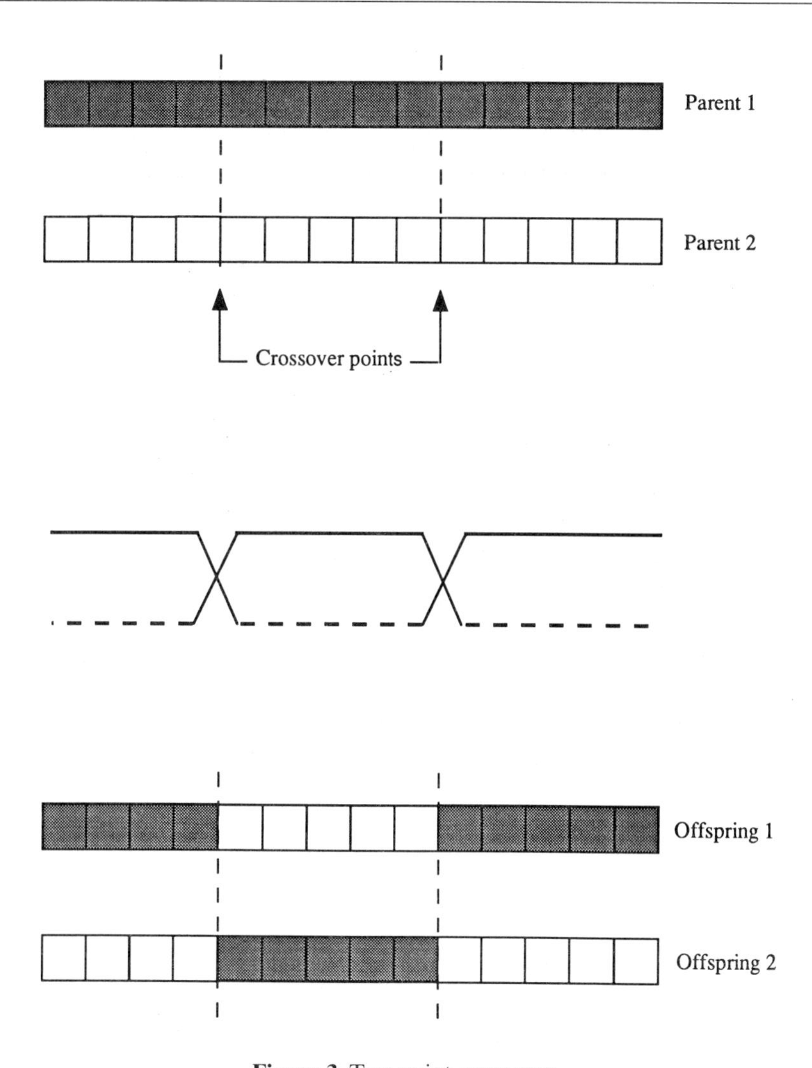

Figure 3 Two-point crossover.

distributional bias (Eshelman *et al.*, 1989). Multi-point crossover is an extension of the two-point crossover. It consists of an increase in the number of crossover points. This reduces positional bias but introduces some distributional bias (Eshelman *et al.*, 1989). The segmented crossover is a variant of multi-point crossover which allows the number of crossover points to vary. Indeed, instead of selecting in advance a fixed number of unique crossover points, a segment switch rate is determined (Eshelman *et al.*, 1989). Uniform crossover exchanges bits rather than segments (Syswerda, 1989). Since the probability of exchanging two bits in each position is not linked to a position,

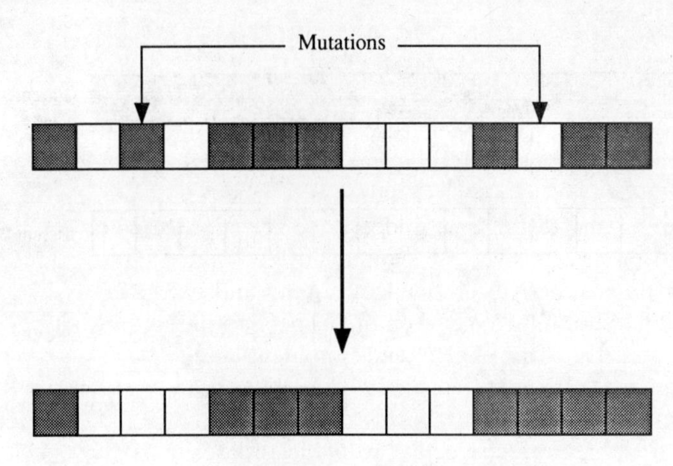

Figure 4 Mutations.

this crossover has no positional bias. However, it presents a distributional bias (Eshelman *et al.*, 1989). Other crossover operators can be found in the genetic algorithm literature (Eshelman *et al.*, 1989; Bui and Moon, 1995; Güvenir, 1995; Höhn and Reeves, 1995; Kapsalis *et al.*, 1995; Robbins, 1995). Some of them are designed for specific applications (e.g. Brown *et al.*, 1994; Djouadi, 1995).

Mutations induce sporadic and random alterations of bit strings (Figure 4). According to Goldberg (1989a, p. 14), mutation plays a secondary role in genetic algorithms in restoring lost genetic material. However, it has recently been shown that it was an important genetic operator (Hinterding *et al.*, 1995). The probability of mutation, which can be held constant or can vary throughout a run of the genetic algorithm (Fogarty, 1989), is generally low (e.g., 0.01). Basically, this probability depends on the choice of encoding (Bäck, 1993; Tate and Smith, 1993). The mutation operator can be improved in several ways. The variants can be problem-dependent (e.g. Goldberg, 1989a; Davis, 1991; Djouadi, 1995; Kapsalis *et al.*, 1995; Montana, 1995; Robbins, 1995).

The offspring created by the genetic manipulation process constitute the next population to be evaluated. The genetic algorithm can replace the whole population or only its less fitted members. The former strategy is termed the generational approach and the latter the steady-state approach.

The genetic algorithm cycle is repeated until a satisfactory solution to the problem is found or some other termination criteria are met (Sutton and Boyden, 1994).

APPLICATIONS OF GENETIC ALGORITHMS IN QSAR AND DRUG DESIGN

Genetic algorithms have been successfully used in all the different steps required for deriving, analyzing and validating QSAR models. Thus, the construction of QSAR models requires in a first step the design of representative training and testing sets. Indeed, the selection of optimal test series is crucial in drug design, since the synthesis of new compounds and the analysis of their biological activity is time-consuming and extremely costly. Most of the selection methods are based on the inspection of graphical displays which are usually derived from linear and nonlinear multivariate analyses (Pleiss and Unger, 1990; Domine *et al.*, 1994a,b, 1996; Devillers, 1995). Recently, Putavy and coworkers have shown that genetic algorithms represented an attractive alternative for selecting valuable test series (Putavy *et al.*, 1996). Their study was performed using a data matrix of 166 aromatic substituents described by means of six substituent constants encoding their hydrophobic, steric and electronic effects. These parameters were the π constant, the H-bonding acceptor (HBA) and donor (HBD) abilities, the molar refractivity (MR) and the inductive and resonance parameters (F and R) of Swain and Lupton (1968). The data were retrieved from the literature (Hansch and Leo, 1979). Different fitness functions based on the calculation of correlation coefficients or Euclidean distances were tested. The different test series proposed by the genetic algorithm were compared from displays on nonlinear maps and calculation of variance coefficients. This showed that genetic algorithms were able to propose sets of aromatic substituents which presented a high information content. It also stressed the complementarity of the genetic algorithms with the graphical displays provided by nonlinear mapping for a better analysis of the results.

In QSAR studies of large data sets, variable selection and model building are also difficult and time-consuming tasks. Different approaches have been proposed for proper model selection. Thus, for example, McFarland and Gans (1993, 1994) used a cluster significance analysis, and Wikel and Dow (1993) applied a backpropagation neural network. Rogers (Rogers and Hopfinger, 1994; Rogers, 1995, 1996) opened an interesting line of research in the domain by using a hybrid system consisting of the Holland's genetic algorithm and Friedman's multivariate adaptive regression splines (MARS) algorithm. After this pioneering investigation, attempts were made by Luke (1994) and Kubinyi (1994a,b) to test the usefulness of evolutionary algorithms.

Basically, a QSAR model allows the prediction of a biological activity from topological and/or physicochemical descriptors. In the activity space, genetic algorithms can be used for their ability to detect outliers (Leardi, 1994; Crawford and Wainwright, 1995). In computer-aided molecular design, genetic algorithms can also be employed for the identification of appropriate molecular structures given the desired physicochemical property or biological activity (e.g., Venkatasubramanian *et al.*, 1994, 1995a,b; Devillers and Putavy, 1996). Sometimes, the genetic algorithms can be used only to guide compound selection during a molecular diversity experiment (Hopfinger and Patel, 1996). Generally, genetic algorithms are linked with a backpropagation neural network in order to produce an intercommunicating hybrid system (Goonatilake and Khebbal, 1995), working into a co-operative environment for generating an optimal modeling solution. Thus, Venkatasubramanian and coworkers proposed a backpropagation neural network-based approach for solving the forward problem of property prediction of polymers based on the structural characteristics of their molecular subunits, and a genetic algorithm-based approach for the inverse problem of constructing a molecular structure given a set of desired macroscopic properties (Venkatasubramanian *et al.,* 1996). In their CAMD approach, the standard genetic algorithm framework needed to be adapted. A similar hybrid system was proposed by Devillers and Putavy (1996) for the design of organic molecules presenting a specific biodegradability. Different constraints were tested in order to estimate the limits of the system for proposing candidate biodegradable molecules presenting specific structural features. A more sophisticated intercommunicating hybrid system constituted of a genetic function approximation (GFA), and a backpropagation neural network was proposed by van Helden and coworkers (van Helden *et al.*, 1996) for predicting the progesterone receptor-binding affinity of steroids.

Progress in medicine and pharmacology depends on our ability to understand the interactions of drugs with their biological targets. Molecular docking is the process which allows recognition between molecules through complementarity of molecular surface structures and energetics. It includes not only the size and shape of molecular surfaces, but also charge–charge interaction, hydrogen bonding and van der Waals interaction. Molecular docking is therefore a difficult problem, in terms of both understanding and modeling. Genetic algorithms have been successfully used to solve numerous problems dealing with this particular aspect of molecular modeling. Thus, Payne and Glen (1993) used a genetic algorithm to fit a series of *N*-methyl-D-aspartate (NMDA) antagonists to a putative NMDA pharmacophore composed of the distance from the amine nitrogen to a phosphonate sp^2 oxygen and the distance from the carboxylic acid oxygen to the same phosphonate sp^2 oxygen. These distances were defined for three molecules. The molecules were generated from arbitrary starting points in rotational and conformational space. Payne and Glen (1993) showed that the algorithm achieved reasonable conformations. They also showed, on the same series of NMDA antagonists, that a genetic algorithm could be used for the elucidation of a pharmacophore. In order to assess the ability of genetic algorithms to find optimal orientations and conformations of flexible molecules, Payne and Glen (1993) carried out different self-fitting experiments. Thus, the monosaccharide 2-deoxyribose in the D-configuration was fitted on to itself using varying constraints (i.e. shape, electrostatic potential), population sizes and mutation rates. The self-fitting of trimethoprim and maltose under various constraints was also studied. The usefulness of genetic algorithms in the more relevant exercise of fitting very dissimilar molecules together (i.e. benzodiazepine receptor ligands on to β-carboline, leu-enkephalin on to hybrid morphine) was also investigated. Finally, 17 GABA analogs were built using SYBYL (Tripos Associates, St Louis, MO, USA) in random conformations and rotations. These were conformationally and spatially restricted by fitting to 4,5,6,7-tetrahydroisoxazolo[5,4-*c*]pyridin-3-ol (THIP), using a genetic algorithm to match molecular properties generated from the test molecules and THIP on a molecular surface. The conformations and spatial orientations of the test structures resulting from the genetic algorithm fit to THIP were used to calculate relevant molecular properties. From the values of these properties, a nonlinear map was produced for the display of the molecules encoded as agonist, weak agonist or inactive. In a more recent publication (Glen and Payne, 1995), the authors have proposed a genetic algorithm using a series of rules allowing the production of realistic molecules in three dimensions. These molecules could then be evaluated using constraints based on calculated molecular properties which are of use in the prediction of biological activity. According to Glen and Payne (1995), the structure–generation algorithm can also produce very large and diverse sets of reasonable chemical structures for searching by 3D database programs. In a recent overview article including new examples of applications, Jones *et al.* (1996) presented different aspects of the above chemical structure handling and molecular recognition problems. Indeed, they clearly demonstrated the usefulness of genetic algorithms during the conformational analysis stage in the searching of databases of flexible 3D molecules to find candidate drug structures that fit a query pharmacophore, the estimation of the binding affinities and the most energetically favorable combination of interactions between a receptor and a flexible ligand, and the superimposition of flexible molecules with the use of the resulting overlays to suggest possible pharmacophoric patterns. Walters and coworkers (Walters and Hinds, 1994; Walters and Muhammad, 1996) proposed a program called Genetically Evolved Receptor Models (GERM) for producing atomic-level models of receptor binding sites, based on a set of known structure–activity relationships. The generation of these models requires no *a priori* information about a real receptor, other than a reasonable assurance that the SAR data are for a set of chemicals acting upon a common site. The receptors generated by the genetic algorithm can be used for correlating calculated binding with measured bioactivity, for predicting the activity of new compounds, by docking the chemicals, calculating their binding energies and their biological activity.

SOFTWARE AVAILABILITY

A large number of software packages which propose genetic algorithms and/or related methods are available commercially or in the public domain. Some examples are given below.

AGAC
Year first available: 1992
Hardware required: anything
Program language: ANSI C
Program size: 58 blocks of source code
Source code availability: yes
User's guide availability: no. Just a README file
Price: free
Contact address: Web: http://www.aic.nrl.navy.mil/~spears, FTP: ftp.aic.nrl.navy.mil under pub/spears, E-mail: spears@aic.nrl.navy.mil

DPGA (Distributed Parallel Genetic Algorithm)
Year first available: 1996
Current version: 1.0
Hardware required: UNIX workstation cluster
Program language: C++, PVM (Parallel Virtual Machine)
Program size: 1.2 MB (compiled)
Source code availability: yes
User's guide availability: yes
Price: on request
Discount for University: free for scientific purposes
Contact address: M. Schwehm, University of Erlangen – IMMD VII, Martensstr. 3, D-91058 Erlangen, Germany. Tel.: 49 9131 85–7617, Fax.: 49 9131 85–7409, E-mail: schwehm@immd7.informatik.uni-erlangen.de

EvoFrame (on PC)
EvoFrame is an object-oriented implemented framework for optimization by using the evolution strategy (Turbo-Pascal 6.0, Turbo-Vision). A prototyping module 'REALizer' is available. REALizer is a completely prepared evolution strategic optimization tool excluding the quality function (vector of reals).
Year first available: 1993
Current version: 2.0
Hardware required: DOS Machine, DPMI allows use of 16 MB under DOS
Program language: Borland Pascal 7.0, Turbo-Vision
Program size: 250 kB (for framework only). Final size depends on implementation of optimization problem
Source code availability: yes. Full source code
User's guide availability: yes. Method reference, 'how to' instructions, introductory papers to evolution strategies
Price (excluding legal VAT): EvoFrame: 1850 DM (single licence), 5200 DM (unlimited licence); REALizer: 200 DM (single licence), 500 DM (unlimited licence)

Discount for university: special prices. EvoFrame: 1250 DM (single licence), 3500 DM (unlimited licence); REALizer: 75 DM (single licence), 200 DM (unlimited licence)
Related scientific article(s): Rechenberg (1973, 1989a,b); Schwefel (1981); de Groot and Wurtz (1990); Hoffmeister and Back (1990); Lohmann (1990, 1991); Rudolf (1990); Stebel (1991, 1992); Herdy (1992); Quandt (1992); Trint and Utecht (1992); Ostermeier *et al.* (1993); Herdy and Patone (1994)
Contact address: Optimum Software, Wolfram Stebel, Bahnhofstr. 12, 35576 Wetzlar, Germany. Tel.: (+49)06441/47633

EvoFrame (on Macintosh)
EvoFrame is an object-oriented implemented framework for optimization by using the evolution strategy (Object-Pascal (MacApp 2.0), C++ (MacApp 3.0)). The prototyping module 'REALizer' is also available for Macintosh.
Year first available: 1993
Current version: 1.0
Hardware required: Apple Macintosh
Program language: MPW, C++, MacApp
Program size: 320 kB (for framework only). Final size depends on implementation of optimization problem
Source code availability: yes. Full source code
User's guide availability: yes. Method reference, 'how to' instructions, introductory papers to evolution strategies
Price (excluding legal VAT): EvoFrame: 1850 DM (single licence), 5200 DM (unlimited licence); REALizer: 200 DM (single licence), 500 DM (unlimited licence)
Discount for University: special prices. EvoFrame: 1250 DM (single licence), 3500 DM (unlimited licence); REALizer: 75 DM (single licence), 200 DM (unlimited licence)
Related scientific article(s): Rechenberg (1973, 1989a,b); Schwefel (1981); de Groot and Wurtz (1990); Hoffmeister and Back (1990); Lohmann (1990, 1991); Rudolf (1990); Stebel (1991, 1992); Herdy (1992); Quandt (1992); Trint and Utecht (1992); Ostermeier *et al.* (1993); Herdy and Patone (1994)
Contact address: Optimum Software, Wolfram Stebel, Bahnhofstr. 12, 35576 Wetzlar, Germany. Tel.: (+49)06441/47633, AppleLink: OPTIMUM, E-mail: Optimum@AppleLink.Apple.Com

Evolver
Evolver is a spreadsheet add-in which incorporates a genetic algorithm.
Year first available: 1990
Current version: 3.0
Hardware required: 486/8 MB RAM (+ Microsoft Excel)
Program language: C++, Visual Basic
Program size: 1 MB
Source code availability: no (or US$100 000)
User's guide availability: yes (US$79)
Price: US$349
Discount for University: US$100 off
Related scientific article(s): Antonoff (1991); Mendelsohn (1994); Ribeiro Filho *et al.* (1994); Zeanah (1994); Begley (1995); Lane (1995)

Contact address: Axcelis Inc., 4668 Eastern Avenue North, Seattle, WA 98 103, USA. Tel.: (206) 632–0885, Fax.: (206) 632–3681

GAC
Year first available: 1992
Hardware required: anything
Program language: C
Program size: 58 blocks of source code
Source code availability: yes
User's guide availability: no. Just a README file
Price: free
Contact address: Web: http://www.aic.nrl.navy.mil/~spears, FTP: ftp.aic.nrl.navy.mil under pub/spears, E-mail: spears@aic.nrl.navy.mil

GAGA
Hardware required: any
Program language: C
Program size: 2142 lines of source code
Source code availability: yes
User's guide availability: no
Price: free
Related scientific article(s): Ribeiro Filho *et al.* (1994)
Contact address: Jon Crowcroft, University College London, Gower St, London WC1E 6BT, UK. Tel.: +44 171 387 7050, Fax.: +44 171 387 1397, E-mail: jon@cs.ucl.ac.uk

GAL
Year first available: 1992
Hardware required: enough RAM to run some version of LISP
Program language: LISP
Program size: 39 blocks of source code
Source code availability: yes
User's guide availability: no. Just a README file
Price: free
Contact address: Web: http://www.aic.nrl.navy.mil/~spears, FTP: ftp.aic.nrl.navy.mil under pub/spears, E-mail: spears@aic.nrl.navy.mil

GAME
GAME is an object-oriented environment for programming parallel GA applications and algorithms, and mapping them on to parallel machines.
Year first available: 1992
Current version: 2.01
Hardware required: numerous machines (e.g. Sun workstations, IBM PC-compatible computers)
Program language: C++
Source code availability: yes
User's guide availability: no
Related scientific article(s): Ribeiro Filho *et al.* (1994)
Contact address: J.L. Ribeiro Filho, Computer Science Department, University

College London, Gower St, London WC1E 6BT, UK. Tel.: +44 71 387 7050, Fax.: +44 71 387 1398, E-mail: j.ribeirofilho@cs.ucl.ac.uk

GANNET
Year first available: 1991
Current version: .91
Hardware required: workstation running UNIX with C compiler
Program language: C
Program size: 4500 lines
Source code availability: yes. FTP from fame.gmu.edu /gannet/source directory
User's guide availability: yes
Price: freeware: GNU General Public License
Related scientific article(s): Hintz and Spofford (1990); Spofford and Hintz (1991)
Contact address: K.J. Hintz, Department of Electrical and Computer Engineering, George Mason University, 4400 University Drive, Fairfax, VA 22030, USA. E-mail: khintz@fame.gmu.edu

GAUCSD
Year first available: 1989
Current version: 1.4
Hardware required: UNIX; can be ported to PC or Macintosh
Program language: C
Program size: 210 kB (source and documents)
Source code availability: yes
User's guide availability: yes
Price: free
Contact address: T. Kammeyer, Computer Science and Engineering Department, University of California, San Diego, La Jolla, CA 92093–0114, USA. Fax.: (619) 534–7029, E-mail: GAuscd-request@cs.uscd.edu

GA Workbench
Year first available: 1989
Current version: 1.10
Hardware required: IBM compatible, MS-DOS, EGA/VGA display
Program language: C++
Program size: ca. 100 kB
Source code availability: no
User's guide availability: yes (ASCII and Postscript formats)
Price: voluntary contribution
Related scientific article(s): Ribeiro Filho *et al.* (1994)
Contact address: M. Hughes, E-mail: mrh@iz.co.uk

GENESIS
GENESIS is a general purpose package implementing a genetic algorithm for function optimization. The user must provide an 'evaluation' function which returns a value when given a particular point in the search space.
Year first available: 1981
Current version: 5.0

Hardware required: numerous machines (e.g. Sun workstations, IBM PC-compatible computers). Runs under UNIX and DOS using Turbo C
Program language: C
User's guide availability: yes
Price: freely used for educational and research purposes. Other rights are reserved. For $52.50 ($60 per copy for addresses outside of North America), you can get the version 5.0 of GENESIS (in C) and OOGA (a GA system in Common LISP and CLOS), along with documentation
Related scientific article(s): Davis (1991)
Contact address: J.J. Grefenstette, The Software Partnership, PO Box 991, Melrose, MA 02176, USA. Tel.: (617) 662–8991, E-mail: gref@aic.nrl.navy.mil

Genocop
Year first available: 1993
Current version: 3.0
Hardware required: SUN workstation
Program language: C
Program size: 300 kB
Source code availability: yes
User's guide availability: yes
Price: free
Related scientific article(s): Michalewicz (1994); Michalewicz *et al.* (1994)
Contact address: Z. Michalewicz, Department of Computer Science, University of North Carolina, 9201 University City Boulevard, Charlotte, NC 28223–0001, USA. Tel.: (704) 547–4873, Fax.: (704) 547–3516, E-mail: zbyszek@uncc.edu

GenocopIII
Year first available: 1995
Current version: 1.0
Hardware required: SUN workstation
Program language: C
Program size: 300 kB
Source code availability: yes
User's guide availability: yes
Price: free
Related scientific article(s): Michalewicz and Nazhiyath (1995); Michalewicz (1996)
Contact address: Z. Michalewicz, Department of Computer Science, University of North Carolina, 9201 University City Boulevard, Charlotte, NC 28223-0001, USA. Tel.: (704) 547–4873, Fax.: (704) 547–3516, E-mail: zbyszek@uncc.edu

MPGA (Massively Parallel Genetic Algorithm)
Year first available: 1993
Current version: 2.0
Hardware required: MasPar MP-1 or MP-2 (Array processor with 1024 to 16384 processor elements)
Program language: MPL
Program size: 1.6 MB (compiled)
Source code availability: yes

User's guide availability: yes
Price: on request
Discount for University: free for scientific purposes
Related scientific article(s): Schwehm (1993)
Contact address: M. Schwehm, University of Erlangen – IMMD VII, Martensstr. 3, D-91058 Erlangen, Germany. Tel.: 49 9131 85–7617, Fax.: 49 9131 85–7409, E-mail: schwehm@immd7.informatik.uni-erlangen.de

OMEGA Predictive Modeling System
OMEGA is a behavioral modeling system offering specific facilities for credit, marketing and insurance applications. Several genetic algorithms (evolutionary and genetic engineering) are used to optimize models.
Year first available: 1994
Current version: 2.2
Hardware required: 486 PC or higher
Program language: 32 bit Fortran and C++
Program size: 12 MB (code), 6 MB (help files)
Source code availability: no
User's guide availability: yes. An overview of the system is available as a Windows help file
Price: £17 500 per year. A further £7500 per year for an inferencing module
Discount for University: special terms are available for Application Development Partners
Related scientific article(s): Barrow (1992); Haasdijk (1993); Babovic and Minns (1994); Haasdijk *et al.* (1994); Ribeiro Filho *et al.* (1994); Babovic (1995); Walker *et al.* (1995); Keijzer (1996)
Contact address: David Barrow, KiQ Ltd., Easton Hall, Great Easton, Essex, CM6 2HD, UK. Tel/Fax.: 44 1371 870 254

SAMUEL
SAMUEL is a multi-purpose machine learning program that uses genetic algorithms and other competition-based heuristics to improve its own decision-making rules. The system actively explores alternative behaviors in simulation, and modifies its rules on the basis of this experience. SAMUEL is designed for problems in which the payoff is delayed in the sense that payoff occurs only at the end of an episode that may span several decision steps.
Program language: C
Source code availability: yes
Price: this program is made available for re-use by domestic industries, government agencies, and universities under NASA's Technology Utilization Program through the COSMIC Software Distribution site at the University of Georgia. Programs and documents may be copied without restriction for use by the acquiring institution unless otherwise noted. License fee: $200.
Discount for University: possible educational discounts
Contact address: COSMIC, The University of Georgia, 382 East Broad Street, Athens, GA 30602, USA. Tel.: (706) 542–3265, Fax.: (706) 542–4807, E-mail: service@cossack.cosmic.uga.edu

XpertRule
Year first available: 1993
Current version: 3.1
Hardware required: Windows 3.1 or later / 4 MB RAM
Program language: Pascal
Program size: 1.5 MB
Source code availability: no
User's guide availability: yes
Price: UK£995
Discount for University: yes (UK£175)
Related scientific article(s): Al-Attar (1994)
Contact address: Attar Software, Newlands Road, Leigh, Lancashire, UK. Tel.: +44 1942 608844, Fax.: +44 1942 601991, E-mail: 100166.1547@CompuServe.com

ADVANTAGES AND LIMITATIONS OF GENETIC ALGORITHMS

Genetic algorithms are robust, adaptive search methods that can be immediately tailored to real problems. In addition, genetic algorithms are very easy to parallelize in order to exploit the capabilities of massively parallel computers and distributed systems (Grefenstette and Baker, 1989; Cantú-Paz, 1995). Furthermore, it has been shown that the elaboration of hybrid systems linking genetic algorithms with other optimization algorithms, pattern-recognition methods or other statistical techniques allows the proposal of powerful modeling tools. Thus, there has been considerable interest in creating hybrid systems of genetic algorithms with neural networks. Genetic algorithms are used to design or train the neural networks (Harp *et al.*, 1989; Miller *et al.*, 1989; Schaffer *et al.*, 1990; Jones, 1993; Shamir *et al.*, 1993; Kitano, 1994; Abu-Alola and Gough, 1995; Kussul and Baidyk; 1995; Montana, 1995; Roberts and Turega, 1995). They can also be employed to perform a particular task in the modeling process (Vahidov *et al.*, 1995; Ventura *et al.*, 1995; Vivarelli *et al.*, 1995; Yip and Pao, 1995; Devillers and Putavy, 1996; van Helden *et al.*, 1996; Venkatasubramanian *et al.*, 1996). In addition to these hybridizations, genetic algorithms have also been hybridized with Kohonen network (Polani and Uthmann, 1993; Hämäläinen, 1995; Merelo and Prieto, 1995), fuzzy logic system or fuzzy network (Feldman, 1993; Karr, 1995), fuzzy decision trees (Janikow, 1995), K nearest neighbors classification algorithm (Kelly and Davis, 1991), MARS algorithm (Rogers, 1991), PLS (Dunn and Rogers, 1996), branch and bound techniques (Cotta *et al.*, 1995) and simulated annealing (Ait-Boudaoud *et al.*, 1995; Ghoshray *et al.*, 1995; Kurbel *et al.*, 1995; Varanelli and Cohoon, 1995). Basically, the above hybrid systems allow the limitations of the individual techniques to be overcome (Goonatilake and Khebbal, 1995). Therefore, they are particularly suitable for solving complex modeling problems.

From the above, it appears that genetic algorithms constitute a very attractive new tool in QSAR studies and drug design. However, despite the successful use of genetic algorithms to solve various optimization problems, progress with regard to research on their theoretical foundations is needed. Indeed, even if the schema theory and building-block hypothesis of Holland and Goldberg (Goldberg, 1989a) capture the essence of genetic algorithm mechanics, numerous studies are required in order to gain a deeper understanding of how genetic algorithms work, and for the correct design of valuable applications.

REFERENCES

Abu-Alola, A.H. and Gough, N.E. (1995). Identification and adaptive control of non-linear processes using combined neural networks and genetic algorithms. In, *Artificial Neural Nets and Genetic Algorithms* (D.W. Pearson, N.C. Steele, and R.F. Albrecht, Eds.). Springer-Verlag, Wien, pp. 396–399.

Ait-Boudaoud, D., Cemes, R., and Holloway, S. (1995). Evolutionary adaptive filtering. In, *Artificial Neural Nets and Genetic Algorithms* (D.W. Pearson, N.C. Steele, and R.F. Albrecht, Eds.). Springer-Verlag, Wien, pp. 269–272.

Al-Attar, A. (1994). A hybrid GA-heuristic search strategy. *AI Expert* **September**, 34–37.

Andrey, P. and Tarroux, P. (1994). Unsupervised image segmentation using a distributed genetic algorithm. *Pattern Recogn.* **27**, 659–673.

Angeline, P.J. (1995). Evolution revolution: An introduction to the special track on genetic and evolutionary programming. *IEEE Expert Intell. Syst. Appl.* **10 (June)**, 6–10.

Antonoff, M. (1991). Software by natural selection. *Popular Science* **October** issue.

Babovic, V. (1995). Genetic programming and its application to rainfall–runoff modelling. *J. Hydrology* (in press).

Babovic, V. and Minns, A.W. (1994). Use of computational adaptive methods in hydroinformatics. In, *Proceedings Hydroinformatics.* Balkema, Rotterdam, pp. 201–210.

Bäck, T. (1993). Optimal mutation rates in genetic search. In, *Proceedings of the Fifth International Conference on Genetic Algorithms* (S. Forrest, Ed.). Morgan Kaufmann Publishers, San Mateo, California, pp. 2–8.

Barrow, D. (1992). Making money with genetic algorithms. In, *Proceedings of the Fifth European Seminar on Neural Networks and Genetic Algorithms.* IBC International Services, London.

Begley, S. (1995). Software au naturel. *Newsweek* **May 8**.

Blaney, F. (1990). Molecular modelling in the pharmaceutical industry. *Chem. Indus.* **XII**, 791–794.

Blickle, T. and Thiele, L. (1995). A mathematical analysis of tournament selection. In, *Proceedings of the Sixth International Conference on Genetic Algorithms* (L.J. Eshelman, Ed.). Morgan Kaufmann Publishers, San Francisco, California, pp. 9–16.

Blommers, M.J.J., Lucasius, C.B., Kateman, G., and Kaptein, R. (1992). Conformational analysis of a dinucleotide photodimer with the aid of the genetic algorithm. *Biopolymers* **32**, 45–52.

Bramlette, M.F. and Bouchard, E.E. (1991). Genetic algorithms in parametric design of aircraft. In, *Handbook of Genetic Algorithms* (L. Davis, Ed.). Van Nostrand Reinhold, New York, pp. 109–123.

Brown, R.D., Jones, G., Willett, P., and Glen, R.C. (1994). Matching two-dimensional chemical graphs using genetic algorithms. *J. Chem. Inf. Comput. Sci.* **34**, 63–70.

Bugg, C.E., Carson, W.M., and Montgomery, J.A. (1993). Drugs by design. *Sci. Am.* **December**, 60–66.

Bui, T.N. and Moon, B.R. (1995). On multi-dimensional encoding/crossover. In, *Proceedings of the Sixth International Conference on Genetic Algorithms* (L.J. Eshelman, Ed.). Morgan Kaufmann Publishers, San Francisco, California, pp. 49–56.

Caldwell, C. and Johnston, V.S. (1991). Tracking a criminal suspect through 'face-space' with a genetic algorithm. In, *Proceedings of the Fourth International Conference on Genetic Algorithms* (R.K. Belew and L.B. Booker, Eds.). Morgan Kaufmann Publishers, San Mateo, California, pp. 416–421.

Cantú-Paz, E. (1995). *A Summary of Research on Parallel Genetic Algorithms.* IlliGAL Report No. 95007, University of Illinois at Urbana-Champaign, Department of General Engineering, p. 17.

Chang, G. and Lewis, M. (1994). Using genetic algorithms for solving heavy-atom sites. *Acta Cryst.* **D50**, 667–674.

Cleveland, G.A. and Smith, S.F. (1989). Using genetic algorithms to schedule flow shop releases. In, *Proceedings of the Third International Conference on Genetic Algorithms* (J.D. Schaffer, Ed.). Morgan Kaufmann Publishers, San Mateo, California, pp. 160–169.

Cotta, C., Aldana, J.F., Nebro, A.J., and Troya, J.M. (1995). Hybridizing genetic algorithms with branch and bound techniques for the resolution of the TSP. In, *Artificial Neural Nets and Genetic Algorithms* (D.W. Pearson, N.C. Steele, and R.F. Albrecht, Eds.). Springer-Verlag, Wien, pp. 277–280.

Crawford, K.D. and Wainwright, R.L. (1995). Applying genetic algorithms to outlier detection. In, *Proceedings of the Sixth International Conference on Genetic Algorithms* (L.J. Eshelman, Ed.). Morgan Kaufmann Publishers, San Francisco, California, pp. 546–550.

Dandekar, T. and Argos, P. (1992). Potential of genetic algorithms in protein folding and protein engineering simulations. *Protein Eng.* **5**, 637–645.

Dandekar, T. and Argos, P. (1994). Folding the main chain of small proteins with the genetic algorithm. *J. Mol. Biol.* **236**, 844–861.

Davidor, Y. (1991). A genetic algorithm applied to robot trajectory generation. In, *Handbook of Genetic Algorithms* (L. Davis, Ed.). Van Nostrand Reinhold, New York, pp. 144–165.

Davis, L. (1991). *Handbook of Genetic Algorithms*. Van Nostrand Reinhold, New York, p. 385.

de Groot, D. and Wurtz, D. (1990). Optimizing complex problems by nature's algorithms: Simulated annealing and evolution strategies – A comparative study. *PPSN, First International Workshop on Parallel Problem Solving from Nature*. October 1–3, 1990, Dortmund, Germany.

de Jong, K. and Spears, W. (1993). On the state of evolutionary computation. In, *Proceedings of the Fifth International Conference on Genetic Algorithms* (S. Forrest, Ed.). Morgan Kaufmann Publishers, San Mateo, California, pp. 618–623.

Devillers, J. (1995). Display of multivariate data using non-linear mapping. In, *Chemometric Methods in Molecular Design* (H. van de Waterbeemd, Ed.). VCH, Weinheim, pp. 255–263.

Devillers, J. and Putavy, C. (1996). Designing biodegradable molecules from the combined use of a backpropagation neural network and a genetic algorithm. In, *Genetic Algorithms in Molecular Modeling* (J. Devillers, Ed.). Academic Press, London, pp. 303–314.

Djouadi, Y. (1995). Resolution of cartographic layout problem by means of improved genetic algorithms. In, *Artificial Neural Nets and Genetic Algorithms* (D.W. Pearson, N.C. Steele, and R.F. Albrecht, Eds.). Springer-Verlag, Wien, pp. 33–36.

Domine, D., Devillers, J., and Chastrette, M. (1994a). A nonlinear map of substituent constants for selecting test series and deriving structure–activity relationships. 1. Aromatic series. *J. Med. Chem.* **37**, 973–980.

Domine, D., Devillers, J., and Chastrette, M. (1994b). A nonlinear map of substituent constants for selecting test series and deriving structure–activity relationships. 2. Aliphatic series. *J. Med. Chem.* **37**, 981–987.

Domine, D., Wienke, D., Devillers, J., and Buydens, L. (1996). A new nonlinear neural mapping technique for visual exploration of QSAR data. In, *Neural Networks in QSAR and Drug Design* (J. Devillers, Ed.). Academic Press, London (in press).

Dunn, W.J. and Rogers, D. (1996). Genetic partial least squares in QSAR. In, *Genetic Algorithms in Molecular Modeling* (J. Devillers, Ed.). Academic Press, London, pp. 109–130.

Easton, F.F. and Mansour, N. (1993). A distributed genetic algorithm for employee staffing and scheduling problems. In, *Proceedings of the Fifth International Conference on Genetic Algorithms* (S. Forrest, Ed.). Morgan Kaufmann Publishers, San Mateo, California, pp. 360–367.

Eshelman, L.J., Caruana, R.A., and Schaffer, J.D. (1989). Biases in the crossover landscape. In, *Proceedings of the Third International Conference on Genetic Algorithms* (J.D. Schaffer, Ed.). Morgan Kaufmann Publishers, San Mateo, California, pp. 10–19.

Feldman, D.S. (1993). Fuzzy network synthesis with genetic algorithms. In, *Proceedings of the Fifth International Conference on Genetic Algorithms* (S. Forrest, Ed.). Morgan Kaufmann Publishers, San Mateo, California, pp. 312–317.

Fogarty, T.C. (1989). Varying the probability of mutation in the genetic algorithm. In, *Proceedings of the Third International Conference on Genetic Algorithms* (J.D. Schaffer, Ed.). Morgan Kaufmann Publishers, San Mateo, California, pp. 104–109.

Fogarty, T.C., Vavak, F., and Cheng, P. (1995). Use of the genetic algorithm for load balancing of sugar beet presses. In, *Proceedings of the Sixth International Conference on Genetic Algorithms* (L.J. Eshelman, Ed.). Morgan Kaufmann Publishers, San Francisco, California, pp. 617–624.

Fogel, D.B. (1995). *Evolutionary Computation. Toward a New Philosophy of Machine Intelligence*. IEEE Press, Piscataway, p. 272.

Fontain, E. (1992a). The problem of atom-to-atom mapping. An application of genetic algorithms. *Anal. Chim. Acta* **265**, 227–232.

Fontain, E. (1992b). Application of genetic algorithms in the field of constitutional similarity. *J. Chem. Inf. Comput.* **32**, 748–752.

Forrest, S. (1993). Genetic algorithms: Principles of natural selection applied to computation. *Science* **261**, 872–878.

Gabbert, P.S., Markowicz, B.P., Brown, D.E., Sappington, D.E., and Huntley, C.L. (1991). A system for learning routes and schedules with genetic algorithms. In, *Proceedings of the Fourth International Conference on Genetic Algorithms* (R.K. Belew and L.B. Booker, Eds.). Morgan Kaufmann Publishers, San Mateo, California, pp. 430–436.

Ghoshray, S., Yen, K.K., and Andrian, J. (1995). Modified genetic algorithms by efficient unification with simulated annealing. In, *Artificial Neural Nets and Genetic Algorithms* (D.W. Pearson, N.C. Steele, and R.F. Albrecht, Eds.). Springer-Verlag, Wien, pp. 487–490.

Glen, R.C. and Payne, A.W.R. (1995). A genetic algorithm for the automated generation of molecules within constraints. *J. Comput.-Aided Mol. Design* **9**, 181–202.

Goldberg, D.E. (1989a). *Genetic Algorithms in Search, Optimization & Machine Learning*. Addison-Wesley Publishing Company, Reading, p. 412.

Goldberg, D.E. (1989b). Sizing populations for serial and parallel genetic algorithms. In, *Proceedings of the Third International Conference on Genetic Algorithms* (J.D. Schaffer, Ed.). Morgan Kaufmann Publishers, San Mateo, California, pp. 70–79.

Goonatilake, S. and Khebbal, S. (1995). *Intelligent Hybrid Systems*. John Wiley & Sons, Chichester, p. 325.

Grefenstette, J.J. and Baker, J.E. (1989). How genetic algorithms work: A critical look at implicit parallelism. In, *Proceedings of the Third International Conference on Genetic Algorithms* (J.D. Schaffer, Ed.). Morgan Kaufmann Publishers, San Mateo, California, pp. 20–27.

Güvenir, H.A. (1995). A genetic algorithm for multicriteria inventory classification. In, *Artificial Neural Nets and Genetic Algorithms* (D.W. Pearson, N.C. Steele, and R.F. Albrecht, Eds.). Springer-Verlag, Wien, pp. 6–9.

Haasdijk, E.W. (1993). Sex between models: On using genetic algorithms for inductive modelling. Master's thesis, Department of Computer Science, University of Amsterdam.

Haasdijk, E.W., Walker, R.F., Barrow, D., and Gerrets, M.C. (1994). Genetic algorithms in business. In, *Genetic Algorithms in Optimisation, Simulation and Modelling* (J. Stender, E. Hillebrand, and J. Kingdon, Eds.). IOS Press, Amsterdam, pp. 157–184.

Hämäläinen, A. (1995). Using genetic algorithm in self-organizing map design. In, *Artificial Neural Nets and Genetic Algorithms* (D.W. Pearson, N.C. Steele, and R.F. Albrecht, Eds.). Springer-Verlag, Wien, pp. 364–367.

Hansch, C. and Leo, A. (1979). *Substituent Constants for Correlation Analysis in Chemistry and Biology.* John Wiley & Sons, New York.

Harp, S.A., Samad, T., and Guha, A. (1989). Towards the genetic synthesis of neural networks. In, *Proceedings of the Third International Conference on Genetic Algorithms* (J.D. Schaffer, Ed.). Morgan Kaufmann Publishers, San Mateo, California, pp. 360–369.

Hartke, B. (1993). Global geometry optimization of clusters using genetic algorithms. *J. Phys. Chem.* **97**, 9973–9976.

Herdy, M. (1992). *Reproductive Isolation as Strategy Parameter in Hierarchically Organized Evolution Strategies.* FB Verfahrenstechnik, FG-Bionik & Evolutionsstrategie, September, TU-Berlin, Germany.

Herdy, M. and Patone, G. (1994). Evolution strategy in action, 10 ES-demonstrations. *International Conference on Evolutionary Computing, The Third Parallel Problem Solving from Nature (PPSN III).* October 9–14, 1994, Jerusalem, Israel.

Hibbert, D.B. (1993a). Genetic algorithms in chemistry. *Chemom. Intell. Lab. Syst.* **19**, 277–293.

Hibbert, D.B. (1993b). Generation and display of chemical structures by genetic algorithms. *Chemom. Intell. Lab. Syst.* **20**, 35–43.

Hightower, R.R., Forrest, S., and Perelson, A.S. (1995). The evolution of emergent organization in immune system gene libraries. In, *Proceedings of the Sixth International Conference on Genetic Algorithms* (L.J. Eshelman, Ed.). Morgan Kaufmann Publishers, San Francisco, California, pp. 344–350.

Hinterding, R., Gielewski, H., and Peachey, T.C. (1995). The nature of mutation in genetic algorithms. In, *Proceedings of the Sixth International Conference on Genetic Algorithms* (L.J. Eshelman, Ed.). Morgan Kaufmann Publishers, San Francisco, California, pp. 65–72.

Hintz, K.J. and Spofford, J.J. (1990). Evolving a neural network. In, *5th IEEE International Symposium on Intelligent Control* (A. Meystel, J. Herath, and S. Gray, Eds.). IEEE Computer Society Press, Philadelphia, pp. 479–483.

Hoffmeister, F. and Back, T. (1990). Genetic algorithms and evolution strategies: Similarities and differences. *PPSN, First International Workshop on Parallel Problem Solving from Nature*. October 1–3, 1990, Dortmund, Germany.

Höhn, C. and Reeves, C. (1995). Incorporating neighbourhood search operators into genetic algorithms. In, *Artificial Neural Nets and Genetic Algorithms* (D.W. Pearson, N.C. Steele, and R.F. Albrecht, Eds.). Springer-Verlag, Wien, pp. 214–217.

Holland, J. (1992). Les algorithmes génétiques. *Pour la Science* **179**, 44–51.

Hopfinger, A.J. and Patel, H.C. (1996). Application of genetic algorithms to the general QSAR problem and to guiding molecular diversity experiments. In, *Genetic Algorithms in Molecular Modeling* (J. Devillers, Ed.). Academic Press, London, pp. 131–157.

Horner, A. and Goldberg, D.E. (1991). Genetic algorithms and computer-assisted music composition. In, *Proceedings of the Fourth International Conference on Genetic Algorithms* (R.K. Belew and L.B. Booker, Eds.). Morgan Kaufmann Publishers, San Mateo, California, pp. 437–441.

Jaeger, E.P., Pevear, D.C., Felock, P.J., Russo, G.R., and Treasurywala, A.M. (1995). Genetic algorithm based method to design a primary screen for antirhinovirus agents. In, *Computer-Aided Molecular Design. Applications in Agrochemicals, Materials, and Pharmaceuticals* (C.H. Reynolds, M.K. Holloway, and H.K. Cox, Eds.). ACS Symposium Series 589, American Chemical Society, Washington, D.C., pp. 139–155.

Janikow, C.Z. (1995). A genetic algorithm for optimizing fuzzy decision trees. In, *Proceedings of the Sixth International Conference on Genetic Algorithms* (L.J. Eshelman, Ed.). Morgan Kaufmann Publishers, San Francisco, California, pp. 421–428.

Jones, A.J. (1993). Genetic algorithms and their applications to the design of neural networks. *Neural Comput. Applic.* **1**, 32–45.

Jones, T. (1995). Crossover, macromutation, and population-based search. In, *Proceedings of the Sixth International Conference on Genetic Algorithms* (L.J. Eshelman, Ed.). Morgan Kaufmann Publishers, San Francisco, California, pp. 73–80.

Jones, G., Willett, P., and Glen, R.C. (1996). Genetic algorithms for chemical structure handling and molecular recognition. In, *Genetic Algorithms in Molecular Modeling* (J. Devillers, Ed.). Academic Press, London, pp. 211–242.

Jones, G., Brown, R.D., Clark, D.E., Willett, P., and Glen, R.C. (1993). Searching databases of two-dimensional and three-dimensional chemical structures using genetic algorithms. In, *Proceedings of the Fifth International Conference on Genetic Algorithms* (S. Forrest, Ed.). Morgan Kaufmann Publishers, San Mateo, California, pp. 597–602.

Judson, R.S. (1992). Teaching polymers to fold. *J. Phys. Chem.* **96**, 10102–10104.

Judson, R.S., Colvin, M.E., Meza, J.C., Huffer, A., and Gutierrez, D. (1992). Do intelligent configuration search techniques outperform random search for large molecules? *Int. J. Quant. Chem.* **44**, 277–290.

Judson, R.S., Jaeger, E.P., Treasurywala, A.M., and Peterson, M.L. (1993). Conformational searching methods for small molecules. II. Genetic algorithm approach. *J. Comput. Chem.* **14**, 1407–1414.

Kapsalis, A., Rayward-Smith, V.J., and Smith, G.D. (1995). Using genetic algorithms to solve the radio link frequency assignment problem. In, *Artificial Neural Nets and Genetic Algorithms* (D.W. Pearson, N.C. Steele, and R.F. Albrecht, Eds.). Springer-Verlag, Wien, pp. 37–40.

Karr, C.L. (1991). Air-injected hydrocyclone optimization via genetic algorithm. In, *Handbook of Genetic Algorithms* (L. Davis, Ed.). Van Nostrand Reinhold, New York, pp. 222–236.

Karr, C.L. (1995). Genetic algorithms and fuzzy logic for adaptive process control. In, *Intelligent Hybrid System* (S. Goonatilake and S. Khebbal, Eds.). John Wiley & Sons, Chichester, pp. 63–83.

Keijzer, M.A. (1996). Representing computer programs in genetic programming. In, *Advances in Genetic Programming II* (K.E. Kinnear and P.J. Angeline, Eds.). The MIT Press, Cambridge (in press).

Kelly, J.D.Jr and Davis, L. (1991). Hybridizing the genetic algorithm and the K nearest neighbors classification algorithm. In, *Proceedings of the Fourth International Conference on Genetic Algorithms* (R.K. Belew and L.B. Booker, Eds.). Morgan Kaufmann Publishers, San Mateo, California, pp. 377–383.

Kidwell, M.D. (1993). Using genetic algorithms to schedule distributed tasks on a bus-based system. In, *Proceedings of the Fifth International Conference on Genetic Algorithms* (S. Forrest, Ed.). Morgan Kaufmann Publishers, San Mateo, California, pp. 368–374.

Kinnear, K.E.Jr (1994). *Advances in Genetic Programming*. The MIT Press, Cambridge, p. 518.

Kirkpatrick, S., Gelatt, C.D., and Vecchi, M.P. (1983). Optimization by simulated annealing. *Science* **220**, 671–680.

Kitano, H. (1994). Neurogenetic learning: An integrated method of designing and training neural networks using genetic algorithms. *Physica D* **75**, 225–238.

Kobayashi, S., Ono, I., and Yamamura, M. (1995). An efficient genetic algorithm for job shop scheduling problems. In, *Proceedings of the Sixth International Conference on Genetic Algorithms* (L.J. Eshelman, Ed.). Morgan Kaufmann Publishers, San Francisco, California, pp. 506–511.

Koza, J.R. (1992). *Genetic Programming. On the Programming of Computers by Means of Natural Selection*. The MIT Press, Cambridge, p. 819.

Kubinyi, H. (1994a). Variable selection in QSAR studies. I. An evolutionary algorithm. *Quant. Struct.-Act. Relat.* **13**, 285–294.

Kubinyi, H. (1994b). Variable selection in QSAR studies. II. A highly efficient combination of systematic search and evolution. *Quant. Struct.-Act. Relat.* **13**, 393–401.

Kurbel, K., Schneider, B., and Singh, K. (1995). VLSI standard cell placement by parallel hybrid simulated-annealing and genetic algorithm. In, *Artificial Neural Nets and Genetic Algorithms* (D.W. Pearson, N.C. Steele, and R.F. Albrecht, Eds.). Springer-Verlag, Wien, pp. 491–494.

Kussul, E.M. and Baidyk, T.N. (1995). Genetic algorithm for neurocomputer image recognition. In, *Artificial Neural Nets and Genetic Algorithms* (D.W. Pearson, N.C. Steele, and R.F. Albrecht, Eds.). Springer-Verlag, Wien, pp. 120–123.

Lane, A. (1995). The GA edge in analyzing data. *AI Expert* **June** issue.

Leardi, R. (1994). Application of a genetic algorithm to feature selection under full validation conditions and to outlier detection. *J. Chemometrics* **8**, 65–79.

Leardi, R., Boggia, R., and Terrile, M. (1992). Genetic algorithms as a strategy for feature selection. *J. Chemometrics* **6**, 267–281.

Li, T.H., Lucasius, C.B., and Kateman, G. (1992). Optimization of calibration data with the dynamic genetic algorithm. *Anal. Chim. Acta* **268**, 123–134.

Lohmann, R. (1990). Application of evolution strategy in parallel populations. *PPSN, First International Workshop on Parallel Problem Solving from Nature*. October 1–3, 1990, Dortmund, Germany.

Lohmann, R. (1991). Dissertation: Bionische Verfahren zur Entwicklung visueller Systeme. FB Verfahrenstechnik, FG-Bionik & Evolutionsstrategie, TU-Berlin, Germany.

Lucasius, C.B. and Kateman, G. (1991). Genetic algorithms for large-scale optimization in chemometrics: An application. *Trends Anal. Chem.* **10**, 254–261.

Lucasius, C.B. and Kateman, G. (1993). Understanding and using genetic algorithms. Part 1. Concepts, properties and context. *Chemom. Intell. Lab. Syst.* **19**, 1–33.

Lucasius, C.B. and Kateman, G. (1994). Understanding and using genetic algorithms. Part 2. Representation, configuration and hybridization. *Chemom. Intell. Lab. Syst.* **25**, 99–145.

Lucasius, C.B., Beckers, M.L.M., and Kateman, G. (1994). Genetic algorithms in wavelength selection: A comparative study. *Anal. Chim. Acta* **286**, 135–153.

Luke, B.T. (1994). Evolutionary programming applied to the development of quantitative structure–activity relationships and quantitative structure–property relationships. *J. Chem. Inf. Comput. Sci.* **34**, 1279–1287.

McFarland, J.W. and Gans, D.J. (1993). Hyperdimensionality in QSAR: Cutting problems to workable sizes. In, *Trends in QSAR and Molecular Modelling 92. Proceedings of the 9th European Symposium on Structure–Activity Relationships: QSAR and Molecular Modelling. September 7–11, 1992, Strasbourg, France* (C.G. Wermuth, Ed.). ESCOM, Leiden, pp. 317–318.

McFarland, J.W. and Gans, D.J. (1994). On identifying likely determinants of biological activity in high dimensional QSAR problems. *Quant. Struct.-Act. Relat.* **13**, 11–17.

McGarrah, D.B. and Judson, R.S. (1993). Analysis of the genetic algorithm method of molecular conformation determination. *J. Comput. Chem.* **14**, 1385–1395.

May, A.C.W. and Johnson, M.S. (1994). Protein structure comparisons using a combination of a genetic algorithm, dynamic programming and least-squares minimization. *Protein Eng.* **7**, 475–485.

Mendelsohn, L. (1994). Evolver review. *Technical Analysis of Stocks and Commodities Magazine* **December** issue.

Merelo, J.J. and Prieto, A. (1995). G-LVQ, a combination of genetic algorithms and LVQ. In, *Artificial Neural Nets and Genetic Algorithms* (D.W. Pearson, N.C. Steele, and R.F. Albrecht, Eds.). Springer-Verlag, Wien, pp. 92–95.

Mestres, J. and Scuseria, G.E. (1995). Genetic algorithms: A robust scheme for geometry optimizations and global minimum structure problems. *J. Comput. Chem.* **16**, 729–742.

Metropolis, N., Rosenbluth, A.W., Rosenbluth, M.N., Teller, A.H., and Teller, E. (1953). Equations of state calculations by fast computing machines. *J. Chem. Phys.* **21**, 1087–1092.

Michalewicz, Z. (1992). *Genetic Algorithms + Data Structures = Evolution Programs.* Springer-Verlag, Berlin, p. 250.

Michalewicz, Z. (1994). *Genetic Algorithms + Data Structures = Evolution Programs*, 2nd (extended) edition. Springer-Verlag, New York.

Michalewicz, Z. (1996). *Genetic Algorithms + Data Structures = Evolution Programs*, 3rd (extended) edition. Springer-Verlag, New York.

Michalewicz, Z. and Nazhiyath, G. (1995). Genocop III: A co–evolutionary algorithm for numerical optimization problems with nonlinear constraints. In, *Proceedings of the 2nd IEEE International Conference on Evolutionary Computation*, Perth, 29 November–1 December 1995.

Michalewicz, Z., Logan, T.D., and Swaminathan, S. (1994). Evolutionary operators for continuous convex parameter spaces. In, *Proceedings of the 3rd Annual Conference on Evolutionary Programming* (A.V. Sebald and L.J. Fogel, Eds.). World Scientific Publishing, River Edge, NJ, pp. 84–97.

Miller, B.L. and Goldberg, D.E. (1995). *Genetic Algorithms, Tournament Selection, and the Effects of Noise.* IlliGAL Report No. 95006, University of Illinois at Urbana-Champaign, Department of General Engineering, p. 13.

Miller, G.F., Todd, P.M., and Hegde, S.U. (1989). Designing neural networks using genetic algorithms. In, *Proceedings of the Third International Conference on Genetic Algorithms* (J.D. Schaffer, Ed.). Morgan Kaufmann Publishers, San Mateo, California, pp. 379–384.

Montana, D.J. (1995). Neural network weight selection using genetic algorithms. In, *Intelligent Hybrid Systems* (S. Goonatilake and S. Khebbal, Eds.). John Wiley & Sons, Chichester, pp. 85–104.

Nordvik, J.-P. and Renders, J.-M. (1991). Genetic algorithms and their potential for use in process control: A case study. In, *Proceedings of the Fourth International Conference on Genetic Algorithms* (R.K. Belew and L.B. Booker, Eds.). Morgan Kaufmann Publishers, San Mateo, California, pp. 480–486.

Ostermeier, A., Gawelczyk, A., and Hansen, N. (1993). *A Derandomized Approach to Self Adaptation of Evolution Strategies*. FB Verfahrenstechnik, FG-Bionik & Evolutionsstrategie, July 1993, TU-Berlin, Germany.

Payne, A.W.R. and Glen, R.C. (1993). Molecular recognition using a binary genetic search algorithm. *J. Mol. Graphics* **11**, 74–91.

Perrin, E., Bicking, F., Fonteix, C., Marc, I., Ruhaut, L., and Shingleton, M. (1993). Identification paramétrique d'un modèle hydrodynamique d'un bioréacteur à l'aide d'un algorithme de type génétique. In, *Conduite et Commande des Procédés*, *Vol. 7*. Grenoble 93, pp. 57–62.

Pleiss, M.A. and Unger, S.H. (1990). The design of test series and the significance of QSAR relationships. In, *Comprehensive Medicinal Chemistry*, *Vol. 4* (C.A. Ramsden, Ed.). Pergamon Press, Oxford, pp. 561–587.

Polani, D. and Uthmann, T. (1993). Training Kohonen feature maps in different topologies: An analysis using genetic algorithms. In, *Proceedings of the Fifth International Conference on Genetic Algorithms* (S. Forrest, Ed.). Morgan Kaufmann Publishers, San Mateo, California, pp. 326–333.

Porto, V.W., Fogel, D.B., and Fogel, L.J. (1995). Alternative neural network training methods. *IEEE Expert Intell. Syst. Appl.* **10 (June)**, 16–22.

Putavy, C., Devillers, J., and Domine, D. (1996). Genetic selection of aromatic substituents for designing test series. In, *Genetic Algorithms in Molecular Modeling* (J. Devillers, Ed.). Academic Press, London, pp. 243–269.

Qi, X. and Palmieri, F. (1993). The diversification role of crossover in the genetic algorithms. In, *Proceedings of the Fifth International Conference on Genetic Algorithms* (S. Forrest, Ed.). Morgan Kaufmann Publishers, San Mateo, California, pp. 132–137.

Quandt, S. (1992). *Evolutionsstrategien mit konkurrierenden Populationen*. Diplomarbeit, FB-Bionik & Evolutionsstrategie, FB Verfahrenstechnik, FG-Bionik & Evolutionsstrategie, November 1992, TU-Berlin, Germany.

Rawlins, G.J.E. (1991). *Foundations of Genetic Algorithms*. Morgan Kaufmann Publishers, San Mateo, California, p. 341.

Rechenberg, I. (1973). *Evolutionsstrategie: Optimierung technischer Systeme nach Prinzipien der biologischen Evolution*. Frommann–Holzboog Verlag, Stuttgart.

Rechenberg, I. (1989a). Evolution strategie: Nature's way of optimization. In, *Optimization: Methods and Applications, Possibilities and Limitations*. Lecture Notes in Engineering 47.

Rechenberg, I. (1989b). Artificial evolution and artificial intelligence. In, *Machine Learning: Principles and Techniques* (R. Forsyth, Ed.). Chapman and Hall, London.

Renders, J.M. (1995). *Algorithmes Génétiques et Réseaux de Neurones.* Hermès, Paris, p. 334.

Ribeiro Filho, J.L., Treleaven, P.C., and Alippi, C. (1994). Genetic–algorithm programming environments. *IEEE Computer* **27 (June)**, 28–43.

Ring, C.S. and Cohen, F.E. (1994). Conformational sampling of loop structures using genetic algorithms. *Israel J. Chem.* **34**, 245–252.

Robbins, P. (1995). The use of a variable length chromosome for permutation manipulation in genetic algorithms. In, *Artificial Neural Nets and Genetic Algorithms* (D.W. Pearson, N.C. Steele, and R.F. Albrecht, Eds.). Springer-Verlag, Wien, pp. 144–147.

Roberts, S.G. and Turega, M. (1995). Evolving neural network structures: An evaluation of encoding techniques. In, *Artificial Neural Nets and Genetic Algorithms* (D.W. Pearson, N.C. Steele, and R.F. Albrecht, Eds.). Springer-Verlag, Wien, pp. 96–99.

Rogers, D. (1991). G/SPLINES: A hybrid of Friedman's multivariate adaptive regression splines (MARS) algorithm with Holland's genetic algorithm. In, *Proceedings of the Fourth International Conference on Genetic Algorithms* (R.K. Belew and L.B. Booker, Eds.). Morgan Kaufmann Publishers, San Mateo, California, pp. 384–391.

Rogers, D. (1995). Development of the genetic function approximation algorithm. In, *Proceedings of the Sixth International Conference on Genetic Algorithms* (L.J. Eshelman, Ed.). Morgan Kaufmann Publishers, San Francisco, California, pp. 589–596.

Rogers, D. (1996). Some theory and examples of genetic function approximation with comparison to evolutionary techniques. In, *Genetic Algorithms in Molecular Modeling* (J. Devillers, Ed.). Academic Press, London, pp. 87–107.

Rogers, D. and Hopfinger, A.J. (1994). Application of genetic function approximation to quantitative structure–activity relationships and quantitative structure–property relationships. *J. Chem. Inf. Comput. Sci.* **34**, 854–866.

Rossi, I. and Truhlar, D.G. (1995). Parameterization of NDDO wavefunctions using genetic algorithms. An evolutionary approach to parameterizing potential energy surfaces and direct dynamics calculations for organic reactions. *Chem. Phys. Lett.* **233**, 231–236.

Rudolf, G. (1990). Global optimization by means of distributed evolution strategies. *PPSN, First International Workshop on Parallel Problem Solving from Nature*, October 1–3, 1990, Dortmund, Germany.

Schaffer, J.D. and Eshelman, L.J. (1991). On crossover as an evolutionarily viable strategy. In, *Proceedings of the Fourth International Conference on Genetic Algorithms* (R.K. Belew and L.B. Booker, Eds.). Morgan Kaufmann Publishers, San Mateo, California, pp. 61–68.

Schaffer, J.D., Caruana, A., and Eshelman, L.J. (1990). Using genetic search to exploit the emergent behavior of neural networks. *Physica D* **42**, 244–248.

Schaffer, J.D., Caruana, R.A., Eshelman, L.J., and Das, R. (1989). A study of control parameters affecting online performance of genetic algorithms for function optimization. In, *Proceedings of the Third International Conference on Genetic Algorithms* (J.D. Schaffer, Ed.). Morgan Kaufmann Publishers, San Mateo, California, pp. 51–60.

Schwefel, H.P. (1981). *Numerical Optimization of Computer Models*. John Wiley & Sons, Chichester.

Schwehm, M. (1993). A massively parallel genetic algorithm on the MasPar MP-1.

In, *Artificial Neural Nets and Genetic Algorithms* (R.F. Albrecht *et al.*, Eds.). Springer-Verlag, Wien, pp. 502–507.

Shamir, N., Saad, D., and Marom, E. (1993). Using the functional behavior of neurons for genetic recombination in neural nets training. *Complex Syst.* **7**, 445–467.

Sheridan, R.P. and Kearsley, S.K. (1995). Using a genetic algorithm to suggest combinatorial libraries. *J. Chem. Inf. Comput. Sci.* **35**, 310–320.

Spears, W.M. (1993). Crossover or mutation? In, *Foundations of Genetic Algorithms. 2.* (L.D. Whitley, Ed.). Morgan Kaufmann Publishers, San Mateo, California, pp. 221–237.

Spears, W.M. and de Jong, K.A. (1991). An analysis of multi-point crossover. In, *Foundations of Genetic Algorithms* (G.J.E. Rawlins, Ed.). Morgan Kaufmann Publishers, San Mateo, California, pp. 301–315.

Spofford, J.J. and Hintz, K.J. (1991). Evolving sequential machines in amorphous neural networks. In, *Artificial Neural Networks* (T. Kohonen, K. Makisara, O. Simula, and J. Kangas, Eds.). Elsevier Science Publishers B.V., North Holland, pp. 973–978.

Srinivas, M. and Patnaik, L.M. (1994). Genetic algorithms: A survey. *IEEE Computer* **27 (June)**, 17–26.

Stebel, W. (1991). *Studienarbeit Experimentelle Untersuchung von Evolutionsstrategie-Algorithmen im Rahmen der Optimierung optischer Systeme.* FB Verfahrenstechnik, FG-Bionik & Evolutionsstrategie, December 1991, TU-Berlin, Germany.

Stebel, W. (1992). *Diplomarbeit Entwicklung und Implementation von Methoden zur Strukturevolution, angewandt auf die Optimierung einfacher optischer Systeme.* FB Verfahrenstechnik, FG-Bionik & Evolutionsstrategie, December, TU-Berlin, Germany.

Sutton, P. and Boyden, S. (1994). Genetic algorithms: A general search procedure. *Am. J. Phys.* **62**, 549–552.

Swain, C.G. and Lupton, E.C. (1968). Field and resonance components of substituent effects. *J. Am. Chem. Soc.* **90**, 4328–4337.

Syswerda, G. (1989). Uniform crossover in genetic algorithms. In, *Proceedings of the Third International Conference on Genetic Algorithms* (J.D. Schaffer, Ed.). Morgan Kaufmann Publishers, San Mateo, California, pp. 2–9.

Syswerda, G. (1991). Schedule optimization using genetic algorithms. In, *Handbook of Genetic Algorithms* (L. Davis, Ed.). Van Nostrand Reinhold, New York, pp. 332–349.

Syswerda, G. (1993). Simulated crossover in genetic algorithms. In, *Foundations of Genetic Algorithms. 2.* (L.D. Whitley, Ed.). Morgan Kaufmann Publishers, San Mateo, California, pp. 239–255.

Syswerda, G. and Palmucci, J. (1991). The application of genetic algorithms to resource scheduling. In, *Proceedings of the Fourth International Conference on Genetic Algorithms* (R.K. Belew and L.B. Booker, Eds.). Morgan Kaufmann Publishers, San Mateo, California, pp. 502–508.

Tate, D.M. and Smith, A.E. (1993). Expected allele coverage and the role of mutation in genetic algorithms. In, *Proceedings of the Fifth International Conference on Genetic Algorithms* (S. Forrest, Ed.). Morgan Kaufmann Publishers, San Mateo, California, pp. 31–37.

Trint, K. and Utecht, U. (1992). *Methodik der Strukturevolution.* FB Verfahrenstechnik, FG-Bionik & Evolutionsstrategie, Seminar *"Ausgewahlte Kapitel zur Bionik und Evolutionsstrategie"*, January 31, 1992, TU-Berlin, Germany.

Tuffery, P., Etchebest, C., Hazout, S., and Lavery, R. (1991). A new approach to the rapid determination of protein side chain conformations. *J. Biomol. Struct. Dynam.* **8**, 1267–1289.

Unger, R. and Moult, J. (1993a). A genetic algorithm for 3D protein folding simulations. In, *Proceedings of the Fifth International Conference on Genetic Algorithms* (S. Forrest, Ed.). Morgan Kaufmann Publishers, San Mateo, California, pp. 581–588.

Unger, R. and Moult, J. (1993b). Genetic algorithms for protein folding simulations. *J. Mol. Biol.* **231**, 75–81.

Vahidov, R.M., Vahidov, M.A., and Eyvazova, Z.E. (1995). Use of genetic and neural technologies in oil equipment computer-aided design. In, *Artificial Neural Nets and Genetic Algorithms* (D.W. Pearson, N.C. Steele, and R.F. Albrecht, Eds). Springer-Verlag, Wien, pp. 317–320.

van Helden, S.P., Hamersma, H., and van Geerestein, V.J. (1996). Prediction of the progesterone receptor binding of steroids using a combination of genetic algorithms and neural networks. In, *Genetic Algorithms in Molecular Modeling* (J. Devillers, Ed.). Academic Press, London, pp. 159–192.

Varanelli, J.M. and Cohoon, J.P. (1995). Population-oriented simulated annealing: A genetic/thermodynamic hybrid approach to optimization. In, *Proceedings of the Sixth International Conference on Genetic Algorithms* (L.J. Eshelman, Ed.). Morgan Kaufmann Publishers, San Francisco, California, pp. 174–181.

Venkatasubramanian, V., Chan, K., and Caruthers, J.M. (1994). Computed-aided molecular design using genetic algorithms. *Comput. Chem. Engng* **18**, 833–844.

Venkatasubramanian, V., Chan, K., and Caruthers, J.M. (1995a). Evolutionary design of molecules with desired properties using the genetic algorithm. *J. Chem. Inf. Comput. Sci.* **35**, 188–195.

Venkatasubramanian, V., Chan, K., and Caruthers, J.M. (1995b). Genetic algorithmic approach for computer-aided molecular design. In, *Computer-Aided Molecular Design. Applications in Agrochemicals, Materials, and Pharmaceuticals* (C.H. Reynolds, M.K. Holloway and H.K. Cox, Eds.). ACS Symposium Series 589, American Chemical Society, Washington, D.C., pp. 396–414.

Venkatasubramanian, V., Sundaram, A., Chan, K., and Caruthers, J.M. (1996). Computer-aided molecular design using neural networks and genetic algorithms. In, *Genetic Algorithms in Molecular Modeling* (J. Devillers, Ed.). Academic Press, London, pp. 271–302.

Ventura, D., Andersen, T., and Martinez, T.R. (1995). Using evolutionary computation to generate training set data for neural networks. In, *Artificial Neural Nets and Genetic Algorithms* (D.W. Pearson, N.C. Steele, and R.F. Albrecht, Eds.). Springer-Verlag, Wien, pp. 468–471.

Vivarelli, F., Giusti, G., Villani, M., Campanini, R., Fariselli, P., Compiani, M., and Casadio, R. (1995). LGANN: A parallel system combining a local genetic algorithm and neural networks for the prediction of secondary structure of proteins. *CABIOS* **11**, 253–260.

Walker, R.F., Haasdijk, E.W., and Gerrets, M.C. (1995). Credit evaluation using a genetic algorithm. In, *Intelligent Systems for Finance and Business* (S. Goonatilake and P. Treleaven, Eds.). John Wiley & Sons, Chichester, pp. 39–59.

Walters, D.E. and Hinds, R.M. (1994). Genetically evolved receptor models: A computational approach to construction of receptor models. *J. Med. Chem.* **37**, 2527–2536.

Walters, D.E. and Muhammad, T.D. (1996). Genetically evolved receptor models (GERM): A procedure for construction of atomic-level receptor site models in the absence of a receptor crystal structure. In, *Genetic Algorithms in Molecular Modeling* (J. Devillers, Ed.). Academic Press, London, pp. 193–210.

Wehrens, R., Lucasius, C., Buydens, L., and Kateman, G. (1993). Sequential assignment of 2D-NMR spectra of proteins using genetic algorithms. *J. Chem. Inf. Comput. Sci.* **33**, 245–251.

Whitley, L.D (1993). *Foundations of Genetic Algorithms. 2.* Morgan Kaufmann Publishers, San Mateo, California, p. 322.

Whitley, L.D and Vose, M.D. (1995). *Foundations of Genetic Algorithms. 3.* Morgan Kaufmann Publishers, San Francisco, California, p. 336.

Wikel, J.H. and Dow, E.R. (1993). The use of neural networks for variable selection in QSAR. *Bio. Med. Chem. Lett.* **3**, 645–651.

Xiao, Y. and Williams, D.E. (1993). Genetic algorithm: A new approach to the prediction of the structure of molecular clusters. *Chem. Phys. Lett.* **215**, 17–24.

Xiao, Y.L. and Williams, D.E. (1994). Genetic algorithms for docking of actinomycin D and deoxyguanosine molecules with comparison to the crystal structure of actinomycin D-deoxyguanosine complex. *J. Phys. Chem.* **98**, 7191–7200.

Yip, P.P.C. and Pao, Y.H. (1995). A perfect integration of neural networks and evolutionary algorithms. In, *Artificial Neural Nets and Genetic Algorithms* (D.W. Pearson, N.C. Steele, and R.F. Albrecht, Eds.). Springer-Verlag, Wien, pp. 88–91.

Zeanah, J. (1994). Naturally selective. Axcelis Evolver 2.1. *AI Expert* **September**, 22–23.

Zeiri, Y., Fattal, E., and Kosloff, R. (1995). Application of genetic algorithm to the calculation of bound states and local density approximations. *J. Chem. Phys.* **102**, 1859–1862.

2 An Overview of Genetic Methods

B.T. LUKE
International Business Machines Corporation, 522 South Road, Poughkeepsie, NY 12601, USA

Genetic methods (genetic algorithms, evolutionary strategies, evolutionary programming, and their variants) represent powerful procedures that can be used to optimize many problems associated with drug design. As opposed to many other optimization methods, these genetic methods store several solutions (i.e., a population) and use them to generate new solutions. This chapter outlines these procedures and describes various options that can be used to customize the algorithm for a particular problem. These options include how the solution is described, which mating, mutation, and/or maturation operators to use, and how new solutions (or offspring) are included in the population. A brief discussion of various published algorithms is also given.

KEYWORDS: *genetic methods; genetic algorithm; evolutionary strategy; evolutionary programming; overview.*

INTRODUCTION

Genetic methods are increasingly becoming popular as procedures to solve difficult combinatorial problems. They can be used to develop one or more quantitative structure–activity relationship (QSAR) or quantitative structure–property relationship (QSPR) models, search for pharmacophores in a database of structures, plan synthesis routes, perform conformational searches and docking simulations, and many other problems involved with drug development. These methods differ from other search strategies in that they use a collection of intermediate solutions. These solutions are then used to construct new, and hopefully improved, solutions to the problem. In biological terms, individual solutions are likened to *animals* in a *population*. Various processes are used to generate *offspring* that increase the *fitness* of the population.

In, *Genetic Algorithms in Molecular Modeling* (J. Devillers, Ed.)
Academic Press, London, 1996, pp. 35–66.
ISBN 0-12-213810-4

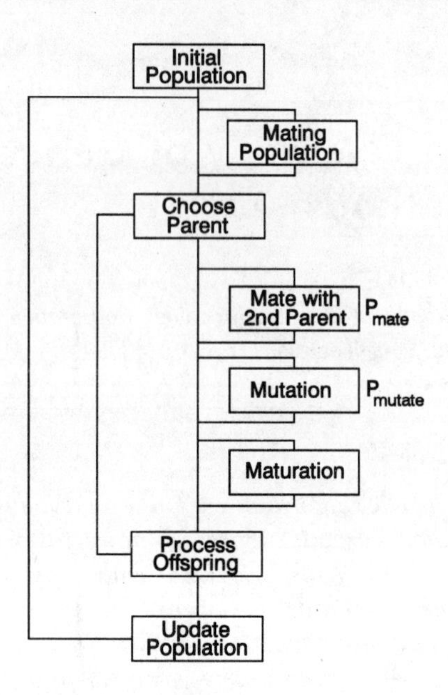

Figure 1 A flowchart of operations that can be used in a genetic method algorithm.

All of the genetic methods described here contain some of the operators depicted in the flowchart shown in Figure 1. Much of what follows will concentrate on describing different options that can be envoked for each of the boxes in this flowchart. Before this can be done, it is necessary to determine how an individual solution, or animal, is described. In general, a solution and how well it solves the problem (e.g., an animal and its fitness) is determined by an n-dimensional set of values. The possible choices for these values represent the *genetic alphabet*, and the collection of these values are the *genes*. The next section deals with the possible forms of the genetic alphabet and how different choices can affect the structure, and difficulty, of the problem. Once this is done, a detailed discussion of each operation presented in Figure 1 is given. In order to understand the possible consequences of each choice, it will first be necessary to describe two terms that will be used throughout this chapter; *focusing* and *similarity*. These terms will be defined before the discussion of each operation since they will be used to compare and contrast different options.

Two general classifications of genetic methods have emerged. The first is genetic algorithms (GAs), and its variants, and the second is evolutionary programming (EP). The major difference between them is that GAs rely on the mating of two solutions to create an offspring while evolutionary programming produces an offspring by mutation of a single solution. Using the notation of Figure 1, GAs use a high mating probability, P_{mate}, and a very low mutation probability, P_{mutate}, while evolutionary programming sets P_{mate} to 0.0 and P_{mutate} to 1.0. When the discussion of each possible operation is concluded, a brief review of Figure 1 and various published algorithms will be presented. This includes a discussion of implementations that can effectively use parallel computers.

Throughout this chapter, the term *NPOP* will denote the number of solutions present in the current population. In addition, each time *NPOP* offspring are generated and added to the population or stored in a new population, a *generation* will have occurred.

The following points about the use of genetic methods will be continually stressed.

1. Not all problems can be efficiently solved using genetic methods. This overview does not want to imply that genetic methods are 'the method of choice'. In particular, if a solution to the particular problem cannot be represented as a string of values, genetic methods cannot be used. In addition, there are many problems that already possess very specialized solution strategies, such as minimizing a relatively simple function. In this case, a specialized strategy will be more effective. Genetic methods can be used in conjunction with other strategies to either produce alternative solutions, and hopefully new insights, to problems, or verify solutions obtained by the other methods.
2. Genetic methods, as depicted in Figure 1, represent a computational methodology, not a computational algorithm. Each of the operations shown in Figure 1 can be performed in many ways. In addition, there can be different coding schemes and extensions of the method beyond what is shown in this figure. Each choice can change the computational speed and ultimate solution of the strategy. Therefore, it can never be said that a genetic method is better, or worse, than another method, only that a particular application of a genetic method is better or worse.
3. Because of all the possible choices, the problem should dictate the particular solution algorithm, not *vice versa*. This means that applying existing computer programs to problems associated with drug design is not recommended, though they do serve a useful function as instructional material and a possible starting point for a problem-specific algorithm. The problem should dictate the coding scheme for a solution; the problem should not be forced to accept a solution in the form dictated by the computer program. In addition, the problem should determine the form, or existence, of the mating, mutation and maturation operators and all other choices needed to uniquely determine the algorithm.

Genetic methods represent an extremely useful computational method. This chapter is unable to enumerate all possible options for each operator and the hope is to convince the reader that the only limit is our imaginations. An operator not discussed here may be more useful for a particular problem. Though it may not yield the best search algorithm, it may provide additional insight into the nature of the problem and may suggest other operators.

GENETIC ALPHABET AND GENES

As stated in the introduction, genetic methods work on a population of solutions and each solution is represented by a string of values. The coding scheme, and the representation of the values, have an effect on the form of the mating and mutation operators. Three coding schemes are possible, and will be denoted *gene based*, *node based* and *delta coding*. In a gene based scheme, each value in the string represents an independent variable. This means that each value can take on any allowed element of the genetic alphabet, independently of the elements used for the other values. This is the most common coding scheme, but is limited to only certain types of problems.

In a node based scheme, the string represents a path or route. This coding scheme is necessary for the traveling salesman problem (TSP) and in other problems associated with drug design, as discussed below and in other chapters. In the TSP, each city can only be included once in a route. This means that the values of the genetic string are not independent.

Delta coding can only be used for problems that can also accept gene based coding. In applications that use delta coding, the first iteration of the genetic method should use a gene based coding scheme. After the first iteration, the best solution to date is used as a template. A new initial population is created, but this time each value in the genetic string represents a change, or delta, relative to the template. The fitness of a solution is determined by applying the string of deltas to the template values, obtaining a solution, and calculating the accuracy of that solution.

Four different forms, or genetic alphabets, can be used to represent each value in a gene based coding scheme. The two most obvious forms for the values in the genetic string are integers and real numbers. Here, the genetic vector, or genetic array, is simply an n-dimensional array of integers or real numbers, respectively.

The other two forms represent these numbers as binary strings. The first uses standard binary coding, while the second uses Gray coding. Gray coding is a one-to-one mapping of a binary representation such that consecutive numbers differ by a single bit. This mapping is not unique (Mathias and Whitley, 1994), but the Gray code, g(i), most often used in genetic methods can be obtained from the standard binary representation, b(i), by numbering the bit positions from 1 to k, where k is the length of the bit string, with b(1) being the highest order bit and using the following rules.

g(1) = b(1)
for i = 2 through k
g(i) = b(i) if b(i – 1) = 0
g(i) = COMPLEMENT(b(i)) if b(i – 1) = 1

Similarly, a Gray coded number can be converted to its standard representation by

b(1) = g(1)
for i = 2 through k
b(i) = g(i) if b(i – 1) = 0
b(i) = COMPLEMENT(g(i)) if b(i – 1) = 1

Each representation for the values in a genetic vector has an effect on the 'landscape' of the function being to minimize or maximize. As an example, a gene based coding scheme can be used to examine the preferred conformation of a molecule. This could be the preferred conformation of an isolated molecule (i.e. finding the global minimum structure of a molecule), the conformation of a molecule that reproduces internuclear distances found from a 2D NMR experiment, or the conformation of a putative substrate that interacts best with a known binding site.

The structure of the molecule can be uniquely defined by setting the torsion angle for each 'rotatable bond'. Therefore, the dimensionality of the problem is equal to the number of rotatable bonds in the molecule. The objective function is the internal energy of the molecule, the level of agreement with the internuclear distances, or the energy of the substrate-binding site complex. A solution (i.e., conformation) is more fit if it makes the objective function smaller.

If the values of the dihedral angles are represented as real numbers, the solution can be exact, but the genetic alphabet contains an infinite number of entries in the range [0.0,360.0°]. If the genetic vector is a string of integers, the genetic alphabet contains exactly 360 members, [0,359], but the solution will be less precise. The same argument holds for binary coding (either form), but the degree of precision can be controlled by the number of bits used to encode a number. If N is the number that is encoded by k bits, the value of this dihedral angle is given by $360.0 \times N/2^k$. You can either say that the dimensionality of the problem is still the number of rotatable bonds and the genetic alphabet contains 2^k elements, or that the dimensionality has been increased to k times the number of rotatable bonds and the genetic alphabet only contains 2 elements (0,1). By increasing k the solution will be more precise, but either the size of the genetic alphabet or the dimensionality of the problem will be increased. Studies have shown that large binary strings can slow down the resulting optimization (Whitley *et al.*, 1990; Alba *et al.*, 1993).

The difference between binary and integer or real coding can be shown by assuming that there is a single minimum in the vicinity of 0° for a particular dihedral angle. For integer and real number coding, 359° and 0° are only 1° apart and will represent the same well. If an 8-bit standard binary coding are used, the minimum and maximum values (0 and 358.6°, respectively) are given by 00000000 and 11111111, respectively. These two representations have the largest possible Hamming distance, and will make the problem look as though it has two minima. One would be approached as more 0s are added from right-to-left, and the other as more 1s are added. Gray coding represents the smallest and largest 8-bit numbers as 00000000 and 10000000, respectively. Since these strings differ in only one position, their Hamming distance is smaller and they, once again, make the problem appear to have a single minimum.

An example counter to the one given above would be a situation where the molecule has minima in the vicinity of 67.5° and 291°. If either integer or real number coding is used, these minima are very well separated and will remain distinct. If 8-bit binary coding is used, these numbers correspond to the binary representations for 48 and 207, respectively. Their standard binary codings are 00110000 and 11001111, respectively. These strings differ in all 8 places, maximizing the distance between these minima. The Gray coding for these numbers is 00101000 and 10101000, respectively. These numbers differ in only a single bit position. Since seven contiguous bits are identical, these two values are right next to each other in search space. The effect of this will be to merge two distinct minima into a single minima and may severely hamper a conformational search.

This discussion should not suggest that binary coding should not be used; only that changing the genetic alphabet from real or integer coding to binary, either standard or Gray, changes the landscape of the objective function being to minimize or maximize. This changing of the landscape with different genetic alphabets may increase the applicability of genetic methods since, if the method does not perform well with a particular genetic alphabet, it may perform much better if a different alphabet is used. In addition, it has been shown that Gray coding can also affect the performance of other optimization methods (Mathias and Whitley, 1994).

Along with conformational studies outlined above, gene based coding can be used in QSAR/QSPR/decision support studies. Given 16 descriptors and the activity for a series of compounds, the string (0,0,1,0,0,0,0,1,2,0,0,1,0,0,0,0) codes for a predictor that uses 4 descriptors (descriptors 3, 8, 9, and 12), with a linear dependence on descriptors 3, 8, and 12 and a quadratic dependence on descriptor 9.

A gene based (integer) coding can also be used in a location–allocation problem. An example of this would be a drug company that contains multiple chemical stockrooms and several geographically separate laboratories or research groups, either dispersed across a single site, in various regions of a country, or world wide. The object would be to store needed chemicals as close as possible to each end-user, accounting for limitations of physical space in the stockroom and any regulatory restrictions. Though this may be outside the topics presented in the rest of this book, it still is an area that may affect the overall drug discovery process. For such a problem, a possible solution can be coded as a 2-dimensional integer array, S(i,j). Each row could correspond to a particular end-user, while each column to a particular chemical. S(3,7) = 4 would mean that end-user 3 would obtain chemical 7 from stockroom 4. The distance from stockroom to end-user for each required chemical, at each required interval, can be used as the metric to determine the cost of a particular solution, and therefore its fitness.

An integer genetic alphabet appears to be the only choice for a node based coding scheme. One problem where node based coding can be used is trying to map one molecule or fragment on another. This can be either a search through a database for a particular substructure or synthesis planning by finding certain substructures within a desired molecule. In a substructure search through a database, for example, the string (3,9,18,5,...) states that atoms 3, 9, 18, and 5 of the compound in the database are mapped to the first 4 atoms of the template structure. The connectivity of the chosen atom can be compared to that of the template for a 2-dimensional search, or its spacial orientation can be compared to the desired one in a 3-dimensional search.

As stated above, delta coding can be used for any problem that can use a gene based coding scheme. The difference between gene based and delta coding is the landscape of the objective function. In a gene based coding problem, the desired optimum value can be anywhere in search space. If delta coding is used, the optimum solution should eventually have a delta vector that is totally composed of zeros. This would mean that the template solution represents the best solution. Another difference between delta and gene based coding is the range of each search dimension that can be examined. K.E. Mathias and L.D. Whitley (1995, private communication) suggest that the maximum of each delta value be decreased each time the template structure changes, and increases if it stays the same from one iteration to the next. The first will concentrate the search if a new region is found, while the second makes sure that the template structure is not one that is in a suboptimal solution but is held there because the search region is not large enough. Since delta coding can be used in any gene based coding problem, it is also able to use any of the genetic alphabets described earlier.

One last example will be presented here, mainly to show that the same problem can be coded in more than one way (as well as using more than one genetic alphabet). This would be a QSAR-type study where the activity may be very well measured, but the other descriptors are very imprecise. This imprecision may be due to experimental or theoretical uncertainties in the measurement, or because the available descriptors are obtained from patient questionnaires and are subjective measures. In this case, obtaining a quantitative relationship may be impractical and the different compounds should simply be placed into groups. Possible solutions to a grouping problem can be coded in many different ways. One way is to use a node based coding scheme. For nine compounds, the string (4,3,5,0,1,6,7,0,8,2,0,9) means that compounds 3, 4, and 5 are placed in one group; 1, 6, and 7 in a second; 2 and 8 in a third and 9 in a fourth group (a 0 delineated one group from the next). The fitness of this grouping would be determined from how well the descriptors are able to form these groups and whether or not compounds that act similarly are placed in the same group. A completely different representation of a solution can be based upon the descriptors. One possibility would be to use a mixed genetic alphabet, gene based coding scheme. A solution could be represented by the string [2,(3,2,5.0),(5,3,B,D)]. Here, the initial 2 means that 2 descriptors will be used to divide the compounds into groups. The first is descriptor 3 and the compounds will be grouped by those that have a 'score' of 5.0 or less (the 2 means that there are 2 regions, so only one number follows) and those with a score of more than 5.0. The second descriptor is 5 and it will be composed of 3 regions; those with a score of B or less (i.e., A or B), those greater than B and less than or equal to D (i.e., C or D), and those greater than D. It is not necessary that one or more compounds lie in each of the 6 possible groups, only that similar compounds lie in the same group.

This section hopefully makes the following points.

1. The problem should dictate which genetic alphabet and which coding scheme can be used. It is possible to treat the same problem using more than one coding scheme (as shown by the example above), especially since delta coding can be used in all situations where gene based coding is appropriate.
2. Changing the genetic alphabet, and coding scheme, changes the landscape of the objective function being examining. If one alphabet/coding does not do well, another combination may.

FOCUSING AND SIMILARITY

As outlined in the preceding section, solutions currently present in a population are composed of a string of values that have a length equal to the dimensionality of the problem (in binary coding, the length is the number of bits per value times the dimensionality). As stated in the introduction, the major action of GAs is to combine two solutions to make new solutions. If, for example, all solutions in the population, or mating population (see Figure 1), contain the same value in a particular position of the genetic string, no combination of solutions can possibly introduce another value into that position. This effectively reduces the dimensionality of the search problem by one, since that value would forever be fixed. This reduction of the dimensionality of the problem will be referred to as *focusing*.

Focusing is at the heart of the strength of GAs, but can also cause convergence to suboptimal solutions. As an example, we will take a situation where the solutions are a 6-bit binary string. If all members of the population have a 1 in the third and fourth positions, and a 0 in the sixth, they all belong to the (**11*0) *schema*, where the * positions can be either a 0 or 1. It is hoped that this schema also contains the optimal solution and effectively reduces the problem from 6-dimensional to 3-dimensional. Building good schema is linked to the effectiveness of GAs in searching high-dimensional space for optimal solutions.

Assume, for example, that this binary coding causes the problem to possess only two minima; a deep, narrow minima at (111111) and a less deep, wide minima at (000000). If at any time in the search process all solutions in the population contain a 0 at a particular position, it will be impossible to find the optimal solution by simply combining existing ones. The best that can be hoped for is to find the suboptimal solution at (000000).

As in the preceding example, focusing is a property that is related to the population as a whole, and may help or hinder the overall results. Instead, it is possible to structure the algorithm such that the total population contains several, potentially overlapping, regions of focus. Creating multiple regions of focus is refered to as *niching*, and may represent an excellent compromise. Niching can be promoted by constructing separate sub-populations that are able to independently focus and occasionally share information with one another, or by carefully choosing mating partners and/or deciding how the offspring are incorporated into the population.

Though focusing is a process that acts on the population as a whole, *similarity* is a measure of how different an offspring is from its parent(s). If the offspring is generated by combining parts of two solutions (as is done in a GA), the offspring should statistically be like either of its parents in about 50% of the values. Inherent in this statement is the assumption that each parent is completely dissimilar from the other. As the number of values present in the schema that is common to all solutions in the population increases, the parents become more similar, and the offspring will become more similar to the parents.

EP produces an offspring by taking a single solution and changing some of the values. The similarity between parent and offspring depends totally upon the number of values selected to change and the maximum allowed size of each change. In the limit of changing all values by an amount that would allow searching the full space of each, EP becomes a random search. This allows for the maximum possible dissimilarity between parent and offspring, but leads to very slow convergence to a good solution and is not recommended. The only point that is being made is that the user has control over the 'search step size'. Using a large step size (making the offspring very dissimilar to the parent) will increase the probability of searching new regions of solution space, and slow convergence, while a small step size (making the offspring quite similar to the parent) increases the convergence, but limits the regions of solution space that is examined. If the step size is too small, convergence to a suboptimal solution is quite possible.

From the above discussion, the user should construct a solution strategy that balances the need to explore all of solution space while allowing for convergence on an optimal solution. It is also necessary for the user to determine whether the process should result in a single optimal solution or a collection of good solutions. If multiple good solutions are desired, using a process that promotes schema formation over the entire population may not be advisable.

CREATING AN INITIAL POPULATION

From the discussion presented in the preceding section, it should be clear that the overall structure of the genetic method should be chosen to effectively solve the problem at hand, and that no single algorithm should be effective in all combinatorial problems associated with drug design (Wolpert and Macready, 1995). In creating an initial population, two decisions need to be made; the size of the population and where the initial guesses at the solution come from.

If the genetic operators have a strong focusing effect, it may be advisable to use a rather large population size. This will improve the chances of at least one initial solution having a relatively high fitness and having many values in common with the optimal solution. If the initial population is too small, a relatively good initial solution may only have many values in common with a suboptimal solution and chances will be high that this solution, and not the optimal one, will be the one found. Conversely, if the size of the population is too large, an initial solution with many values in common with the optimal one may be overwhelmed by the shear number of other solutions and may not be able to have an impact on the overall population. In other words, studies have shown that if good solutions are too 'diluted' by less fit ones, the good solutions may actually disappear after several generations (Cecconi and Parisi, 1994).

If the genetic operator has a nonfocusing effect, a smaller population may suffice. This will also depend upon how similar the offspring is to the parent(s). If this similarity is high, and the offspring displace other solutions in the population, it is possible that the population will converge on a solution that is close (in search space) to an initial solution. Here, the population size should also be large so that there is a better chance that one of the initial solutions will be in the vicinity of the optimal one.

In general, these initial solutions can come from a purely random guess, random guesses that satisfy some minimal fitness criteria, or from the results of other calculations or studies. Just as for population size, there is no single best method for generating initial populations for all problems associated with drug design. Different methods will have to be tried with each initial population size and set of genetic operators to see which method yields the best results.

One point that should be made is that better initial populations will not necessarily yield better final solutions. If the initial solutions lie in pronounced suboptimal wells of the search space in a minimization problem and the genetic operators are focusing and/or yield offspring with a high similarity to their parent(s), the simulation may not be able to escape the wells and find the optimal solution. The more fit the initial solutions, the deeper within the wells they are, and the harder it may be to get them out. On the other hand, the previous section shows how this solution landscape can be changed by simply changing the representation of the solutions, and simply using a different genetic alphabet and/or coding scheme may produce an optimal solution.

BUILDING A MATING POPULATION

In many applications of genetic methods presented in the literature, a *selection* process occurs where members of the current population are placed in a mating population, with very fit members being placed more often than less fit ones. This mating population is then used to create offspring. In the following examples, the mating population is chosen to be the same size as the current population, *NPOP*. Though this is the dominant choice, it isn't the only one possible. If the mating population has a size of *NPOP*, certain solutions in the current population (usually those with a low fitness) will never make it to the mating population and never transfer their genetic information to offspring. Instead, it is possible to copy all solutions in the current population to the mating population and then add extra copies of fit solutions.

Several selection strategies are available, and some will be described here. The first is to simply use some linear ramping function. For example, if the population size was selected to be 100, the best solution can be copied to the mating population 6 times, the second best 5 times, the third best 4 times, the fourth best 3 times and the fifth best 2 times. Since the mating population now contains 20 solutions, members 6 through 85 (assuming they are pre-ordered by fitness) can be copied once to the mating population. This ramping can be more severe such that fewer different solutions are placed in the mating population and better ones are placed more often.

A second selection procedure has been presented by Whitley (1989). Here, the population is ordered from best to worst and the number of the animal selected is determined by the formula

$$I = NPOP \times (\alpha - (\alpha^2 - 4.0 \times (\alpha - 1.0) \times \mathrm{RAN})^{1/2})/(2.0 \times (\alpha - 1.0))$$

where *NPOP* is the number of solutions in the population, RAN is a random number between 0.0 and 1.0, and α is a user defined value that is greater than 1.0. As α increases in magnitude, the resulting index I, is more skewed towards 1.

Another selection procedure that is heavily used is called *tournament selection*. In this procedure, the user has to select a tournament size. If this size is 3, for example, three members of the current population are selected at random. The solution with the highest fitness of the three is placed in the mating population. If the tournament size is increased from 1 to the population size, the mating population will change from one that is likely to resemble the original population to one that it completely composed of the most fit, or *dominant*, solution.

The main thing to realize is that selecting a mating population is a focusing process. Good individual values of otherwise bad strings may be removed from the population by this process before they have a chance to be combined with other good values through mating or mutation. This should really increase the rate of convergence of the population, but may eventually yield a suboptimal solution. If selection is used, care should be taken to ensure that the dominant solution in the initial population does not contribute too heavily to all future offspring. This can be done by not biasing the selection process too heavily towards the most fit solutions and/or allowing mutations to play a large role in generating offspring.

CHOOSING A PARENT

In most applications of genetic methods, the first, or only, parent is chosen at random. Naturally, if a mating population is used, this parent is not completely random. If no mating population is used, this first parent can still be chosen using any of the biased selection procedures presented in the preceding section. In a standard EP algorithm, each parent is sequentially chosen and allowed to produce a single offspring through mutation. Other options are available and some will be discussed when various published algorithms are presented.

MATING

Mating is an operation where two solutions are combined to generate one or more new solutions. There are two steps in this operation; the first is choosing the second parent and the other is combining the solutions. In principle, any of the processes used to choose the first parent can be used to choose the second. In practice, constraints are placed on how the second parent is chosen. The set of constraints most often used is to choose the second parent based upon its fitness; the more fit the solution, the greater its chances of being chosen as the second parent (which can be a random choice if a mating population is used). Another option is to choose the second parent based upon its similarity to the first. This latter option will be discussed more thoroughly when niching algorithms are described.

The form of the mating operator will be different depending upon whether a gene based or a node based coding scheme is used. For a gene based coding, a crossover operator is used. The simplest operator is called a *one-point crossover*. With this operator, a cut point is chosen along the string of values, and all values to the right of the cut point in each solution are exchanged. For example

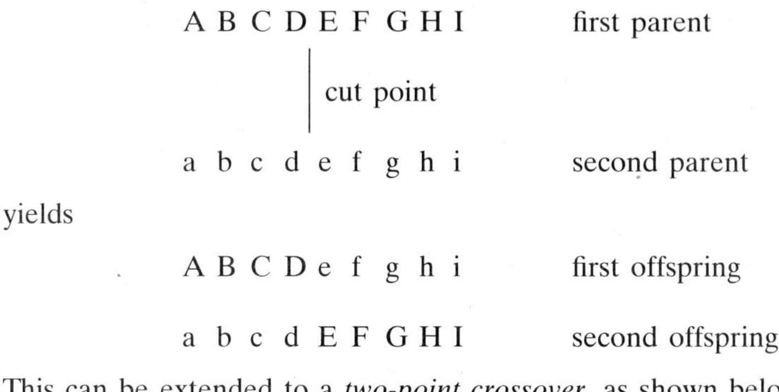

This can be extended to a *two-point crossover*, as shown below.

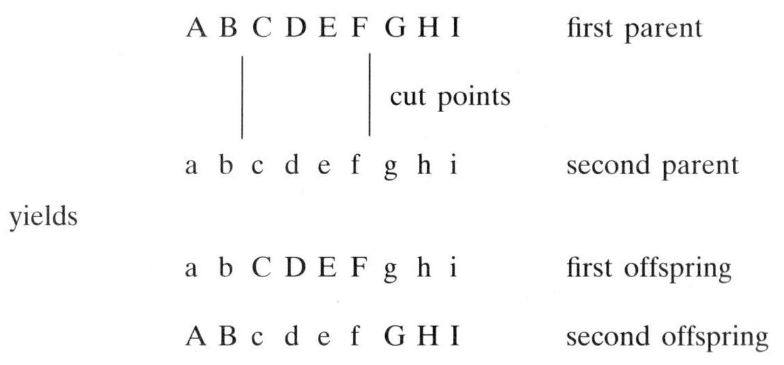

The number of cut points can be continually increased until each value for a given offspring can be independently chosen from one parent or the other, the other offspring simply receives the complementary value. This is called a *uniform crossover*. If there is an even probability that a given value will come from one parent or the other, it is called an *unbiased uniform crossover*. If there is a greater probability of the values coming from one parent rather than the other, it is a *biased uniform crossover* (J. Culberson, 1994, private communication).

If the two parents have different values in k positions of the genetic vector, a k-dimensional hyper-rectangle can be drawn in search space where the diagonal is given by the parents. The crossover operator simply places the offspring at one of the vertices of this hyper-rectangle. If a uniform crossover is used, an offspring can be any of the 2^k vertices. At the other extreme, a one-point crossover only allows an offspring, or its complement, to be one of $2(k - 1)$ possible vertices.

The type of crossover (e.g., the number of crossover points allowed) selected should be determined by the problem since it affects the search. Assume, for example, the goal is to choose a set of 11 descriptors out of a collection of 100 that accurately predicts six different properties or activities. Also assume that the best, or dominant, set of descriptors found to date is

$$(3,12,26,33,39,51,67,72,77,83,97)$$

If a one-point crossover is used, there is a 90% probability that adjacent descriptors will be transferred to the same offspring, and a 0% probability that the first and last descriptors will be transferred to the same offspring from this parent. Similarly, if 3 and 26 are good descriptors and 12 is not, there is no way to only place 3 and 26 into an offspring from this parent. In this case, a multi-point or uniform crossover may be preferable (note that a uniform crossover is mathematically equivalent to randomizing the descriptor labels and then performing a one-point, or any-point, crossover).

There are other cases where correlating the placement of adjacent values into an offspring may improve the results. One such application would be conformational searching or finding the best interaction between a putative substrate and a binding site. Here, the values would represent the dihedral angles of the rotatable bonds in the substrate and a one-point crossover would preserve much of the local geometry in favorable conformations and only affect the longer-range interactions within the molecule (assuming the dihedral angles were stored left-to-right, or right-to-left, in the genetic string and not randomly placed).

Another mating operator that can be used in a gene based (non-binary) coding problem is called *intermediate recombination*. If the value at string position i is given by a_i and b_i for the two parents, the value for an offspring, c_i, is given by either

$$c_i = a_i + u(b_i - a_i)$$

or

$$c_i = a_i + u_i(b_i - a_i)$$

In the first equation, a single random number u, between 0 and 1 is used to place all values of the offspring between those parents while in the second, a different random number is used for each value. This operator has the effect of allowing the offspring to occupy any point in the interior of the

k-dimensional hyper-rectangle denoted by the parents, and not just one of the vertices.

This can be extended to an *interval crossover* operator, where the offspring's values are determined from either

$$c_i = a_i + u(b_i - a_i + 2\delta) - \delta$$

or

$$c_i = a_i + u_i(b_i - a_i + 2\delta_i) - \delta_i$$

where it is assumed that $a_i < b_i$. Here, δ (or δ_i) is usually a small number, though this is not required, and allows the offspring to move outside the range of values defined by the parents. This increases the dimensionality of the hyper-rectangle to the full dimensionality of the search space since c_i can be different from a_i even if $a_i = b_i$. The line connecting the two parents no longer represents the diagonal, but the center of the hyper-rectangle is still the average of the two parents. The 'width' of the rectangle in a particular direction is determined by the magnitude of the difference between the parents in that value and the size of δ_i.

For a node based coding problem, the crossover operator is slightly different. Since each value in the genetic alphabet must be represented once, a check has to be done at the end to remove any duplicates. There are many different possibilities, and only some of them will be presented here. In each example, the following two parents will be used.

A B C D E F G H I J	first parent
B D E J H G A F C I	second parent

1. Cut the first parent at a random point and throw away all node values to the right. Append the second parent to the end of the first parent fragment and remove the duplicates from the second parent.

A B C D | E F G H I J	cut point
A B C D B D E J H G A F C I	discard and append
A B C D E J H G F I	remove duplicates

2. Choose two random cut points in the first parent and only keep the node values between the points. Fill in the missing values by examining the second parent from left to right.

A B | C D E F G | H I J	cut points
* * C D E F G * * *	discard
B J C D E F G H A I	add from second

3. Choose one cut point in the first parent and two in the second. Insert the node values between the cut points of the second at the cut point in the first, and delete all duplicates found when reading from left-to-right.

A B C D | E F G H I J	first parent cut point
B D E | J H G A | F C I	second parent cut points
A B C D J H G A E F G H I J	insert
A B C D J H G E F I	remove duplicates

Each of the procedures outlined above, or any others, can be done again, reversing the first and second parent, to create a second offspring. Which operator to use in a particular algorithm depends upon how much continuity in node ordering is to be transferred from parent to offspring, and if this order is kept from the start of the string or in a central region of the string.

Since at least one of the parents is chosen based upon its fitness, the 'genes' from a relatively small number of solutions will represent approximately 50% of the genes in all offspring generated from that population (depending upon the mating operator that is used). The object is to seed the offspring with favorable genes. The net result is to focus the population into favorable, and hopefully the optimal, regions of search space by forming schema and reducing the dimensionality of the search space. The way the parents are chosen and the percentage of good genes placed in the offspring will control the focusing rate for the population as a whole. For example, if 80% of the genes in all offspring are obtained from the best solution in the current population, focusing will be very fast. In contrast, if both parents are chosen randomly from the full population (i.e., not chosen based upon their fitness), focusing will either be very slow or will not occur at all.

MUTATION OPERATOR

A mutation operator simply modifies a single solution. In a standard genetic algorithm, the mutation probability (P_{mutate} in Figure 1) is very small and its action is to try to prevent premature convergence to a suboptimal solution. In evolutionary programming, an offspring is created from a single parent through mutation, so the effect of the mutation is larger and $P_{mutate} = 1.0$.

As with the mating operator, the form of the mutation operator depends upon the coding scheme. For a gene based coding scheme, a mutation operator will simply change one or more of the values in the genetic string. If real or integer coding is used, this operator can simply add or subtract a random amount from the value. In binary coding, one or more of the bits are flipped to its complement. If normal coding is used, the magnitude of this change depends upon which bit happens to be changed and it can be very large. With Gray coding, the magnitude of the change will be smaller for a single bit flip. As with the crossover operator, it is possible to define a *k-point mutation*, where *k* values are chosen one at a time and changed. Since some values may be chosen more than once, up to *k* values will be changed. In addition, a *uniform mutation* is possible, where each value, or bit, is changed with some fixed probability (J. Culberson, 1994, private communication).

Since any value, or any bit encoding a value, can be changed by a mutation operator, this operator is nonfocusing. The number of values, or bits, changed and the magnitude of the change can be set by the user to control how similar the final solution is to the one before mutation. If many values are changed by a large amount the similarity will be very small, and the convergence of the population to the optimal value may be extremely slow.

It has been argued that P_{mutate} should be large at the beginning of a GA, so that more of solution space can be examined for the first few generations, and then reduced so that the mating operator can focus the population into one or more good regions. Again, this may be promising for some problems but should not be considered *the* way to go for all problems.

All possible results of the mutation operator are contained in a hyper-rectangle that has the full dimensionality of the search space. The 'width' of the rectangle in any dimension is simply twice the maximum change allowed for that value and the center of the hyper-rectangle is the solution prior to mutation.

Though mating and mutation operators are generally treated as very different, the preceding discussion shows that there is a smooth path between a mating-based algorithm (e.g. a GA) and a mutation-based one (e.g. EP). At one extreme is the standard crossover operator where the offspring represent vertices of a, possibly reduced-dimension, hyper-rectangle centered at the average of the parents. The intermediate recombination operator allows the offspring to become any point within this hyper-rectangle, while the interval crossover operator ensures that the dimensionality of the hyper-rectangle is always the full dimensionality of the search space and the center of this rectangle is again the average of the parents. Niching will increase similarity between the parents, which means that their distance in search space decreases. Therefore, the mutation-based EP can simply be viewed as the extreme case of a niching GA (both parents are the same) that uses an interval crossover operator. Here δ_i simply represents the maximum allowed change in that particular value (or one-half the width of the hyper-rectangle in that direction).

For node based coding problems, the mutation operator will also cause effects similar to the mating operators discussed above. Again, only a few of the many possible mutation operators will be presented here. The initial solution (before mutation) will simply be

A B C D E F G H I J

1. Randomly pick two points on the string and switch their values.

A B C D E F G H I J initial

A B C D E F G H I J final

2. Cut the string at two places and invert the center region.

A B C | D E F G | H I J initial

A B C G E F D H I J final

3. Cut the string at two places, move the first element of the center region to the end and all others up one position on the string.

A B C | D E F G | H I J initial

A B C E F G D H I J final

Again, the main point is that mutation is nonfocusing, and the similarity between the intial and final solution can be controlled by setting the number of values changed and the magnitude of this change. It is also possible to have the number and magnitude of the changes decrease during the simulation, or to have it dynamically change as is done with delta coding.

MATURATION OPERATOR

Up to this point, a general paradigm has been presented that searches for the optimal solution of a problem by using a population of solutions. Methods for generating the initial population and creating offspring from either one or two parents have been presented. This general construct can be used to 'drive' many existing computational methods.

For example, if the optimum conformation of a molecule is desired, with or without a binding site and/or solvent, a local minimization, a dynamics or Monte Carlo simulation, or a simulated annealing run can be performed. The final, or average, energy and its corresponding structure can be used to determine the fitness and genetic values of the offspring.

If the object is to determine the maximum common substructure of two compounds, an initial offspring can be generated using mating and/or mutation operators. This offspring will be an incomplete one-to-one mapping of one compound's atoms to the other compound's atoms. A local search procedure can be used to remove 'bad' atoms from this list and/or add atoms that improve the overlap.

In a QSAR/QSPR investigation, a single position in the array of chosen descriptors can be selected and sequentially replaced with all descriptors that haven't been chosen; keeping the best one.

Each of the operations described in this section are denoted *maturation operators*. They are similar to mutation operators in that they modify a single solution, but differ in the fact that they are guided, as opposed to random, modifications. A maturation operator will either leave the initial offspring alone or improve its fitness. There is no guarantee that using a maturation operator will increase the probability of locating the optimum solution; it is only presented as another option available to this computational method.

The maturation operator may be useful whenever a real number genetic alphabet is used, since it may help simplify problems associated with the infinite number of members in the alphabet. It is important to realize that the maturation operator will result in a focusing of the overall population. If a local optimization is used, all initial offspring that lie within the attractive basin of this local minimum will all converge to exactly the same structure and fitness. Many offspring generated from these solutions may also lie in this well and the percentage of the population represented by this structure may grow with time, unless some other minimum has a deeper attractive well.

PROCESS OFFSPRING

From the preceding discussion, it may be possible to imply that, using a mating operator, two parents will produce two offspring, and that these offspring can be subjected to mutation and maturation operators. Similarly, one could imply that a process that does not use a mating operator produces a single offspring from a single parent, which may be acted upon by a maturation operator. There is no reason to require that the number of offspring must equal the number of parents, or that all offspring should be placed into the population. Again, there are any number of options, and only some of them will be described here.

1. Use mating (with or without mutation and maturation) to create as many offspring as desired from two parents. From this group of offspring, keep the best one or more. Please realize that if only the best solution is kept, or if a solution and its complement are not kept together, this mating will be strongly focusing. If a solution and its complement are kept together, this pair spans the same search space dimensionality as their parents and will not result in immediate focusing. Conversely, if a large population size is used, it may be advantageous to focus onto good areas as quickly as possible. Keeping only the best offspring will help this.
2. Use a mutation operator to generate multiple offspring from a single parent (with or without maturation) and only keep the best one. This

is nonfocusing, and the similarity between parent and offspring can be controlled by the number of values changed and the magnitude of the change caused by mutation.

3. Use a mutation operator (with or without maturation) to generate 100/N offspring from N% of the parents. For example, a new population can be created by generating 5 offspring from the best 20% of the solutions through mutation (Nolfi and Parisi, 1991; Parisi *et al.*, 1991; Nolfi *et al.*, 1994a,b; Calabretta *et al.*, 1995). This process has interesting consequences since it is a mutation based reproduction that has the ability to focus the population. Since each offspring has many values in common with its parents, the descendents of a single parent can take over the entire population in only a few generations (three generations for the example above). If this happens, certain values may be the same for all members of the population. Obviously, the degree of focusing depends upon how similar each offspring is to its parent. In addition, this process will not guarantee that the fitness of the population increases from generation to generation. In fact, the maximum, minimum and average fitness of the population can actually decrease from one generation to the next.

When one or more offspring are selected, they can either be added to the current population by displacing an existing member, or can be placed in a new population. If an offspring is placed into the current population, it can either

1. Displace the weakest member of the population.
2. Displace a parent. This could be the only parent if a single parent generates a single selected offspring through mutation, or the weaker parent if two parents generate a single selected offspring through mating.
3. Displace the most similar member of the current population, or a subset of this population.

Though the first option is the most popular, it is also the most focusing. Since the offspring will have several values in common with its parent(s), and it and both parents will be in the population after the offspring is selected, there will be an increase in the number of solutions with a given schema after each offspring is produced. This also has the disadvantage that potentially valuable genetic information contained in the least fit member of the population will be lost before it has a chance to pass this information on to an offspring. If this offspring is created by a mating operation and one or both of the parents are chosen based upon their fitness, chances are good that this offspring will have a fitness that is higher than the worst solution in the population. Therefore, this option has the effect of increasing the average fitness of the population and reducing the population's minimum fitness.

The second and third options slow down the focusing process. If the second option is used in conjunction with a mating operator, chances are reasonably good that it will improve the average fitness of the population, especially if it is the most fit offspring from several. It may have values in common with the remaining parent, but this is less focusing than values in common with both surviving parents. The third option is much less focusing for the population as a whole since the member it replaces may have just as many values in common with a given schema as the offspring.

The next section discusses the options that are available if the offspring are placed in a separate population.

UPDATING THE POPULATION

When the selected offspring are placed in a separate population and the size of this population reaches a predetermined size, which will be called *KPOP* (usually *KPOP* = *NPOP*, but this doesn't have to be the case) the generation of offspring stops. At this point, some mechanism has to be implemented to create a new 'current' population. As with all the other operators discussed so far, there are many options as to how this can be done.

1. Take the *NPOP* best solutions from the new population and make this the current population.
2. Combine the *NPOP* – 1 best solutions from the new population and the best solution from the current population to make the next current population. Including the best current solution in the next generation is called the *elitist strategy*.
3. Combine the current and new populations to make a population of size *NPOP* + *KPOP*. Take the *NPOP* best solutions from this combined population.
4. Combine the current and new populations to make a population of size *NPOP* + *KPOP*. At this point, use a selection strategy to select *NPOP* members of this combined population. This can be done with or without forcing the elitist strategy.
5. Combine the current and new populations and select the most fit n unique solutions to produce focus points. Then, in a cyclic fashion, choose the solution from the combined population that is most similar to a focus point.

The only possible disadvantage of the first choice is that the overall best solution may be lost from one generation to the next.

If the fourth choice is used, a slight variation on the tournament selection process could be implemented, or any of the selection procedures outlined in the section on building a mating population. In this variant, a tournament size of q is selected. Each member of the combined population is compared against q randomly chosen members and receives a point each time its fitness is better than a randomly selected one. When completed, each solution in the combined population will have a score between 0 and q. These solutions are then ranked in order of decreasing score and the first *NPOP* solutions comprise the new population. Note that this procedure forces the best overall solution into the new population. In addition, the user has the option of ordering solutions with the same score by decreasing fitness or using a random order.

The fifth choice would be used to promote niching. In addition, choices 3 to 5 can be modified to place as many unique solutions as possible into the new generation's population. Promoting uniqueness among the solutions may be useful if the object is to find multiple good descriptors from a single run in a QSAR/QSPR investigation (Luke, 1994).

As a final point in this section, the choices presented above can be cast in the (μ, λ) and $(\mu + \lambda)$ notation used in much of the literature on genetic methods. This will not lead to any new insights; it simply relates this discussion to other works. In this notation, μ parents (i.e., *NPOP*) create λ offspring (i.e., *KPOP*). In a (μ, λ) method, λ must be greater than or equal to μ, and the μ individuals needed to build the new population are taken solely from the λ offspring. In a $(\mu + \lambda)$ method, there are no restrictions on the size of either parameter. The μ solutions are simply chosen from the combined population, $\mu + \lambda$, using a selection procedure. Therefore, choices 1 and 2 are (μ, λ) methods (without and with an elitist strategy) and choices 3 to 5 are $(\mu + \lambda)$ methods.

SUMMARY

At this point, many different choices have been presented for each of the operations presented in Figure 1. In the hopes of keeping a good overall view, and not getting lost in all of the details, a brief review of the processes outlined in Figure 1 will be presented. Before choices are made for each of the operations, a coding scheme, genetic alphabet, and fitness function will have to be determined. The coding scheme and genetic alphabet have already been described. The fitness function has been referred to many times so far, but no real discussion of its possibilities has been presented. This will be done now.

In general, the fitness is simply a measure of how well a solution fits the problem, and is really a relative term since an ordering of the solutions is generally used. On the other hand, a selection scheme can be devised that uses the normalized fitness of an individual as a measure of its probability for selection. In this case, the form of the fitness function is important (e.g. is the value of the function being maximized or the exponent of this value used).

If a function is being minimized, such as the conformational energy of a molecule, the *cost* of a solution is being measured, not its fitness. For a relative measure, ordering the solutions from lowest to highest cost is the same as from highest to lowest fitness. For an absolute measure, the fitness can be some maximum value minus the cost, the inverse of the cost, or some other function.

The fitness, or cost, of a solution can be more than the value being optimized. For example, in Luke (1994) evolutionary programming is used in QSAR/QSPR investigations. To compare with previous studies, the major value to minimize is the root-mean-square error (*RMS_ERROR*) between the predicted and observed activities/properties. A solution is represented as an *NDESCR* string of integers, where *NDESCR* is the number of available descriptors. A given element is 0 if that descriptor is not used, 1 if the predictor is linear in that descriptor, 2 if the predictor is quadratic in that descriptor, –1 if it depends upon the inverse of that descriptor, and so on. The overall cost is given by

$$COST = (RMS\_ERROR)\,(XVAL)^{|ITERM\,-\,nterm|}\,\Pi WEIGHT(n_i)$$

Here, *ITERM* is a user-supplied number of terms that this descriptor should have and *nterm* is the number of terms present (i.e., the number of nonzeros) in the string representing the solution. *XVAL* is a user-supplied penalty value, and in Luke (1994) it is set to 1.2. This means that if 3-term descriptors are desired, a 4-term predictor would yield a lower overall cost if it has an *RMS_ERROR* that is lower by more than a factor of 1.2 (assuming the *WEIGHT* term yields the same value). *WEIGHT* is present to control the allowed exponents. *WEIGHT*(1) = 1.0, which means that using linear descriptors will not add to the cost. If *WEIGHT*(2) = 2.0, a relation that is quadratic in a descriptor yields a lower overall cost only if the *RMS_ERROR* decreases by more than a factor of 2.

The key point is that the fitness, or cost, function need not simply represent the value of the function being optimized; it can also include terms that will guide the solution into a desired form, or structure. The final point is that the fitness function need not be the same over the course of the search, or for all members of the population. For example, the search may start with a subset of the function to optimize, and terms can be increased in magnitude, or added, from generation to generation. Conversely, different populations may evolve using pieces of the overall problem. The best solutions of each piece can be combined to have them evolve on the full problem, and then redistribute the evolved solutions to the individual nodes, and pieces of the problem. This procedure was used in Kauffman *et al.* (1994, private communication) using a different optimization procedure. Finally, the weights of different terms in the function may be stored as values in the genetic string of each solution (Lund and Parisi, 1994). Here the weights, and therefore the fitness function, could evolve along with the solution values.

Once a coding scheme, genetic alphabet, and fitness function (or fitness scheme) has been determined, the actual algorithm used to solve the problem can be built. The first step is to determine a population size, *NPOP*, and generate an initial population. The next step is to determine if the initial population will be used with either a random or biased selection method, or if a mating population is created (to jump ahead, if the offspring replace members of the population containing the parents, a mating population can only be created once since this population will continually change with each offspring added).

After this, the method of generating offspring must be determined. This procedure can use a mating operator with a given probability (P_{mate}), a mutation operator with a given probability (P_{mutate}), and/or a maturation operator. Finally, these offspring can either replace members of the current population, or be placed into a new population. If they replace members of the current population, the flowchart in Figure 1 will only cycle over the inner loop. Such a scheme is denoted *non-generational* since no partial or total change in the population occurs at a particular time, though it is still valid to say that a generation occurs each time *NPOP* offspring have been added to the current population. The difference between a non-generational and a generational scheme is that in the first case an offspring can become a parent in the very next offspring-creation step while in the latter, there must be a population turnover before any of the offspring can become parents. If the offspring are placed into a new population, it is necessary to decide how they will be used (with or without the current population) to create a new population. Since there is a turnover in the population, it is possible to re-run the selection step and create a new mating population each generation.

The point of emphasis is that there are a great number of choices to be made, and these choices should be guided by an understanding of the complexity and landscape of the problem being solved. No single combination will optimally work for all problems.

REVIEW OF VARIOUS PUBLISHED ALGORITHMS

In no way is the following intended to be a full review of all previously published algorithms that use genetic methods. The purpose is mainly to spark the reader's imagination by listing some of the combinations described above that have been tried. This section starts with three independently developed algorithms that represent the first applications of genetic methods; the standard GA, evolution strategies (ES), and EP.

Standard genetic algorithm (SGA)

The 'standard' GA (Bledsoe, 1961; Holland, 1975) uses a gene based, binary coding scheme. There is a high mating probability, a very low mutation probability, and no maturation process. At least one of the parents is chosen based upon its fitness, a one-point crossover is used, and the offspring replaces the weakest member of the population. Since members of the current population are replaced, this is a non-generational algorithm; the offspring can become parents in the next mating.

Evolution strategies

Though this method was originally a mutation-only procedure with a population size of one (Rechenberg, 1973; Schwefel, 1977), it has been expanded to allow for multimember populations and mating operators (T. Back, G. Rudolph and H.P. Schwefel, 1995, private communication). In the ES method, mating operators are denoted *recombination operators*. The offspring are placed into a new population and the *NPOP* best solutions are selected from either the new population or the combined old and new populations. In other words, this can either be a (μ, λ) or a $(\mu + \lambda)$ method. One aspect that makes this procedure unique is that the *strategy parameters* are included with the values stored in the genetic string and can be modified during the search by mating/mutation operators. These strategy parameters are used to determine the magnitude of the mutation.

Evolutionary programming

EP (Fogel *et al.*, 1966, 1991; Fogel, 1993) differs from GA in that it does not use a mating operator. In general, *NPOP* parents create *NPOP* offspring by mutation/maturation. These offspring are placed into a new population. The next generation's population is formed by choosing *NPOP* solutions from the combined parent + offspring pool. In other words, this is a $(\mu + \lambda)$ procedure. As originally proposed, this procedure uses a *probabilistic replacement* scheme as opposed to the *deterministic replacement* scheme used in the ES method. This means that any of the selection procedures outlined above can be used to determine the new population from the combined one, though there is no reason why a deterministic replacement cannot be used.

SGA, ES and EP

As a quick review, the SGA is non-generational in that the offspring replace members of the current population, while both the ES and EP methods are generational. The ES method can either be (μ, λ) or $(\mu+ \lambda)$, while the EP method is always $(\mu + \lambda)$. The SGA method places an offspring into the population whether or not it is more fit than the solution it is replacing. The ES method uses a deterministic replacement where it only keeps the most fit solutions, while the EP method uses a probabilistic replacement where a less-fit solution has a chance of surviving to the next generation. The SGA method always uses a mating operator (and the mutation operator is of minor importance), the ES method can use a mating operator, and the EP method never uses this operator. Finally, the ES method is unique in that the strategy parameters are stored in the genetic string and modified along with the values being optimized. Please realize that these are not 'all or nothing' methods and that parts can be taken from one or the other to create a different algorithm. The choices listed above are only those that were originally used in these standard methods.

Crowding genetic algorithm

This is very similar to a standard GA, the only difference is on how the offspring are placed into the current population. Here (de Jong, 1975), both offspring obtained by mating and mutation are kept. A user-defined number of solutions (denoted C, and is referred to as the *crowding factor*) are randomly chosen from the mating population. The first offspring is compared to these solutions and replaces the member that is most similar. The same procedure is followed for the second offspring. This means that there is a small, but finite, probability that the second offspring will replace the first. In a *deterministic crowding GA* (Mahfound, 1992), the parents are chosen randomly, but the replacement is done by comparing offspring to parents. In particular, each offspring is compared to the most similar parent, and the higher fitness solution is kept; the other is discarded. This replacement could also be done using a Boltzmann selection probability, as is done in a Metropolis Monte Carlo simulation, or with a variable temperature Boltzmann probability, as in simulated annealing.

Multi-niche crowding genetic algorithm

In this algorithm (Cedeno *et al.*, 1995), a *crowding selection* is used. The first parent is either chosen randomly or sequentially from the population. The second parent is chosen by randomly choosing C_s solutions and selecting the one that is most similar.

This promotes mating between solutions of the same niche, but does not require it. A *worst among the most similar* replacement policy is used. Here, C_f groups of s solutions are chosen at random and the one most similar to the offspring is selected from each group. This yields C_f solutions and the one with the lowest fitness is replaced by the offspring. This replacement method is slightly different from the *enhanced crowding* technique (Goldberg, 1989) where the offspring replaces the most similar solution out of a group of least fit candidates. The multi-niche crowding algorithm uses crowding for both selection of the second parent and replacement of the offspring.

Gene invariant genetic algorithm (GIGA)

This is a GA that also uses binary, gene based coding (Culberson, 1992a,b,c, 1994, private communication). It uses crossover and no mutation. It selects a pair of parents, usually a pair that is close in fitness value, and produces a family of pairs of offspring through crossover. The best of these pairs replaces the parents in the population. In this way, the multiset of binary values in any column of the population matrix remains unchanged over time. This property is referred to as *genetic invariance*. The best pair is the set of offspring that contains the best individual, independent upon how good the complementary solution is. This is non-focusing for the population as a whole, but does focus the selection by choosing parents that are close in fitness. In addition, this method does not require a relatively high similarity between parent and offspring. *Local elitism* can be used whereby the parents are also included in the set of offspring pairs to determine which will be put into the population.

Delta coding genetic algorithm

This is also a GA without mutation. A standard GA is run until a population convergence is obtained. The best solution at this point is denoted the *interim solution* (K.E. Mathias and L.D. Whitley, 1995, private communication). At this point, *delta iterations* are begun. Here, each solution in the population represents a set of *delta values* that are applied to the interim solution to generate a new solution whose fitness is evaluated. Crossover is used on the population of delta values to generate new offspring. In the first delta iteration, the maximum size of each delta value is set so that the full space can still be searched (only the representation of a solution is different). For subsequent iterations, the magnitude of the maximum delta is adjusted depending upon the result of the previous iteration. If the previous iteration converged to the same solution, the maximum delta is increased; if a different solution is obtained, the maximum delta is decreased. Obviously, there are limits on the largest and smallest maximum delta. After each iteration, the best solution becomes the new interim solution and a new population of delta values is randomly created. This algorithm is non-focusing in that the delta values represent a uniform mutation operation with a probability of 1.0. Because the maximum delta is constantly changing, the similarity between an offspring and its parents will also change. The search space will always occupy a hyper-rectangle centered at the interim solution, and it will always have the full dimensionality of the problem (non-focusing). The 'width' of the rectangle in any direction is twice the maximum delta for that variable. If an interim + delta value is greater than the maximum allowed for that variable, the value should be folded from the beginning. In other words, all variables are assumed cyclic such that $max + x = min + x$. The random restarts that occur at the beginning of each iteration sustain the search and preclude having to use large population sizes or high rates of mutation.

CHC adaptive search algorithm

This method is a GA that uses *heterogeneous recombination* (Eshelman, 1991). In other words, there is an unbiased selection of parents. The offspring are stored in a temporary population. The best *NPOP* solutions from the parent and temporary populations are used to form a new population. In other words, this is a ($\mu + \lambda$) method that uses deterministic replacement. If the new population is equal to the old (no offspring survive to the next generation), a restart mechanism is used that is known as a *cataclysmic mutation*. Here, the best solution is copied to the new population and is used to create all other members by mutation. This creates a new population that is biased towards a good solution, but with new diversity.

Parallel genetic algorithm (PGA)

In this algorithm (Muhlenbein *et al.*, 1991), a different population of initial solutions is placed on different nodes of a parallel computer, or network of workstations. Each population evolves using a GA for a given number of generations. At this point, the best solution of a local population is copied to neighboring populations. This is one method to promote niching, and the user is free to determine exactly how the most fit solution of one population is sent to one or more other populations. One option is to 'order' the nodes in a ring and copy the solution to the neighbor in a clockwise fashion. Another is to randomly pick one or more other populations each time a solution is to be copied; and a third is to copy the best solution to all other populations. Care must be taken to ensure that the population size on each node is not too small, or else each population will be trapped in the same suboptimal solution (Dymek, 1992).

Adaptive parallel genetic algorithm (apGA)

This is similar to a parallel GA, with one added feature (Liepins and Baluja, 1991). After a given number of generations, the best solution from each population is placed in a 'central population'. This population can be increased by adding randomly generated solutions, or other solutions from each population. The central population evolves using a GA and, after a given number of generations, the best solution is copied to the populations on each node. At this point, the independent evolutions proceed.

Massively distributed parallel genetic algorithm (mdpGA)

This is a GA that can be run on a large array of CPUs, or nodes (Baluja, 1992). Each node contains two solutions, and the parents are chosen from these two and solutions present on adjacent nodes. If the nodes are arranged in a ring, ten solutions can be selected by randomly choosing one of the solutions from each of the four nodes to the left and one from each of the four nodes to the right, plus the two already present. From these ten, two solutions are chosen based upon their fitness, and the other eight are discarded. These parents produce a complementary pair of offspring by mating/mutation/maturation, and these offspring become the solutions stored on this node. The population of ten can be formed in many other ways (three from the left, three from the right, and two copies of the two solutions present on that node, or arrange the nodes to be on the surface of a torus and take one solution from each of the eight nearest neighbors). Since both parents are chosen based upon their fitness, a very good solution may migrate over the nodes. It will take many matings for this to happen, as opposed to being immediately available for every mating as is found in a standard GA.

Meta-genetic algorithm

This algorithm is very similar to a parallel GA, the only difference being that the genetic operators can be different from one population to another (Mansfield, 1990). Since all populations are using the same fitness function, one of the objectives is to find a combination of mating/mutation/maturation operators that perform well on this function. After a given number of generations, the best solution is copied to a neighboring population. If this population performed better during these generations (e.g. the average fitness rose more, or became higher; or the population did better by any metric the user wants to impose) than the population on the 'receiving end', some or all of the genetic operators are transferred also. Care should be taken since it has been argued (Wolpert and Macready, 1995) that an algorithm's performance to date (i.e., a particular choice of operators and application probabilities) does not correlate at all with the quality of its final result.

Messy genetic algorithm (mGA)

In a mGA (Goldberg *et al.*, 1989; Goldberg and Kerzic, 1990), the values stored in the genetic string may be incomplete. This method can only be used for gene based coding problems. The genetic string for a solution is doubly indexed; the first gives the position along the string and the second gives the value for this position. The solution is applied with respect to a template, similar to the delta coding method. Any missing values in the solution are taken from the template. If the same position is given more than one value in the string, the left-most value on the string is used. Instead of a crossover mating operator, SPLIT and JOIN are used. SPLIT splits the parent at a random position to create two offspring, while JOIN combines two parents into a single offspring. In either case, the original parents are no longer present in the population. In practice, better results are found if the SPLIT probability is larger than the JOIN. The net effect is to increase the population size over time, so a periodic 'pruning' of the least fit individuals has to be done.

CONCLUSION

This chapter has shown that there are a very large number of combinations that can be employed to construct a particular algorithm. Though this may seem to diminish the applicability of this construct, it actually means that the same general computational philosophy can be customized to efficiently solve particular problems.

This customization can also control the rate of niching or focusing of the population. EP does not promote focusing unless the mutation causes a very small change in the offspring (i.e., the offspring is very similar to its parent). Liepins and Baluja (1991) propose something similar for GAs called an *adaptive mutation*. Here, the mutation probability is inversely proportional to the similarity of the parents. If the parents are very similar, any offspring generated by mating must be similar to both of them. If the mutation probability is increased, the similarity between parents and offspring should decrease.

Niching slows down the focusing of the total population. If this is done using a parallel GA, the total population may eventually focus onto a single region of search space since fit members are copied to other populations. This seeds a particular region of search space into multiple, if not all, subpopulations and may promote overall focusing; especially if the size of the subpopulations is small.

Only through further testing can the full power of genetic methods' ability to solve problems associated with drug discovery be determined. At present, it does represent a new methodology that can be used in conjunction with existing techniques, and may yield valuable insights and solutions.

REFERENCES

Alba, E., Aldana, F., and Troya, J.M. (1993). Full automatic ANN design: A genetic approach. In, *Proceedings of the International Workshop on Artificial Neural Networks* (J. Mira, J. Cabentany, and A. Prieto, Eds.). Springer Verlag, New York, pp. 399–401.

Baluja, S. (1992). A Massively Distributed Parallel Genetic Algorithm. *CMU-CS-92-196*, School of Computer Science, Carnegie-Mellon University.

Bledsoe, W.W. (1961). The use of biological concepts in the analytical study of systems. Paper presented at *ORSA-TIMS National Meeting*, San Francisco.

Calabretta, R., Nolfi, S., and Parisi, D. (1995). An artifical life model for predicting the tertiary structure of unknown proteins that emulates the folding process (private communication).

Cecconi, F. and Parisi, D. (1994). Learning During Reproductive Immaturity in Evolving Populations of Neural Networks. *Technical Report PCIA-94-12*, Institute of Psychology, C.N.R.-Rome.

Cedeno, W., Vemuri, V.R., and Slezak, T. (1995). Multi-niche crowding in genetic algorithms and its application to the assembly of DNA restriction-fragments (private communication).

Culberson, J. (1992a). Genetic Invariance: A New Paradigm for Genetic Algorithm Design. *Technical Report TR 92-02*, University of Alberta Department of Computing Science.

Culberson, J. (1992b). GIGA Program and Experiments. University of Alberta Department of Computing Science.

Culberson, J. (1992c). GIGA Program Description and Operation. *Technical Report TR 92-06*, University of Alberta Department of Computing Science.

de Jong, K.A. (1975). An analysis of the behavior of a class of genetic adaptive systems. Ph.D. Thesis, University of Michigan. *Dissertation Abstracts International* **36(10)**, 5140B.

Dymek, Captain A. (1992). An examination of hypercube implementations of genetic algorithms. Thesis, Air Force Institute of Technology *AFIT/GCS/ENG/92M-02*.

Eshelman, L. (1991). The CHC adaptive search algorithm. How to have safe search when engaging in nontraditional genetic recombination. In, *Foundations of Genetic Algorithms* (G. Rawlings, Ed.). Morgan Kaufmann Publishers, San Mateo, CA, pp. 265–283.

Fogel, D.B. (1993). Applying evolutionary programming to selected traveling salesman problems. *Cybern. Syst. (USA)* **24**, 27–36.

Fogel, D.B., Fogel, L.J., and Porto, V.W. (1991). Evolutionary methods for training neural networks. In, *IEEE Conference on Neural Networks for Ocean Engineering*, 91CH3064-3, pp. 317–327.

Fogel, L.J., Owens, A.J., and Walsh, M.J. (1966). *Artificial Intelligence through Simulated Evolutions*. John Wiley, New York.

Goldberg, D.E. (1989). *Genetic Algorithms in Search, Optimization & Machine Learning*. Addison-Wesley Publishing Company, Reading, p. 412.

Goldberg, D.E. and Kerzic, T. (1990). mGA 1.0: A Common LISP Implementation of a Messy Genetic Algorithm. *NASA-CR-187260*, Cooperative Agreement NCC 9-16, Research Activity AI.12, N91-13084.

Goldberg, D.E., Korb, B., and Deb, K. (1989). Messy genetic algorithm: Motivation, analysis, and first results. *Complex Systems* **3**, 493–530.

Holland, J.H. (1975). *Adaptation in Natural and Artificial Systems*. University of Michigan Press, Ann Arbor.

Liepins, G.E. and Baluja, S. (1991). apGA: An adaptive parallel genetic algorithm. *Fourth International Conference on Genetic Algorithms*. San Diego, CONF-910784–2.

Luke, B.T. (1994). Evolutionary programming applied to the development of quantitative structure-activity relationships and quantitative structure-property relationships. *J. Chem. Inf. Comput. Sci.* **34**, 1279–1287.

Lund, H.H. and Parisi, D. (1994). Simulations with an evolvable fitness function (private communication).

Mahfound, S.W. (1992). Crowding and preselection revisited. In, *Proceedings of Parallel Problem Solving from Nature 2* (R. Manner and B. Manderick, Eds.). Elsevier Science B.V., New York, pp. 27–36.

Mansfield, Squadron Leader R.A. (1990). Genetic algorithms. *Dissertation*, University of Wales, College of Cardiff, School of Electrical, Electronic and Systems Engineering, Crown Copyright.

Mathias, K.E. and Whitley, L.D. (1994). Transforming the search space with Gray coding. Proceedings of the First IEEE Conference on Evolutionary Computation. In, *IEEE World Congress on Computational Intelligence*, Vol. 1.

Muhlenbein, H., Schpmisch, M., and Born, J. (1991). The parallel genetic algorithm as a function optimizer. *Parallel Computing* **17**, 619–632.

Nolfi, S. and Parisi, D. (1991). Growing Neural Networks. *Technical Report PCIA-91-15*, Institute of Psychology. C.N.R.-Rome.

Nolfi, S., Miglino, O., and Parisi, D. (1994a). Phenotypic Plasticity in Evolving Neural Networks. *Technical Report PCIA-94-05*, Institute of Psychology. C.N.R.-Rome.

Nolfi, S., Elman, J.L., and Parisi, D. (1994b). Learning and Evolution in Neural Networks. *Technical Report PCIA-94-08*, Institute of Psychology. C.N.R.-Rome.

Parisi, D., Nolfi, S., and Cecconi, F. (1991). Learning, behavior and evolution. In, *Toward a Practice of Autonomous Systems* (F. Varela and P. Bourgine, Eds.). MIT Press.

Rechenberg, I. (1973). *Evolutionsstrategie: Optimierung technischer Systeme nach Prinsipien der biologischen Evolution*. Frommann-Holzboog Verlag, Stuttgart.

Schwefel, H.-P. (1977). Numerische Optimierung von Computer-Modellen mittels der Evolutionstrategie. Volume 26 of *Interdisciplinary Systems Research*, Birkhauser, Basel.

Whitley, D. (1989). The GENITOR algorithm and selection pressure: Why rank-based allocation of reproductive trials is best. In, *Proc. Third International Conference on Genetic Algorithms* (J.D. Schaffer, Ed.). Morgan Kaufmann, Los Altos, CA, pp. 116–121.

Whitley, D., Starkweather, T., and Bogart, C. (1990). Genetic algorithms and neural networks: Optimizing connections and connectivity. *Parallel Computing* **14**, 347–361.

Wolpert, D.H. and Macready, W.G. (1995). No Free Lunch Theorems for Search. *Technical Report TR-95-02-010*, The Santa Fe Institute.

3 Genetic Algorithms in Feature Selection

R. LEARDI
Istituto di Analisi e Tecnologie Farmaceutiche ed Alimentari, Università di Genova, via Brigata Salerno (ponte), I-16147 Genova, Italy

Starting from the original algorithm, several changes made genetic algorithms a powerful tool in feature selection. A full validated version has also been implemented, to check that the selected subset is really predictive and not due only to chance correlations (very easy when working with a large variables/objects ratio).

Hybrid algorithms are conceptually very simple: after a certain number of generations of genetic algorithms, the best experimental condition so far found undergoes a 'classical' method of optimization (in the case of feature selection, stepwise selection); the results thus obtained can enter the population, and then a new genetic algorithm is started with the updated population. This approach allows further improvement of the performance of the genetic algorithm.

The application of genetic algorithms to two quantitative structure–activity relationship data sets will be presented, and the results will be compared with those described in literature.

KEYWORDS: *genetic algorithms; feature selection; regression; full validation.*

ABOUT FEATURE SELECTION

One of the main problems when elaborating large data sets is the detection of the relevant variables (i.e. the variables holding information) and the elimination of the noise. Roughly speaking, each data set can be situated somewhere in between the following two extremes:

(a) data sets in which each variable is obtained by a different measurement (e.g. clinical analyses);
(b) data sets in which all the variables are obtained by a single measurement (e.g. spectral data).

In, *Genetic Algorithms in Molecular Modeling* (J. Devillers, Ed.)
Academic Press, London, 1996, pp. 67–86.
ISBN 0-12-213810-4

It can be easily understood that while for data sets of type (b) the only goal of feature selection is the elimination of noise, together with the simplification of the mathematical model, for data sets of type (a) it is very worthwhile to try to reduce as much as possible the number of variables involved, since this also means shorter analysis time and lower costs.

The procedure of feature selection, apparently so simple, is indeed very dangerous and needs a very careful validation, to avoid the risk of overestimating the predictive ability of the selected model; in such cases, when using it on new data, one can be deceived, discovering that it has no predictive ability at all (Lanteri, 1992).

This is mainly due to the random correlations: if you try to describe 10 hypothetical objects with 100 random X variables and a random response, you will surely have some X variables perfectly modelling your response. This risk is of course higher when the ratio variables–objects is very high: this is the typical case of quantitative structure–activity relationships (QSAR), when only a few molecules are described by several tens of molecular descriptors.

APPLICATION OF GENETIC ALGORITHMS TO FEATURE SELECTION

Genetic algorithms (GAs) can be very easily applied to feature selection; in this paper it will be shown that very good results are obtained with a 'tailor-made' configuration, in which the classical GA has been slightly modified taking into account several peculiarities of this particular problem (Leardi *et al.*, 1992; Leardi, 1994).

In a GA applied to feature selection, the structure of the chromosomes is very simple, since each X variable is a gene coded by a single bit (0 = absent, 1 = present) (Lucasius and Kateman, 1991; Leardi *et al.*, 1992; Lucasius *et al.*, 1994).

The response to be maximized is the percentage of predicted variance (in the case of a regression problem) or the percentage of correct predictions (in the case of a classification problem). As an example, suppose we have 10 variables, and we want to find the best subset of variables to be used with partial least squares (PLS). If the chromosome to be evaluated is 0010011001, then the response will be the variance predicted by the PLS model computed by taking into account variables 3, 6, 7, and 10.

The usual way to evaluate the predictive ability of a model is cross-validation with a predefined number of deletion groups (e.g. 5). When performing a process of feature selection, one has to be aware that such a 'prediction' is only partially validated, since the objects on which the selection is performed are also those on which the prediction ability is computed, and this does not get rid of the problem of random correlation (we will see later on how a full-validated GA has to be). The reliability of this approach is therefore highly dependent on the objects–variables ratio: the results obtained when this is very high are really very near to full-validation; on the other side, when only a few objects are present, the results are almost meaningless.

CLASSICAL METHODS OF FEATURE SELECTION VS GENETIC ALGORITHMS

The only way to be sure of selecting the best subset of variables (of course, without taking into account the problem of random correlations) would be to compute all the possible models. The only limitation to this approach is given by the fact that, with n variables, $2^n - 1$ combinations are possible; as a consequence, this method becomes not applicable when the variables are more than just a few. To give an idea, with 30 variables (a rather small data set) more than 1 billion combinations are possible (computing one model per second, it would take 34 years!).

The most commonly used technique is the stepwise approach, which can run forward or backward.

In the forward version, the variables are selected and added to the model one at a time, until no improvement in the results is obtained. In the backward version, the first model to be computed is the one with all the variables; the variables are then selected and removed one at a time. A mix of these two approaches is also used, in which each addition of a variable by the forward selection is followed by a backward elimination.

The stepwise techniques have two main disadvantages:

- each choice heavily affects the following choices (e.g., in the forward version, once one of the variables has been selected, all the models that don't contain it will never be taken into account);
- the final result is expressed by a single combination, and then no choice is given to the user.

GA, in their favour, always allow the exploration of the whole experimental space: due to the occurrence of the mutations, each possible combination can occur at any moment.

The result obtained by GA is a whole population of solutions: the user can then choose the one he prefers, taking into account at the same time the response and the variables used (the user could, for example, be interested in limiting as much as possible the use of certain variables or in taking into account a model that can also have good theoretical explanations).

From this it is also evident that a relevant feature of GA is the ability to detect several local maxima, and then to give a good idea about the presence of several regions of interest.

Usually the performance of a GA is evaluated only in terms of how often and in how much time it leads to the detection of the global optimum. Apart

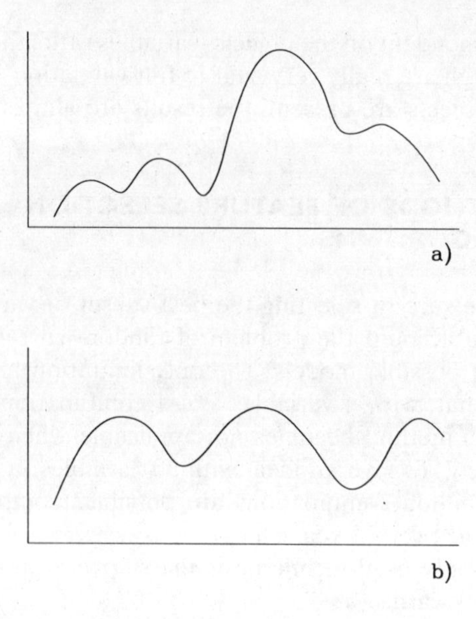

Figure 1 The problem of localizing the global maximum: in (a) the global maximum is by far better than the other local maxima, while in (b) the three maxima are equivalent, and the difference among them is only due to the noise.

from the fact that one should know where it is (people cannot afford to wait 34 years to check which is the best combination in a 30 variables data set), this can be sensible in the case shown in Figure 1a, in which the global maximum is much better than the other local maxima.

In the case of Figure 1b, on the contrary, the three local maxima are totally equivalent, and then it would be absolutely nonsense to look only for the global one. The best algorithm would be the one giving the information that three equivalent optima are present: the user will then be free to choose the most convenient one.

Another very important consideration is linked to the fact that we are always working with experimental data, affected by an experimental error; as a consequence, going back to Figure 1b, another set of measurements would produce a slightly different result, according to which the order of the three local maxima could be changed.

CONFIGURATION OF A GENETIC ALGORITHM FOR FEATURE SELECTION

It is clear that, in comparison with the configuration of GA usually employed for the resolution of other kinds of problems, the configuration of a GA devoted to feature selection must take into account some peculiarities. Some of them are the following:

- The time required by the evaluation of the response is much higher than the time required by GA itself: indeed, to evaluate the response means to perform a full multivariate analysis (e.g. multiple linear regression (MLR), partial least squares or linear discriminant analysis), with different deletion groups to obtain a predicted value (though only in partial validation). In a reasonable amount of time, working on a PC (all the computing times will be referred to a 486 PC with mathematical coprocessor, 33 MHz), only a few thousands of chromosomes can be evaluated, compared with the classical applications in which tens of thousands of generations are produced.

  Furthermore, since we are dealing with experimental data, one has to be aware of the possibility of random correlation and random predictions: the effect of both is higher the more combinations are tested (and then the more chromosomes are evaluated).

  An algorithm allowing a fast improvement of the response function is then required. This is obtained by reducing the population size to about 30 chromosomes (usual sizes range between 50 and 500 elements) and by introducing the highest possible degree of elitism (population k + 1 is formed only by the two offspring of a pair of parents; they can immediately replace the two lowest ranked elements of population k, and they can coexist with their parents).

  There are several advantages deriving from this configuration: once a good chromosome has been found, it immediately enters the population, and then can be immediately picked up as a parent, without having to wait for the evaluation of the other chromosomes of its same generation; a low size population has a lower exploration ability and a higher exploitation ability, since the best chromosomes are more frequently selected as parents; the high elitism allows all the best chromosomes found to be kept.

  Another modification to the original algorithm is the creation of a library of chromosomes in which all the already evaluated chromosomes are listed. A chromosome will be evaluated only if it never had a 'twin', since a library search is by far less time consuming than the evaluation of a response.
- The 'quality' of a subset of variables is related both to the value of the response and to the selected variables (how many and which ones). In many cases it can be important to have solutions containing as few variables as possible, since it will lead to an easier mathematical model and sometimes to lower costs.

  It is therefore interesting to evaluate how the response changes as a function of the number of selected variables. To do that, the following rule has been added: the best chromosome using the same number of variables is protected, regardless of its position in the ordered population, unless a chromosome with a lower number of variables gives a better response. Applying this rule, a protected chromosome can be 'killed' only by another chromosome giving a better response by using the same (or a lower) number of variables.

At the end of the run, the user can see very easily the evolution of the response as a function of the number of selected variables and then decide what is the best compromise between the number of variables and the response function.

- Usually, the best combinations select only a small percentage of the variables, and then in a chromosome the number of 0s is much higher than the number of 1s. This is rather peculiar of this kind of optimization, since in numerical optimization the number of 0s and 1s is on average the same. The usual initialization of the population, in which each bit is given a value of 0 or 1 at random with the same probability, would be almost inapplicable in this case, since the initial population would be formed by chromosomes corresponding to subsets containing as an average half of the variables of the data set.

  Apart from the fact that in some cases these chromosomes could not be evaluated (e.g., in MLR the number of variables must be lower than the number of objects), such a situation would lead to two main disadvantages: the computation of the response for a model containing a higher number of variables requires a much higher time, and then a much lower number of chromosomes can be evaluated in the same time; since the presence of a 'bad' variable amongst several 'good' variables would be almost unnoticed, that chromosome would have a very high response and then that variable would stay undisturbed in it.

  The solution applied to this problem is very simple: at the stage of the creation of the initial population, the probability of having a '1' is much lower than that of having a '0'. As a consequence, the original population is formed by chromosomes corresponding to subsets of only a few variables each (as a guideline, for each gene the probability of being a '1' is set to 5/number of variables; on average, in each chromosome 5 variables will be selected). This means that during the first stages a much higher number of chromosomes is evaluated and that 'bad' variables can be more easily discarded, since one of them can be enough to significantly worsen the response of a chromosome in which only a few 'good' variables are selected. Within each run, the combination of these small, highly informative 'building blocks' will lead to a gradual increase in the number of selected variables, until when it will stabilize around the 'optimal' number.
- One of the advantages of GA is that at the end of a run a population of possible solutions is obtained. As an example, a final population

Table I *Final population of a GA run on the ANTIM data set.*

#	Resp.	Selected variables
1	95.80	2, 7, 13, 17, 20, 50, 51, 52
2	95.01	2, 7, 13, 15, 20, 50, 51, 52
3	94.48	2, 7, 13, 20, 50, 51, 52
4	94.36	2, 9, 13, 20, 50, 51, 52
5	94.13	1, 2, 13, 17, 20, 46, 50, 51
6	93.93	2, 9, 20, 50, 51, 52
7	93.87	1, 7, 13, 20, 47, 50, 51
8	93.83	2, 20, 38, 50, 51
9	93.66	1, 2, 13, 17, 20, 50, 51
10	93.58	2, 17, 20, 38, 50, 51
11	93.24	2, 7, 13, 16, 20, 46, 50, 51, 52
12	93.21	1, 7, 20, 39, 50, 51
13	93.09	2, 7, 14, 17, 20, 50, 51, 52
14	92.92	2, 7, 13, 16, 20, 50, 51, 52
15	92.88	2, 7, 13, 20, 38, 50, 51, 52
16	92.76	2, 9, 13, 17, 20, 50, 51
17	92.73	1, 2, 13, 20, 50, 51, 52
18	92.57	1, 7, 20, 47, 50, 51, 52
19	92.57	1, 7, 13, 20, 39, 50, 51
20	92.55	2, 7, 9, 20, 50, 51, 52
21	92.27	1, 7, 20, 38, 39, 50, 51
22	92.24	2, 13, 20, 50, 51, 52
23	92.18	2, 20, 44, 50, 51, 52
24	91.89	1, 7, 13, 16, 20, 50, 51, 52
25	91.84	2, 7, 13, 20, 46, 50, 51, 52
26	91.78	1, 7, 13, 20, 47, 50, 51, 52
27	89.07	2, 20, 50, 51
28	79.64	50, 51, 52
29	40.74	50, 52
30	30.21	52

obtained by a GA applied to the ANTIM data set (see later in the text) is the one shown in Table I. If we look at chromosome 3, we can see that the variables selected by it are a subset of the variables selected by chromosomes 11, 14, 15, and 25. Since chromosome 3 gives the best result, chromosomes 11, 14, 15, and 25 are useless (they give a less good result by using the same variables as chromosome 3, plus some more). The same thing happens also with other chromosomes, so that we can say that the only 'useful' chromosomes are the ones reported in Table II. Only 21 chromosomes out of 30 bring forward valuable information. In such a case, the presence of 'useless' chromosomes produces bad effects for two main reasons: the information obtained by the final population is reduced since not all the chromosomes are relevant; the 'useless' chromosomes are very similar to some other chromosomes,

Table II *'Useful' chromosomes from the population of Table I.*

#	Resp.	Selected variables
1	95.80	2, 7, 13, 17, 20, 50, 51, 52
2	95.01	2, 7, 13, 15, 20, 50, 51, 52
3	94.48	2, 7, 13, 20, 50, 51, 52
4	94.36	2, 9, 13, 20, 50, 51, 52
5	94.13	1, 2, 13, 17, 20, 46, 50, 51
6	93.93	2, 9, 20, 50, 51, 52
7	93.87	1, 7, 13, 20, 47, 50, 51
8	93.83	2, 20, 38, 50, 51
9	93.66	1, 2, 13, 17, 20, 50, 51
12	93.21	1, 7, 20, 39, 50, 51
13	93.09	2, 7, 14, 17, 20, 50, 51, 52
16	92.76	2, 9, 13, 17, 20, 50, 51
17	92.73	1, 2, 13, 20, 50, 51, 52
18	92.57	1, 7, 20, 47, 50, 51, 52
22	92.24	2, 13, 20, 50, 51, 52
23	92.18	2, 20, 44, 50, 51, 52
24	91.89	1, 7, 13, 16, 20, 50, 51, 52
27	89.07	2, 20, 50, 51
28	79.64	50, 51, 52
29	40.74	50, 52
30	30.21	52

and so the exploration potential is very much reduced; as a consequence, the risk of being stuck on a local maximum is increased.

The following rule has been added: if chromosome A uses a subset of the variables used by chromosome B, and the response of chromosome A is higher than the response of chromosome B, then chromosome B is discarded.

THE HYBRIDIZATION WITH STEPWISE SELECTION

Generally speaking, the classical techniques are characterized by a very high exploitation and a very poor exploration: this means that, given a starting point, they are able to find the nearest local maximum, but, once they have found it, they get stuck on it. On the contrary, techniques such as GA have a very high exploration and a rather low exploitation: this means that they are able to detect several 'hills' leading to different local maxima, but that it is not very easy for them to climb up to the maximum.

It is therefore rather intuitive to think that coupling the two techniques should produce a new strategy having both high exploration and high exploitation, since the application of the classical technique to one of the solutions found by GA should lead to the identification of the local maximum near which the chromosome was lying (Hibbert, 1993).

The most used classical technique is stepwise selection, which can be used in its forward or backward form. The former adds one variable at a time, while the latter removes one variable at a time, and the procedure continues until step n + 1 leads to results not better than step n.

Both forms of stepwise selection can be very useful when combined with GA. Starting from one of the chromosomes of the population, the backward procedure allows 'cleaning' of the model, by eliminating those variables that give no relevant information but that have been selected by chance. On the other hand, the forward procedure allows those variables to enter that give relevant information but that have never been selected. Such a hybrid GA alternates generations of GA with cycles of stepwise selection, in the backward or in the forward form.

Once more, the main risk connected to this step is overfitting: a stepwise selection pushed too hard will very much improve the fitting to the training set, but can lead to a loss of true predictivity. To limit this danger, only those models are accepted that produce a relevant decrease of the prediction error (compared with that of the 'parent' model). A good value for this threshold seems to be 2%, and of course it has to be higher the higher is the risk of overfitting (i.e. with large variables/objects ratios or with a very high noise).

After their evaluation, the new models deriving from the stepwise selection and fulfilling the previous requirement will be considered as new chromosomes, and then undergo the usual steps of check of subsets and insertion into the population. According to this consideration, the sentence 'the procedure continues until step n + 1 leads to results not better than step n' will become 'the procedure continues until step n + 1 leads to results not *significantly* better than step n' (the word 'significantly' has to not be interpreted in a 'statistical' way).

Since the variables present in a chromosome are generally a small subset of the total variables, it is evident that a step in backward selection will be much faster than a step in forward selection. As a practical example, if our data set has 100 variables and our initial solution is formed by 10 variables, a backward step will require the estimation of 10 new models, each one lacking one of the selected variables. A forward step will require the estimation of 90 new models, each one with one of the unselected variables. Furthermore, since our aim is to reduce as much as possible the number of variables used, and since it is highly probable that, due to the randomness connected to the generation of the chromosomes, some non-relevant variables have also been selected, it is evident that the backward strategy should be more frequently used.

Having defined b as the frequency of backward stepwise and f as the frequency of forward stepwise, with f a multiple of b, the general structure of the algorithm is the following:

(1) start with GA;
(2) after the evaluation of b chromosomes perform backward stepwise;

(3) if $b = f$ then perform forward stepwise;
(4) go to 1.

As a general guideline, $b = 100$ and $f = 500$ are values allowing to obtain a good compromise between exploration and exploitation.

Another problem is to decide when to stop with the stepwise selection. The general algorithm for stepwise selection on v variables, starting from a subset of s selected variables, is the following:

(1) evaluate the model with s variables;
(2) for t times ($t = s$ in backward, $t = v - s$ in forward) evaluate the models with p variables ($p = s - 1$ in backward, $p = s + 1$ in forward);
(3) if the best model evaluated in step 2 is better than that evaluated in step 1 (producing a decrease of the prediction error greater than the predefined value), then go to step 1, with the corresponding subset of variables being the starting model; else end.

The stepwise selection is usually performed on the best chromosome. It can happen anyway that since the last cycle of stepwise the GA didn't find any new better chromosome. In this case, the first chromosome has already undergone stepwise selection, and it would then be useless to repeat it. A downward search among the chromosomes of the population would detect the first one which has not yet undergone stepwise, and this one will be the starting chromosome.

Table III shows a comparison among classical methods, GAs and hybrid GAs.

Table III *Comparison between different strategies.*

Classical methods
– Perform local search, finding a local maximum
– At every step the domain in which the search takes place is reduced
– Produce a single result (the 'best' result)

Genetic algorithms
– No local search
– At every moment every point of the experimental domain can be explored
– Produce a series of almost equivalent results (the user can choose the 'best')

Hybrid genetic algorithms
– Perform local search, finding local maxima
– At every moment every point of the experimental domain can be explored
– Produce a series of almost equivalent results (the user can choose the 'best')

THE PROBLEM OF FULL-VALIDATION

The approach so far described has one main drawback: since the prediction and the choice of the variables are performed on the same objects, the results obtained are not fully validated. Therefore, they cannot be considered as expressing the real predictive capability of the model, and this overestimation increases with the ratio variables/objects and with the number of chromosomes evaluated.

As a confirmation, try to create a data set with, say, 10 objects, 100 X variables and 1 response, in which each datum is a random number. With such a dimensionality, whatever the technique of feature selection you choose, you will always find a subset of variables almost perfectly 'predicting' the response. It is anyway evident that the only fact that you could find a good relationship between the X variables and the response doesn't mean that you can rely on that model to predict the response of other objects of the same data set (i.e. objects whose response is a random number).

The only way to be sure of the predictive ability of a model is to test it on objects that never took part to its definition. A good approach could be the following:

- divide the objects into *d* deletion groups;
- for *i* = 1 to *d* outer cycles
  - perform the GA on the objects not constituting deletion group *i*
  - test the chromosomes of the final population on the objects of deletion group *i*;
- next *i*;
- evaluate the predictions in the *d* outer cycles. As a first guess, one can say that the real predictive ability of the model computed with all the objects will be not worse than the prediction obtained in the worst outer cycle. More information will also be given by the difference between the partial and the full validated predictions obtained with the same subset of variables (the lower it is, the higher the stability of the model and then the probability that the results obtained can have the same value also on a new data set coming from the same population).

In the case of a large number of objects, a much simpler approach is the 'single evaluation set'. The objects are split into a training set, on which the model is computed, and a validation set, on which the model will be validated. Apart from the problem deriving from the fact that the two sets must be homogeneous, this approach will never take into account the information brought by the objects of the evaluation set, and therefore it is hardly applicable in the case of QSAR, since the number of molecules under study is usually not so great to allow to renounce *a priori* to the information brought by some of them.

As one can see, the need of a good validation becomes more and more important at the same conditions at which it becomes more and more difficult to perform it!

TWO QSAR EXAMPLES

Two elaborations will be performed on two QSAR data sets published in the literature, and the results obtained after the application of GA will be compared with the conclusions of the authors of the papers.

Data set ANTIM

Selwood and coworkers published a paper (Selwood *et al.*, 1990) in which a set of antifilarial antimycin analogues described by 53 X variables was studied, with the goal of correlating the structure with the antifilarial activity by a MLR model (Table IV).

The authors used objects 1–16, available at a first time, as training set, and objects 17–31, available only at a following time, as evaluation set. They wanted to use a MLR model, without using PLS; to reduce the number of variables they used the stepwise technique, selecting variables 24, 50, and 51. Figure 2 shows the plot of the observed vs. computed (for the training set) or predicted (for the evaluation set) values. Even discarding objects 17, 18, and 24, that are clear outliers, the predictive ability surely cannot be satisfying: the predicted variance is only 29.74%, with a residual mean square error in prediction (RMSEP) of 0.653.

A deeper analysis of the training and of the evaluation sets reveals that one of the reasons of such a poor performance is that some objects to be predicted are out of the domain spanned by the training set. Figure 3 shows the PC1-PC2 plot of the 31 molecules (each described by the 53 X variables). Though this plane explains only 45.69% of total variance, it is very clear that objects 17, 18, 21, 22, 25, and 31 are well outside the space encompassed by the objects of the training set (1–16). Since their leverage is very high, the prediction error on these objects will be very high, whatever the model computed with objects 1–16. As a consequence, these objects too should be discarded from the evaluation set, that is then composed by the following 8 objects: 19, 20, 23, 26, 27, 28, 29, and 30.

On this evaluation set, a much more acceptable value of 58.38% of explained variance (RMSEP = 0.567) has been obtained.

GAs have then been applied, both in the 'partial validated' and in the 'full validated' form, with the goal of comparing the predictive ability of the selected subsets. Each version has been run five times, and then the subset giving the best result on the training set has been checked on the evaluation set.

Table IV *Data set ANTIM.*

Compounds:

	R1	*R2*
1)	3-NHCHO	$NHC_{14}H_{29}$
2)	3-NHCHO	NH-3-Cl-4-(4-ClC_6H_4O)C_6H_3
3)	5-NO_2	NH-3-Cl-4-(4-ClC_6H_4O)C_6H_3
4)	5-SCH_3	NH-3-Cl-4-(4-ClC_6H_4O)C_6H_3
5)	5-$SOCH_3$	NH-3-Cl-4-(4-ClC_6H_4O)C_6H_3
6)	3-NO_2	NH-3-Cl-4-(4-ClC_6H_4O)C_6H_3
7)	5-CN	NH-3-Cl-4-(4-ClC_6H_4O)C_6H_3
8)	5-NO_2	NH-4-(4-$CF_3C_6H_4O$)C_6H_4
9)	3-SCH_3	NH-3-Cl-4-(4-ClC_6H_4O)C_6H_3
10)	5-SO_2CH_3	NH-3-Cl-4-(4-ClC_6H_4O)C_6H_3
11)	5-NO_2	NH-4-(C_6H_5O)C_6H_4
12)	5-NO_2	NH-3-Cl-4-(4-ClC_6H_4CO)C_6H_3
13)	5-NO_2	NH-4-(2-Cl-4-$NO_2C_6H_3O$)C_6H_4
14)	5-NO_2	NH-3-Cl-4-(4-$CH_3OC_6H_4O$)C_6H_3
15)	3-SO_2CH_3	NH-3-Cl-4-(4-ClC_6H_4O)C_6H_3
16)	5-NO_2	NH-3-Cl-4-(4-ClC_6H_4S)C_6H_3
17)	3-NHCHO	NHC_6H_{13}
18)	3-NHCHO	NHC_8H_{17}
19)	3-$NHCOCH_3$	$NHC_{14}H_{29}$
20)	5-NO_2	$NHC_{14}H_{29}$
21)	3-NO_2	$NHC_{14}H_{29}$
22)	3-NO_2-5-Cl	$NHC_{14}H_{29}$
23)	5-NO_2	NH-4-C$(CH_3)_3C_6H_4$
24)	5-NO_2	$NHC_{12}H_{25}$
25)	3-NO_2	$NHC_{16}H_{33}$
26)	5-NO_2	NH-3-Cl-4-(4-ClC_6H_4NH)C_6H_3
27)	5-NO_2	NH-4-(3-$CF_3C_6H_4O$)C_6H_4
28)	5-NO_2	NH-3-Cl-4-(4-$SCF_3C_6H_4O$)C_6H_3
29)	5-NO_2	NH-3-Cl-4-(3-$CF_3C_6H_4O$)C_6H_3
30)	5-NO_2	NH-4-(C_6H_5CHOH)C_6H_4
31)	5-NO_2	4-ClC_6H_4

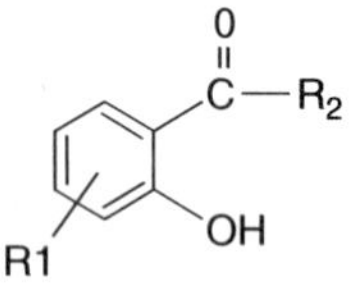

Descriptors:

1–10) partial atomic charges for atoms 1–10
11–13) vectors (X, Y, and Z) of the dipole moment
14) dipole moment
15–24) electrophilic superdelocalizability for atoms 1–10
25–34) nucleophilic superdelocalizability for atoms 1–10
35) van der Waal's volume
36) surface area
37–39) moments of inertia (X, Y, and Z)
40–42) principal ellipsoid axes (1, 2, and 3)
43) molecular weight
44–46) parameters describing substituent dimensions in the X, Y, and Z
47–49) parameters describing directions and the coordinates of the center of the subsituent
50) calculated log P
51) melting point
52–53) sums of the F and R substituent constants

Response: antifilarial activity

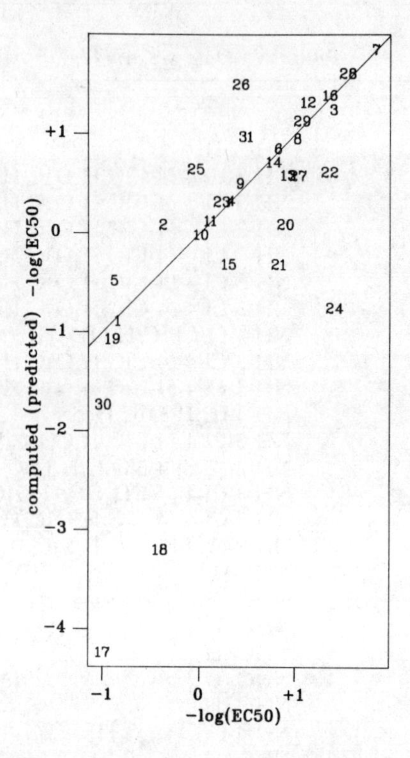

Figure 2 Data set ANTIM: observed vs. computed (obj. 1–16) or predicted (obj. 17–31) values (variables 24, 50, and 51, stepwise selection).

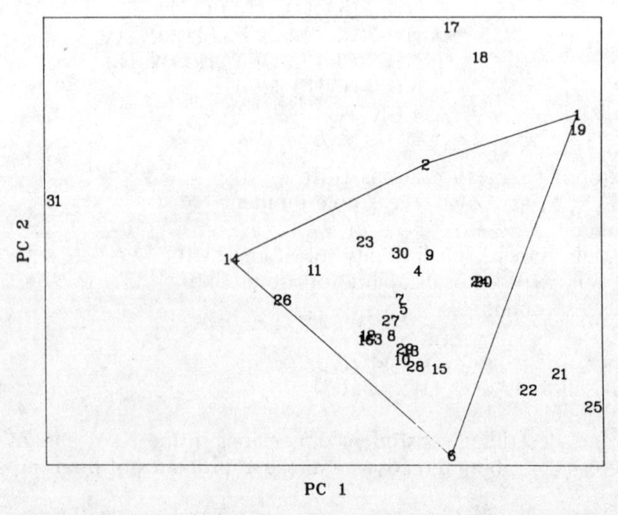

Figure 3 Data set ANTIM: the plot PC1 vs. PC2 shows that some objects of the evaluation set are well outside the domain spanned by the objects of the training set (1–16).

Table V *Configuration of the genetic algorithm.*

	Partial validation	Full validation
size of the population	30	20
probability of initial selection	0.0943	0.0943
probability of cross-over	0.5	0.5
probability of mutation	0.01	0.01
frequency of backward elimination	100	100
frequency of forward selection	500	–
minimum reduction in prediction error to accept stepwise	5%	5%
threshold value	–	70%
outer cycles	–	5
deletion groups	5	5
stop criterion	(a) 1 hour (b) 2' (+ 1 stepwise)	300 ev. (+ 1 stepwise)

Table V shows the configurations of the algorithm. All the parameters are set in such a way that the growth of the response is as fast as possible. The minimum reduction in prediction error to accept a model deriving from stepwise selection has been set a higher value (5%), since the variables–objects ratio is very large and the data are rather noisy.

The GA in partial validation (stop criterion 1h) selected variables 7, 13, 16, 19, 45, 50, and 51, with a response (cross-validated variance on the training set) of 98.67%. When used to predict the objects of the validation set, the predicted variance was only 22.59% (RMSEP 0.773). This is a confirmation that with such a large variables–objects ratio the partial validation used as response criterion can lead to highly overfitted models, with no real predictive ability at all.

When observing the evolution of the response in time, one can see that, after having improved very much in the first few minutes, it reaches a plateau. The interpretation of this phenomenon is very simple: the algorithm, whose configuration is planned to reach in a short time a very high response, can recognize very quickly the general structure of the data; from that moment on, the small improvements are mainly due to the adaptation of the models to that particular data set (random correlations, random predictions, structure of the deletion groups, ...).

Of course, the main problem is to detect when to stop. A great help in taking that decision can come from a randomization test. In it, the responses in the response vector are randomized, so that each molecule will have the activity of another molecule. A GA is run on that data set, and the evolution of the response is recorded. One can consider that these results are noise, and the results obtained with the 'true' data set are information + noise; as a consequence, the stop criterion will be easily detected as the moment in

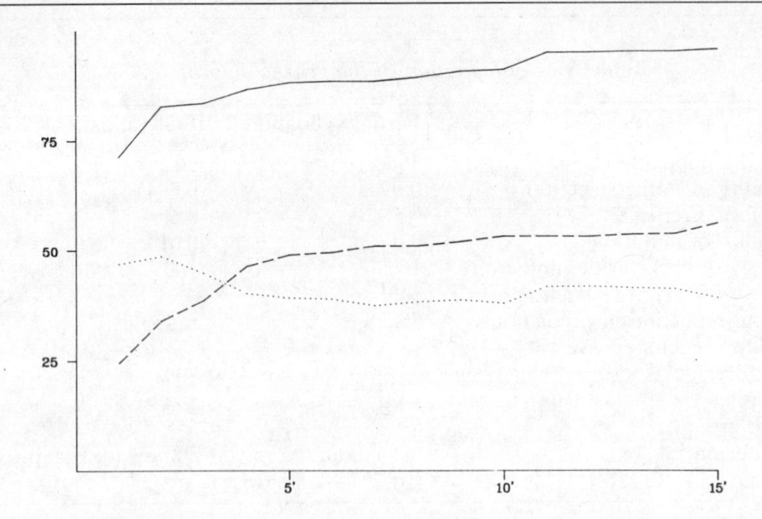

Figure 4 Data set ANTIM: evolution of response (predicted variance in partial validation) vs. time. Solid line: 'true' data set; dashed line: 'randomized' data set; dotted line: difference (each line is the average of 5 runs; in the case of randomization test 5 different randomizations have been performed).

which the difference between the 'true' and the 'random' response is the greatest (Figure 4).

From the data obtained by 5 runs on randomized data and 5 runs on 'true' data, it resulted that the highest difference between the average responses was obtained after 2' (about 300 evaluations). Stopping the elaboration at such an early time reduces very much the overfitting, as shown by the fact that the selected subset (variables 19, 21, 50, and 51) explains 71.50% of the variance of the evaluation set (RMSEP 0.469).

The full-validated version selected variables 16, 19, 50, and 51. With this subset, the variance predicted on the objects of the evaluation set is 79.62% (RMSEP 0.397), significantly better than that obtained with the 3 variables selected by the stepwise technique. Figure 5 shows the plot of the observed vs. computed (for training set) or predicted (for evaluation set) values. In the case of full-validation, the stop criterion is not as critical as in the partially validated version, since the algorithm itself will eliminate the low predicting models.

This example confirms that with a very large variables/objects ratio the partial validation, unless a previous study is performed, leads to highly overfitted models, with no real predictive ability at all; on the contrary, the algorithm based on full validation also in these difficult conditions gives results better than those obtained by classical methods. It is also interesting to notice that the 4 variables selected by the fully validated version are a subset of the 7 variables selected by the partially validated one (stop criterion 1h): it

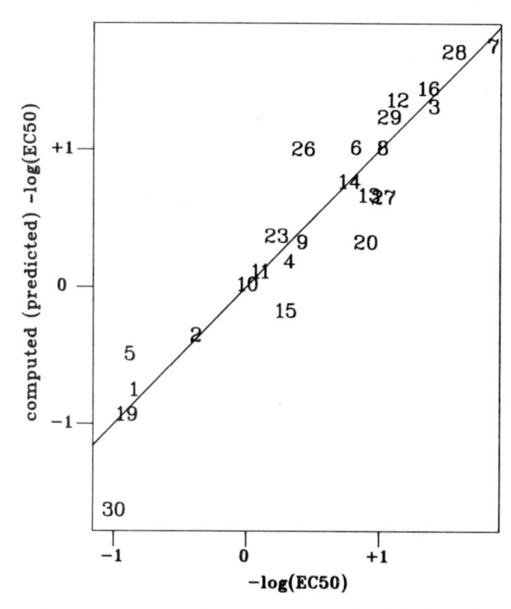

Figure 5 Data set ANTIM: observed vs. computed (obj. 1–16) or predicted values (variables 16, 19, 50, and 51, genetic algorithm).

is therefore very clear that variables 7, 13, and 45 have been selected due to a random correlation with the response.

Data set HEME

Hellberg and coworkers published a paper (Hellberg *et al.*, 1985) in which a data set of halogenated ethyl methyl ethers (HEME) was described by 41 X variables (molecular descriptors) and 2 responses (anesthetic activity and toxicity).

The total number of molecules was 22, but not each of them had both responses: 14 objects had the anesthetic activity, 10 the toxicity, and 5 had none of them. The first response will be studied, and then the data set will be formed by 14 objects, 41 descriptors and 1 response (Table VI).

Due to the very limited number of objects, it is impossible to think to an independent validation set, and then the only possible form of prediction will be cross-validation.

The authors applied PLS to the whole data set, with the 2-component model explaining 86% of the variance in fitting; by cross-validation, the best predictions are obtained with 5 components (predicted variance 75.05%).

A cluster analysis applied to the variables (Figure 6) shows that they form well defined clusters, and that several variables have a very high similarity. At a similarity level of 0.6 the following 9 clusters are identified:

Table VI *Data set HEME.*

Compounds:
1) CH_3-CF_2-O-CF_2Cl
2) CHF_2-CH_2-O-CH_3
3) CH_2F-CF_2-O-CHF_2
4) CH_2Cl-CF_2-O-CF_3
5) CF_2Cl-CH_2-O-CH_3
6) $CHCl_2$-CF_2-O-CF_3
7) CH_2Cl-CF_2-O-CF_2Cl
8) $CFCl_2$-CF_2-O-$CHCl_2$
9) $CFCl_2$-CF_2-O-CCl_3
10) $CHCl_2$-CF_2-O-CHF_2
11) $CHCl_2$-CF_2-O-CH_2F
12) CCl_3-CF_2-O-CHF_2
13) CHFCl-CF_2-O-$CHCl_2$
14) $CHCl_2$-CF_2-O-CH_2Cl

Descriptors:
1) log P
2) $(\log P)^2$
3) molar volume
4) molecular weight
5) mean molecular polarizability
6) $\Sigma\pi$ – C1
7) $(\Sigma\pi)^2$ – C1
8) $\Sigma\pi$ – C3
9) $(\Sigma\pi)^2$ – C3
10) charge of C1
11) electronegativity of C1
12) charge of C2
13) electronegativity of C2
14) charge of C3
15) electronegativity of C3
16) Q of most E halogen on C1
17) E of most E halogen on C1
18) Q of most E halogen on C2
19) E of most E halogen on C2
20) Q of most E halogen on C3
21) E of most E halogen on C3
22) Q of most acidic hydrogen
23) E of most acidic hydrogen
24) Q of the oxygen
25) E of the oxygen
26) ΔQ – HC
27) ΔE – HC
28) E – CH
29) Σ Q – C1
30) Σ Q – C2
31) Σ Q – C3
32) Σ Q – (Q – C1)
33) Σ Q – (Q – C2)
34) Σ Q – (Q – C3)
35) Σ Q * (Q – C1)
36) Σ Q * (Q – C2)
37) Σ Q * (Q – C3)
38) strongest H-bonder
39) 2nd strongest H-bonder
40) (O) strongest H-bonder
41) (O) 2nd strongest H-bonder

Response: anesthetic activity

- Cluster 1: variables 1, 2, 3, 4, 5, 8, 9 (bulk, lipophilicity and polarizability);
- Cluster 2: variables 6, 7 (lipophilicity on C1);
- Cluster 3: variables 12, 13, 30, 33, 36 (description of C2);
- Cluster 4: variables 10, 11, 16, 17, 19, 20, 24, 29, 32, 35 (mainly, description of C1);
- Cluster 5: variables 14, 15, 21, 25, 31, 34, 37 (mainly, description of C3);
- Cluster 6: variables 18, 26;
- Cluster 7: variables 22, 23, 27, 28 (most acidic hydrogen);
- Cluster 8: variables 38, 40 (strongest H-bonder);
- Cluster 9: variables 39, 41 (second strongest H-bonder).

Since one can expect a very high degree of redundancy on the data, a rather important reduction in the number of the variables should be possible.

GA has been applied in the full-validated version, with the same structure

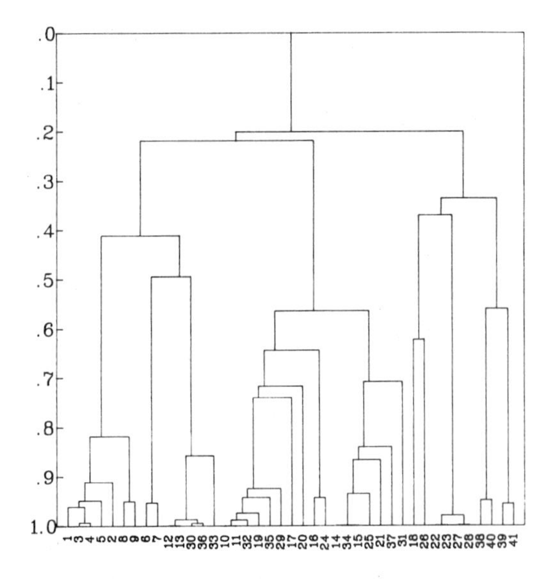

Figure 6 Data set HEME: dendrogram of the variables (distance: absolute value of the correlation coefficient; average linkage, weighted).

of the previous example; the only differences were the initial probability of selection (0.0732) and the threshold (75%). The best result obtained in the five runs selected 6 variables (4, 6, 7, 17, 31, 39); the most predictive model was the one with five components and the outer cycle with the worst prediction in full-validation scored 82.44%. This means that in any of the five outer cycles at least one combination of variables giving a prediction higher than 82% on the independent evaluation set has been found. Of course, since no external evaluation set is present, no direct final comparison of the two models can be performed; anyway, the fact that the 75% threshold has been by far succesfully fulfilled in each outer cycle means that the global predictive ability is at least comparable with that obtained with all the variables.

When taking into account the selected variables, one can see that, except for variables 6 and 7, each of them is in a different cluster; this is another confirmation of the good choice performed by the GA.

ACKNOWLEDGEMENTS

The author wishes to thank Drs Carlos Lucasius and David Rogers for the very useful scientific discussions held during the *Second International Workshop on Neural Networks and Genetic Algorithms Applied to QSAR and Drug Design* (Lyon, France, June 12–14, 1995).

This work received financial support from the Ministry of University and Scientific Research (MURST 40% and 60%) and from the CNR (Italian National Research Council), Comitato Scienze e Tecnologia Informazione.

The source codes of the programs are available from the author upon request.

REFERENCES

Hellberg, S., Wold, S., Dunn III, W.J., Gasteiger, J., and Hutchings, M.G. (1985). The anesthetic activity and toxicity of halogenated ethyl methyl ethers, a multivariate QSAR modelled by PLS. *Quant. Struct.-Act. Relat.* **4**, 1–11.

Hibbert, D.B. (1993). A hybrid genetic algorithm for the estimation of kinetic parameters. *Chemom. Intell. Lab. Syst.* **19**, 319–329.

Lanteri, S. (1992). Full validation procedures for feature selection in classification and regression problems. *Chemom. Intell. Lab. Syst.* **15**, 159–169.

Leardi, R. (1994). Application of a genetic algorithm to feature selection under full validation conditions and to outlier detection. *J. Chemom.* **8**, 65–79.

Leardi, R., Boggia, R., and Terrile, M. (1992). Genetic algorithms as a strategy for feature selection. *J. Chemom.* **6**, 267–281.

Lucasius, C.B. and Kateman, G. (1991). Genetic algorithms for large-scale optimization in chemometrics: An application. *Trends Anal. Chem.* **10**, 254–261.

Lucasius, C.B., Beckers, M.L.M., and Kateman, G. (1994). Genetic algorithms in wavelength selection: A comparative study. *Anal. Chim. Acta* **286**, 135–153.

Selwood, D.L., Livingstone, D.J., Comley, J.C.W., O'Dowd, A.B., Hudson, A.T., Jackson, P., Jandu, K.S., Rose, V.S., and Stables, J.N. (1990). Structure–activity relationships of antifilarial antimycin analogues: A multivariate pattern recognition study. *J. Med. Chem.* **33**, 136–142.

4 Some Theory and Examples of Genetic Function Approximation with Comparison to Evolutionary Techniques

D. ROGERS
Molecular Simulations Incorporated, 9685 Scranton Road, San Diego, CA 92121, USA

Genetic function approximation (GFA) is a statistical modeling algorithm which builds functional models of experimental data. Since its inception, several applications of this algorithm in the area of quantitative structure–activity relationship modeling have been reported, and at least two related modeling techniques based upon evolutionary (mutation-only) search have appeared. This chapter gives an expanded discussion of issues that arise in the application GFA, some additional theory concerning the lack-of-fit score, new experiments, and comparisons with the three evolutionary or genetic modeling techniques: Kubinyi's MUSEUM, Luke's evolutionary programming, and Leardi's feature selection method.

KEYWORDS: *genetic algorithms; genetic function approximation; quantitative structure–activity relationships; evolutionary algorithms; spline modeling.*

INTRODUCTION

'If your experiment needs statistics,
you ought to have done a better experiment.'

Ernest Rutherford

As physicist Ernest Rutherford's comment shows, statistics were hardly a respected branch of mathematics at the turn of the century, not least by the physicists who were determined to resist believing in God's dice-throws as long as possible. Modern physicists, at least, have grudgingly grown to accept the

In, *Genetic Algorithms in Molecular Modeling* (J. Devillers, Ed.)
Academic Press, London, 1996, pp. 87–107.
ISBN 0-12-213810-4

fundamental role of statistics in their search for a mathematics model of the physical world. In their entry into the quantum realm, statistics turned from an unwanted side-product of a poorly-designed experiment to become the property they wished to measure. Rutherford would hardly have been pleased.

Quantitative structure–activity relationship (QSAR) modeling is a branch of chemistry that attempts to use statistical modeling principles to estimate the biological activity of molecules. Often it appears no more wanted or respected by chemists than statistical physics was to many physicists. Chemists, like those physicists, prefer concrete mechanisms to abstract predictions, and absolute knowledge to merely probable outcomes. But much as physicists were forced to change their stance under pressure from quantum theory, chemists face an equally powerful pressure: the explosion of availability of both experimental and computational information. Combinatorial chemistry threatens to present experimental data in huge quantities. Computational models of 3D structures are continuing a steady improvement in their ability to generate realistic estimates of molecular properties. A request for a QSAR model of 100 000 compounds with associated descriptors and activity is a real possibility in the near future. Rather than being left behind in the current surge of change, computational chemistry in general and QSAR in particular will soon be required to perform staggering feats of data analysis. It is the author's hope that this chapter, and many of the other chapters in this book, will stimulate critical discussion and evaluation of methods which may play a key role in that undertaking.

GENETIC FUNCTION APPROXIMATION

The genetic function approximation (GFA) algorithm is a genetic algorithm (GA) (Holland, 1975) derived from the previously-reported G/SPLINES algorithm (Rogers, 1991, 1992), and has been recently applied to the generation of QSAR models (Rogers, 1995; Rogers and Hopfinger, 1994).

In most cases, the QSAR models are represented as sums of linear or nonlinear terms. The form is shown in Eq. (1).

$$\mathrm{F}(X) = a_0 + \sum_{k=1}^{M} a_k \phi_k(X) \tag{1}$$

The terms are called *basis function*, and are denoted $\{\phi_k\}$; these are functions of one or more features, such as $(\mathbf{X}_5 - 10)^2$, $\mathrm{SIN}(\mathbf{X}_1) \times \mathrm{SIN}(\mathbf{X}_2)$, or $\mathbf{X}_{10}$, where the $\mathbf{X}_i$s are the feature measures. The model coefficients $\{a_k\}$ are determined using least-squares regression or another suitable fitting technique. The linear strings of basis functions play the role of the DNA for the application of the GA.

The initial QSAR models are generated by randomly selecting a number of features from the training data set, building basis functions from these features using the user-specified basis function types, and then constructing the genetic models from random sequences of these basis functions. Most commonly, only linear polynomial basis functions are desired, so as to create models which are significant and easy to interpret. Thus, SIN(TANH(log P)) may allow for a close fit to the data samples, but likely would have no reasonable relationship to the underlying mechanism of activity, and so should not be used.

The fitness function used during the evolution is derived from Friedman's LOF (lack-of-fit) scoring function (Friedman, 1988a, b, c), which is a penalized least-squares error measure. This measure balances the decrease in error as more basis functions are added against a penalty related to the number of basis functions and complexity of the model. By insisting on compactness as well as low error on the training set, the LOF-derived fitness function guides the system towards models which predict well and resist overfitting. For a more detailed discussion of the LOF measure and related measures, the reader should consult the next section.

At this point, we repeatedly perform the *genetic recombination* or *crossover* operation:

- Two good models are probabilistically selected as 'parents' proportional to their fitness.
- Each is randomly 'cut' into two sections. The cuts occur between the basis functions; individual basis functions are not cut. A new model is created using the basis functions taken from a section of each parent.
- Optional mutation operators may alter the newly-created model.
- The model with the worst fitness is replaced by this new model.
- The overall process is ended when the average fitness of the models in the population stops improving. For a population of 300 models, 3000 to 10 000 genetic operations are usually sufficient to achieve 'convergence'.

A description of a typical crossover operation is shown in Figure 1. Upon completion, the user can simply select the model from the population with the best fitness score, though it is usually preferable to inspect the different models and select on the basis of the appropriateness of the features, the basis functions, and the feature combinations.

COMMENTS ON THE LACK-OF-FIT MEASURE

One of the novel additions of GFA to the application area of quantitative structure–activity relationships was the use of Friedman's lack-of-fit (LOF) function as an appropriate fitness measure. Given the newness of this function in the QSAR community, it is useful to review the history and development of this measure.

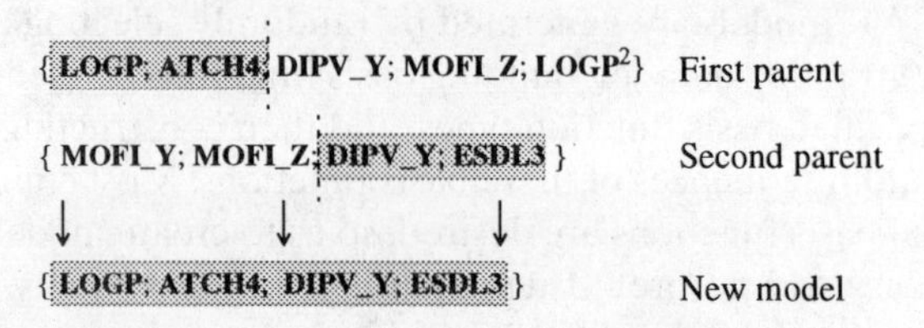

Figure 1 The crossover operation. Each parent is cut at a random point, and a piece from each parent is used to construct the new model which now uses some basis functions from each parent.

Friedman's work on regression splines can be seen as a special case of studies on smoothing splines. These are functions which minimize the relationship shown in Eq. (2).

$$\frac{1}{N}\sum_{i=1}^{N}(y_i - f(x_i))^2 + \lambda \int_a^b (f''(x))^2 \, \mathrm{d}x \tag{2}$$

In this equation, the first term is the well-known least-squares error. The latter term is a measure of the smoothness of the function. The coefficient λ determines the relative weight between fitting and smoothing, and is usually determined through cross-validation. The solution to this problem is in a class of functions known as *natural polynomial splines*. The appropriateness of a given solution at a given value of λ can be shown to be measured by a generalized cross-validation (GCV) criteria shown in Eq. (3), as developed in the work of Craven and Wahba (1979).

$$\mathrm{GCV}\,(\text{model}, \lambda) = \frac{1}{N} \frac{\mathrm{LSE}\,(\text{model})}{\left(1 - \frac{C(\text{model}, \lambda)}{N}\right)^2} \tag{3}$$

The GCV criterion is the average-squared residual of the fit to the data (the numerator) times a penalty (inverse denominator) to account for the increased variance associated with model complexity. N is the number of samples in the data set, and C (model, λ) is a complexity cost function which estimates the cost of the model.

Regression splines are a special case, in which $\lambda = 0$ is assumed. In this case, the complexity cost function C is simply the number of nonconstant basis function 'c' plus one. This assumes that the parameters of the basis functions are determined independently of the dependent variable or variables. The special-case criterion is shown in Eq. (4).

$$\text{GCV (model)} = \frac{1}{N} \frac{\text{LSE (model)}}{\left(1 - \frac{(c+1)}{N}\right)^2} \tag{4}$$

However, in both Friedman's work on multivariate adaptive regression splines (MARS) (Friedman, 1988c) and GFA, the parameters of the basis functions are not determined independently of the dependent variables. Indeed, it is by the use of the responses in constructing the models that these techniques achieve their power and flexibility. However, this flexibility results in the GCV function overestimating the true fit of the models.

Friedman and Silverman (1989) recognized this fact, and proposed the addition of another penalty term, and a user-settable variable, to reflect this bias:

$$\text{LOF(model)} = \frac{1}{N} \frac{\text{LSE (model)}}{\left(1 - \frac{(c + 1 + (d \times p))}{N}\right)^2} \tag{5}$$

In this LOF function, c is the number of nonconstant basis functions, N is the number of samples in the data set, d is a smoothing factor to be set by the user, and p is the total number of parameters in the model (coefficients + knots). Most commonly, cross-validation is used to assign a value for d; Friedman and Silverman (1989) suggest a value of $d = 2$ based upon the expected decrease in the average-squared residual by adding a single knot to make a piecewise linear model. Friedman's models contain only spline basis functions; from experiments with models containing linear terms, I find that a value of $d = 1$ is often better. From experiments with MARS, Friedman notes:

- The actual accuracy of the modeling either in terms of expected squared error is fairly insensitive to the value of d in the range $2 < d < 4$.
- However, the value of the LOF function for the final model does exhibit a moderate dependence on the choice of d.

These two claims, taken together, mean that while how well one is doing is fairly insensitive to the choice of d, how well one thinks one is doing does depend on its value. This is fortunate, as one would not want the underlying quality of the model to be highly sensitive to small changes in the value of d. Friedman suggests a sample re-use technique to obtain an additional estimate of goodness-of-fit of the final model if it needs to be known with more precision.

Note that if a model contains both linear and spline terms, the spline terms 'cost' more than the linear terms. Depending on the value of d, a spline term costs between 50% and 200% more than a linear term. This creates a bias towards linear models which is only overcome by a spline term's ability to decrease LSE significantly more than a competing linear term. The use of GFA to construct nonlinear models is discussed further in the next section.

NONLINEAR MODELING

Splines

Recent work on neural-network-based QSAR modeling is partly motivated by the desire to model data sets which contain nonlinear relationships between descriptors and the activity or activities. GFA can create nonlinear models using spline-based terms. Spline models have several advantages over neural network models for the treatment of nonlinear effect. They are easier to interpret; the location of the discontinuity is presented to the user in an understandable manner. They allow a natural mixing of linear and nonlinear data; adjustments to the LOF function cause an additional penalty to be applied to reflect the added complexity of spline terms. Spline models have been well-studied in the statistics literature, and so the user can utilize the standard statistical methods in evaluating the models. A fairly technical review of the theoretical basis of spline modeling is given in Wahba (1990).

The splines used are *truncated power splines*, and in this work are denoted with angle brackets.* For example, <f(**x**) – a> is equal to zero if the value of (f(**x**) – a) is negative, else it is equal to (f(**x**) – a). For example, <**Energy** –2.765> is zero when **Energy** is less than 2.765, and equal to (**Energy** –2.765) otherwise, as shown in Figure 2. The constant a is called the *knot* of the spline. When a spline term is created, the knot is set using the value of the descriptor in a random data sample.

A spline partitions the data samples into two classes, depending on the value of some feature. The value of the spline is zero for one of the classes, and nonzero for the other class. When a spline basis function is used in a linear sum, the contribution of members of the first class can be adjusted independently of the members of the second class. Linear regression assumes

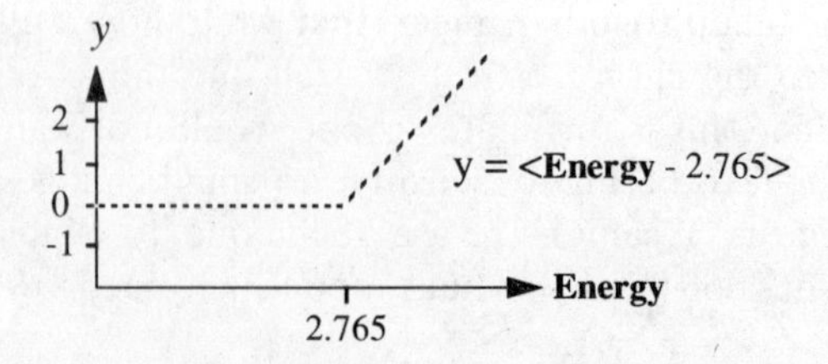

Figure 2 A graph of the truncated power spline <**Energy** – 2.765>.

* This is different from the standard notation which is of the form $(f(\mathbf{x}) - a)_+$. The change was made to allow representation in standard ASCII printouts.

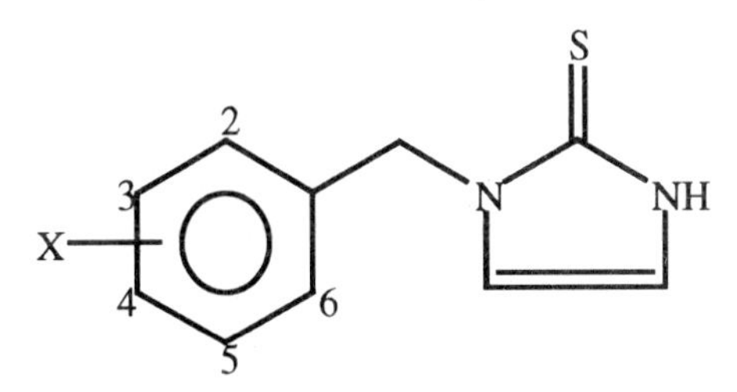

Figure 3 The shared structure of the dopamine β-hydroxylase inhibitors.

that the effect of a feature on the response is linear over its range. Regression with splines allows the incorporation of features which do not have the linear effect over their entire range.

Example of nonlinear modeling with splines: dopamine β-hydroxylase inhibitors

The dopamine β-hydroxylase inhibitor data set is a set of 47 1-(substituted-benzyl)imidazole-2(3H)-thiones with associated inhibitory activities described in the work of Kruse *et al.* (1987). These inhibitors effectively reduce blood pressure and are used for treatment of cardiovascular disorders related to hypertension. The series of analogs are of the general form shown in Figure 3.

We were interested in studying this data set because the compounds had been studied using molecule shape analysis (MSA) by Burke and Hopfinger (1990). The original Kruse *et al.* (1987) study contained 52 compounds though only 25 were used for QSAR generation. The less-active compounds had their activity reported as percent inhibition at fixed concentration, and the remaining compounds had their activity reported as –log (IC_{50}) values. Burke and Hopfinger (1990) chose 47 of the 52 compounds for their study, estimating –log (IC_{50}) for the samples reported with percent inhibition at fixed concentration. Five compounds were considered problematical due to differences in structure, flexibility, and charge from the remaining 47 compounds, and were not used in their study, nor were their –log (IC_{50}) estimates reported. While we realized such removal of samples can bias the resulting QSAR, we wished to compare our work with the results of Burke and Hopfinger (1990), and so we used this same 47 compound subset in our QSAR study. The original compound numbers are retained from the Kruse *et al.* (1987) study.

The –log (IC_{50}) values were generated by Burke and Hopfinger (1990) from two different data sets. The first, which contained most of the more actives, was in molar concentration units. The latter were in percent inhibition at fixed inhibitor concentration in molar units. The latter values were

- $\mathbf{V_o}$: Common overlap steric volume against the most-active compound
- π_0: Molecular lipophilicity
- π_4: Water/octanol fragment constant of the 4-substituent
- $\mathbf{Q_6}$: The partial atomic charge on atom 6
- $\mathbf{Q_{3,4,5}}$: Sum of partial atomic charges on atoms 3, 4, and 5

Figure 4 The QSAR descriptors generated by Burke and Hopfinger (1990).

transformed into the former units using a simple transformation. While the appropriateness of this transformation process may be questioned, for the interests of comparing models built with the same data set it should be satisfactory. However, it is clear that superior descriptors should result in higher-quality models, whatever the subsequent modeling technique.

Linear free energy descriptors were used by Kruse *et al.* (1987) to construct their QSARs. Burke and Hopfinger (1990) constructed QSARs for this data set; they generated 5 descriptors for each of the compounds. The QSAR descriptors generated by Burke and Hopfinger (1990) are shown in Figure 4. For more detailed information on the method of calculation, meanings, and units of these descriptors, the reader is referred to Burke and Hopfinger (1990).

Burke and Hopfinger (1990) proposed two models. The first model contained six terms and a constant, and used the complete set of 47 compounds, and is shown in Eq. (6).

$$\begin{aligned} -\log(IC_{50}) = {} & 52.27 \\ & - 116.9 \times \mathbf{V_o} \\ & + 69.1 \times \mathbf{V_o}^2 \\ & + 2.06 \times \mathbf{Q_{3,4,5}} \\ & - 4.68 \times \mathbf{Q_6} \\ & + 0.0465 \times \pi_0^2 \\ & - 0.578 \times \pi_4 \end{aligned} \qquad (6)$$

N: 47; r^2: 0.828

The second model contained three terms and a constant and used 45 of the compounds (compounds 23 and 39 were identified as outliers and removed). This reduced model is shown in Eq. (7). The interpretation of this model is that the best compounds have the most shape-similarity to compound 51 (as measured by the $\mathbf{V_o}$ descriptor) and have a substituent pattern which make atoms 3, 4, and 5 a positive-charge domain (as reflected in the $\mathbf{Q_{3,4,5}}$ descriptor).

$$\begin{aligned} -\log(IC_{50}) = {} & 52.15 \\ & - 117.5 \times \mathbf{V_o} \\ & + 70.4 \times \mathbf{V_o}^2 \\ & + 2.32 \times \mathbf{Q_{3,4,5}} \end{aligned} \qquad (7)$$

N: 45 (samples 23 and 39 removed); r^2: 0.810

The energetic descriptors calculated by a receptor surface model were tested for their ability to represent 3-dimensional steric and electrostatic information critical to the estimation of activity, and hence their ability to serve as components of QSAR models in situations where 3-dimensional effects must be considered.

A receptor surface model was constructed from a subset of the most-active compounds, their associated activities, and a steric 'tolerance'. This tolerance is the distance from the van der Waals surface of the overlapped molecules that the receptor surface is constructed. The most useful descriptors are derived from the receptor surface model constructed from the three most-active compounds (i.e., 50–52) at a tolerance of 0.1 Å.

We analyzed the dopamine β-hydroxylase inhibitors with GFA using linear polynomials and linear splines. The data set contained 4 receptor surface descriptors generated from the receptor surface model of the 3 most active compounds at a steric tolerance of 0.1 Å. These receptor surface descriptors were combined with the Burke and Hopfinger (1990) QSAR descriptors shown in Figure 4. The population of QSAR models was evolved for 5000 generations. The best QSAR model, as rated by the GFA's LOF score, is shown in Eq. (8).

$$
\begin{aligned}
-\log\,(\mathrm{IC}_{50})_{\text{P-QSAR}} &= 3.762 \\
&+ 0.296 \times \langle -10.203 - \mathbf{E}_{\mathbf{interact}} \rangle \\
&+ 0.089 \times \langle 26.855 - \mathbf{E}_{\mathbf{inside}} \rangle
\end{aligned}
\tag{8}
$$

N: 47; r^2: 0.808; regression-only CV–r^2: 0.788; fully CV–r^2: 0.669

This model explains the entire training set (no outliers removed) nearly as well as either of the models of Burke and Hopfinger (1990), and contains fewer descriptors. If we count each spline-based term as the equivalent of two linear terms, this four-term model performs as well as the six-term Burke and Hopfinger (1990) model. None of the Burke and Hopfinger (1990) descriptors are chosen in the best-rated models, including their molecular shape descriptor $\mathbf{V_0}$. This suggests the information provided by $\mathbf{V_0}$ is replaceable by receptor surface model descriptors. This suggestion is confirmed by calculating the correlation coefficient r^2 between $\mathbf{V_0}$ and $\mathbf{E_{interact}}$. They are highly correlated, with an r^2 of 0.78. That $\mathbf{E_{interact}}$ rather than $\mathbf{V_0}$ is chosen by the GFA procedure implies that $\mathbf{E_{interact}}$ contains additional useful information (likely electrostatic) in its variance noncorrelated with $\mathbf{V_0}$. This demonstrates the ability of receptor surface model descriptors to provide a compact representation of 3-dimensional information for QSAR modeling of binding data, incorporating both steric and electrostatic components. The conclusion of the QSAR study is that steric and electrostatic effects from a model of 3D binding can explain much of the variance in activity.

Perhaps surprisingly, even though GFA is parameterized to favor linear terms over spline terms, the best model contained only spline terms. Splines are preferred if the data contains nonlinear relationships. This is understandable if we look at a scatterplot of $\mathbf{E_{interact}}$ and $\mathbf{E_{inside}}$ versus $\mathbf{-log\ (IC_{50})}$. These scatterplots are shown in Figure 5. The dotted lines show the location of the spline knot, which is where the spline separates the linear region from the nonlinear region. The spline term in each case has discovered a linear relationship between the points to the left of the knot and the activity. Samples to the right of the knot cause the spline term to return 0.0. The QSAR model uses spline terms to expose, for the most active compounds, the linear relationship between the descriptors and activity. Without splines, these relationships could have been obscured or missed entirely.

There are other techniques available for nonlinear studies; see, for example, the review of Sekulic *et al.* (1993). One possibility is the direct introduction of nonlinearity into the partial least squares (PLS) process ('nonlinear-PLS') (Berglund *et al.*, 1995). Minimally, the user can visually inspect the scatterplots of the descriptors versus activity, and create new descriptors (such as quadratic terms) from the nonlinear descriptors. Notwithstanding, there are advantages to our approach: the nonlinearities are discovered automatically, which may be difficult to do visually as the number of descriptors increases; the spline terms are easy to interpret, unlike quadratic terms or PLS latent variables; and the technique allows but does not require the construction of nonlinear models, since the spline terms are eliminated if they do not significantly increase the performance of the model.

The above results confirm that the receptor surface models allow us to model the dopamine β-hydroxylase inhibitor training compounds better, but in most cases what we are interested in is predictiveness: how well can we estimate the activity of compounds outside the training set? This is commonly estimated using cross-validation.

Cross-validation is done by dividing the training set into some number of groups, called cross-validation groups. Each group is left out in turn, and the remaining groups used to build a model of activity. The samples in the left-out group are then predicted using this model. At the end of the process, all samples have been predicted. Commonly, each cross-validation group may contain only one sample; this is leave-one-out cross-validation. In any case, the final estimate is usually expressed as a *cross-validated*-r^2.

For the dopamine β-hydroxylase data set, we divided the samples into 16 cross-validation groups. Leaving out each group in turn, we used PLS on the data set containing the ten 2-dimensional QSAR descriptors and the four 3-dimensional receptor surface model based descriptors. The nonlinear GFA analysis of the data gave a cross-validated-r^2 of 0.669; the PLS analysis performed much worse, with a cross-validated-r^2 of 0.471.

The relatively poor result from PLS is due to the nonlinearities in the data variables. A linear technique such as PLS cannot discover these values automatically; GFA was able to discover them and so create superior models and a superior cross-validation score.

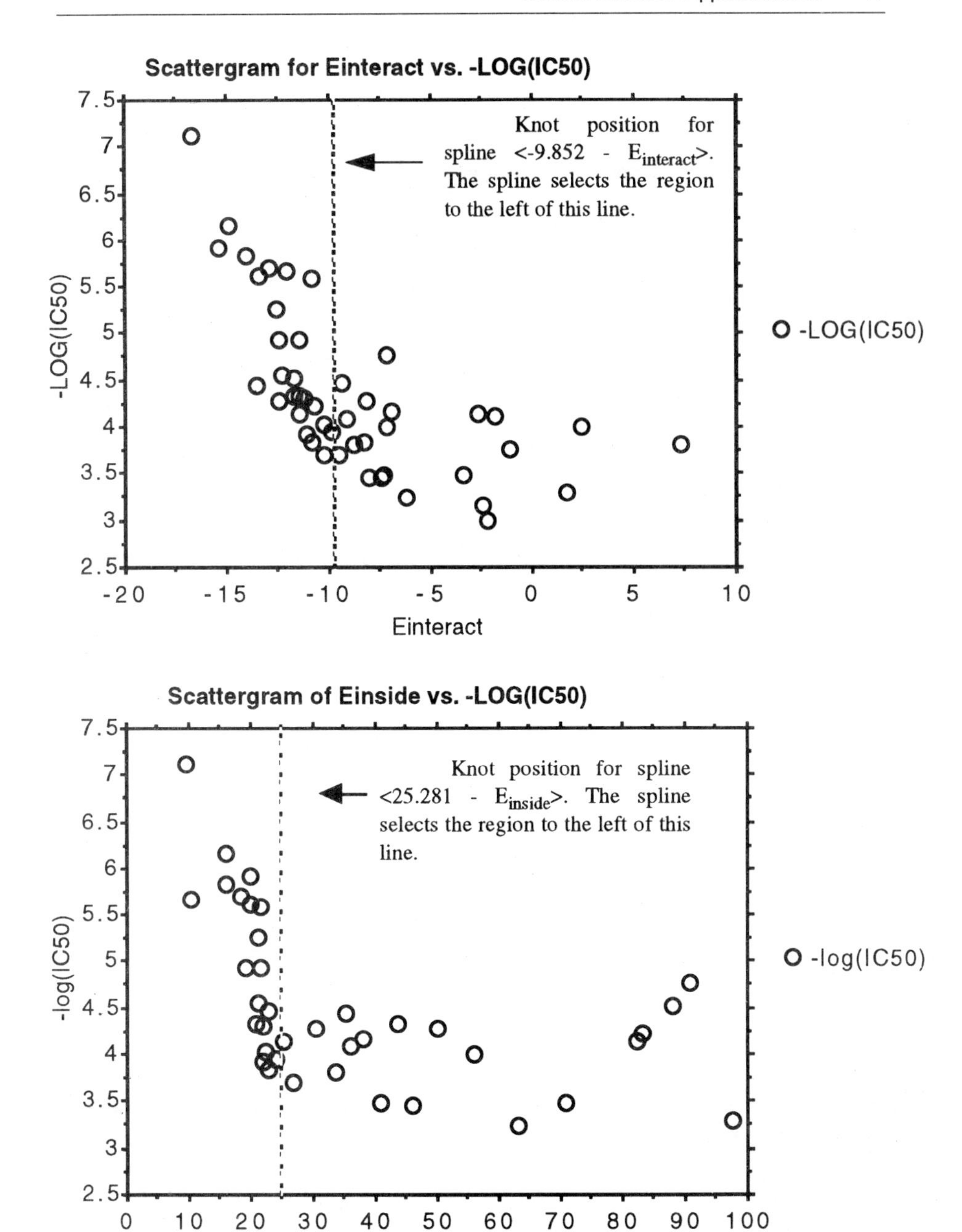

Figure 5 Scatterplots of the descriptors $\mathbf{E}_{\mathbf{interact}}$ and $\mathbf{E}_{\mathbf{inside}}$ versus –log (IC_{50}). The dotted lines show the location of the knot; the spline term in each case has discovered a linear relationship between the points to the left of the knot and the activity. Samples to the right of the knot cause the spline term to return 0.0.

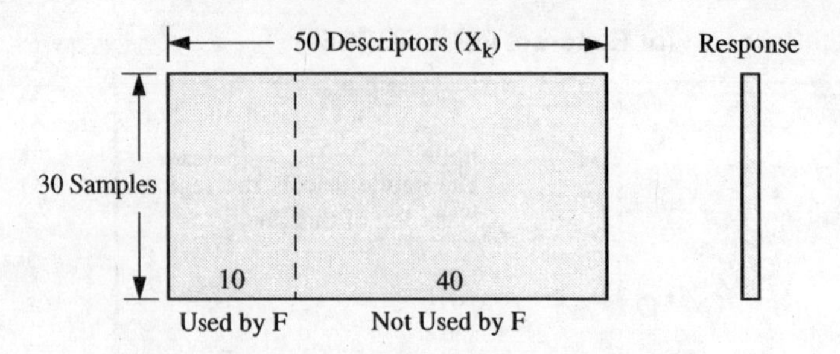

Figure 6 The structure of the generated data set.

GFA VERSUS PLS MODELING

To illustrate the use of GFA, and illustrate a situation in which it has an advantage over PLS, a specific test case was constructed. While this is not experimental data, I would claim that the structure of the useful information in this data set is not unlike the structure of many QSAR data sets.

The data set was constructed as follows:

- The underlying function is defined as: $F(X) = \mathbf{X}_1 + 2 \times \mathbf{X}_2 + 3 \times \mathbf{X}_3 + \ldots + 10 \times \mathbf{X}_{10}$.
- 30 random samples of 50 descriptors were created using random values in the range [0.0 . . . 1.0].
- The response was calculated as: Response = $F(X)$ + (*constant* × noise). The structure of the function F means that the response truly only depends on the first ten variables in the data set; the remaining variables have only a chance relationship. The *constant* is set so 30% of the response variance is from noise.

The structure of the data set is illustrated in Figure 6.

GFA was applied to the generated data set. 100 random models were generated using linear polynomial basis functions. 5000 genetic crossover operations were performed. The best three models are shown in Figure 7. This demonstrates that in circumstances where the information is contained in a small subset of variables, GFA discovers the salient descriptors to generate small, interpretable models. By 'interpretable', we mean that the reduced set of variables more directly exposes the underlying equation which generated the data, and the model coefficients better reflect the relative weight that each variable contributes towards activity.

PLS was applied against the generated data set. The resulting model is shown in Figure 8. We see one advantage of GFA models: they are smaller and easier to interpret. The PLS model does not expose the underlying

$$\begin{aligned} \text{Response}_1 = {} & 3.919 \\ & + 4.458 * \mathbf{X_4} \\ & + 3.353 * \mathbf{X_5} \\ & + 6.947 * \mathbf{X_6} \\ & + 8.642 * \mathbf{X_7} \\ & + 8.401 * \mathbf{X_8} \\ & + 10.615 * \mathbf{X_9} \\ & + 8.218 * \mathbf{X_{10}} \\ & - 4.251 * \mathbf{X_{20}} \end{aligned}$$

r: 0.973
xvalid-r: 0.943

$$\begin{aligned} \text{Response}_2 = {} & 1.541 \\ & + 4.512 * \mathbf{X_4} \\ & + 3.110 * \mathbf{X_5} \\ & + 5.167 * \mathbf{X_6} \\ & + 9.024 * \mathbf{X_7} \\ & + 7.233 * \mathbf{X_8} \\ & + 10.610 * \mathbf{X_9} \\ & + 7.670 * \mathbf{X_{10}} \\ & + 3.684 * \mathbf{X_{47}} \end{aligned}$$

r: 0.971
xvalid-r: 0.941

$$\begin{aligned} \text{Response}_3 = {} & 0.350 \\ & + 3.252 * \mathbf{X_3} \\ & + 3.899 * \mathbf{X_4} \\ & + 4.877 * \mathbf{X_5} \\ & + 5.401 * \mathbf{X_6} \\ & + 9.104 * \mathbf{X_7} \\ & + 6.987 * \mathbf{X_8} \\ & + 9.990 * \mathbf{X_9} \\ & + 10.633 * \mathbf{X_{10}} \end{aligned}$$

r: 0.970
xvalid-r: 0.934

Figure 7 The three top models discovered by GFA on the generated data set.

$$\begin{aligned} \text{Response}_{PLS} = {} & 18.679 \\ & + 1.302 * \mathbf{X_1} \\ & + 0.304 * \mathbf{X_2} \\ & + 0.044 * \mathbf{X_3} \\ & + 2.032 * \mathbf{X_4} \\ & + 1.617 * \mathbf{X_5} \\ & + 4.506 * \mathbf{X_6} \\ & + 4.822 * \mathbf{X_7} \\ & + 4.353 * \mathbf{X_8} \\ & + 4.056 * \mathbf{X_9} \\ & + 4.110 * \mathbf{X_{10}} \\ & + 0.824 * \mathbf{X_{11}} \\ & - 2.312 * \mathbf{X_{12}} \\ & - 1.094 * \mathbf{X_{13}} \\ & + 1.273 * \mathbf{X_{14}} \\ & + 0.257 * \mathbf{X_{15}} \\ & - 0.719 * \mathbf{X_{16}} \\ & + 1.495 * \mathbf{X_{17}} \\ & + 3.248 * \mathbf{X_{18}} \\ & - 0.052 * \mathbf{X_{19}} \\ & - 1.074 * \mathbf{X_{20}} \\ & - 0.483 * \mathbf{X_{21}} \\ & + 0.426 * \mathbf{X_{22}} \\ & - 2.976 * \mathbf{X_{23}} \\ & + 1.066 * \mathbf{X_{24}} \\ & - 0.351 * \mathbf{X_{25}} \\ & + 1.894 * \mathbf{X_{26}} \\ & - 1.050 * \mathbf{X_{27}} \\ & - 0.593 * \mathbf{X_{28}} \\ & - 0.535 * \mathbf{X_{29}} \\ & + 1.778 * \mathbf{X_{30}} \\ & - 2.370 * \mathbf{X_{31}} \\ & - 1.461 * \mathbf{X_{32}} \\ & + 2.142 * \mathbf{X_{33}} \\ & - 0.444 * \mathbf{X_{34}} \\ & + 1.077 * \mathbf{X_{35}} \\ & - 2.118 * \mathbf{X_{36}} \\ & + 0.570 * \mathbf{X_{37}} \\ & + 0.743 * \mathbf{X_{38}} \\ & + 0.156 * \mathbf{X_{39}} \\ & - 4.102 * \mathbf{X_{40}} \\ & - 0.031 * \mathbf{X_{41}} \\ & - 2.830 * \mathbf{X_{42}} \\ & + 0.838 * \mathbf{X_{43}} \\ & - 1.172 * \mathbf{X_{44}} \\ & - 0.516 * \mathbf{X_{45}} \\ & + 1.584 * \mathbf{X_{46}} \\ & + 3.347 * \mathbf{X_{47}} \\ & - 0.597 * \mathbf{X_{48}} \\ & - 2.121 * \mathbf{X_{49}} \\ & - 0.816 * \mathbf{X_{50}} \end{aligned}$$

r: 0.992
xvalid-r: 0.619

Figure 8 A 4-component PLS model of the generated data set.

structure of the function F (that it depends only on the first 10 variables) in a clear way.

However, the cross-validation scores are not comparable: the bias introduced in the descriptor selection step is not measured by standard cross-validation. To compare GFA against PLS fairly, we need to perform *full cross-validation*.

For each cross-validation group, a separate population of models is created and evolved. The best model in each group predicts the removed samples. The combined predictions from each group are used to calculate a *fully cross-validated-r*. This fully cross-validated-r is directly comparable to the cross-validated-r from PLS.

When this form of a cross-validation is performed, we can make a valid comparison. With PLS on the full data set, the cross-validated-r was: 0.619. With GFA and using only the best model, the fully cross-validated-r was: 0.874. Thus, not only are the GFA models more interpretable, in this instance they are more predictive than the PLS model.

Of course, no technique is preferable in all situations for all data sets. What this example is intended to illustrate is that for data sets which reflect a given underlying structure (only a few variables have nonrandom relationships to the response), GFA is a highly-competitive method versus PLS.

COMPARISON OF GFA WITH OTHER GENETIC AND EVOLUTIONARY METHODS

MUSEUM is an evolutionary algorithm recently reported by Kubinyi (1994 a, b). Evolutionary programming (EP) is an evolutionary algorithm recently reported by Luke (1994). In this paper, I will use *evolutionary algorithm* to mean an algorithm that derives new models by stochastic (that is, random) mutations on a *single* parent model. This is distinct from a *genetic algorithm*, which derives new models by using crossover on *two* parent models, followed by a possible mutation step. Leardi reports a feature selection algorithm using a genetic algorithm (Leardi *et al.*, 1992). This section describes, compares, and contrasts these methods for QSAR model construction.

The MUSEUM procedure consists of three phases: a *random* phase, a *systematic* phase, and a *cleanup* phase. A simplified description of the procedure follows. The process starts with a randomly-constructed model; a number of trials are conducted where one or a few variables are added, removed or changed. If a better model is discovered, it replaces the parent, and the procedure restarts with the new model. If no better model is discovered after a given number of steps, the program systematically tries to add or remove all known variables in turn searching for a better model. Again, if a better model is discovered, the procedure restarts with that model. If no better model is found, a final cleanup removes any variable with less than a 95% confidence score, and the remaining model is returned as the answer. However, the system does keep a record of all intermediate models for presentation to the user and to avoid generation of duplicate models.

Luke's EP is similar. It starts with a population of N randomly-created models, in which the basis function are each variables raised to some power. The coefficients of the models are constructed with least-squares regression. Mutation is conducted on each model, and the child added to the population. The models are sorted by fitness, and the bottom N models in the new population are removed. The process repeats for a user-defined number of cycles. The fitness function is a product of least-squares error (LSE), a term to favor models of a given length, and a final term which drives up the error for models which have values for variable exponents outside of a user-defined range. Three mutation operators are used: the first mutation reduces the exponent value for some variable in the model, eliminating it if the exponent falls to zero. The second mutation adds a currently non-used variable to the model. The last mutation randomly replaces one variable in the model with a current non-used variable.

Leardi's method (Leardi *et al.*, 1992) is a genetic algorithm designed for feature selection. It represents a feature set as a bit-string in which each bit position codes for the inclusion or exclusion of a given variable. A fitting technique such as least-squares regression can then be used to create a QSAR, and its fitness measured by its cross-validated-r^2. From a randomly-generated population of initial feature sets (biased to contain a given number of features), both crossover and mutation are used to evolve the population. The best bit-string of each size is 'protected' and guaranteed to survive until a superior bit-string of the same size is discovered. No effort is made to automatically select which size is best; instead, the user is present at the end with a list of models for each size. Leardi's method is similar, though not identical, to GFA when only linear polynomial basis functions are used.

MUSEUM, Luke's evolutionary programming method, Leardi's method, and GFA can be seen as one of a continuum of variable-selection methods which create new models by altering one or more current variables in a procedural step. This continuum is shown in Figure 9. In the figure, the 'step size' is the number of variables which may be altered going from the parent model or models to the child model.

Stepwise regression is the most conservative technique, always making a deterministic single-variable addition or removal starting from a single model. Luke's EP algorithm is less conservative; single-variable addition, deletion, or replacement is allowed in a step. Unlike MUSEUM and stepwise, it keeps a population of models between steps. MUSEUM takes even larger steps, making random changes to several variables starting from a single model. GFA and Leardi's method use two models chosen from a population to construct a child in a step which combines both mutation and crossover; thus, the child model may contain a large number of variable changes from either parent model, and may be able to combine partial solutions discovered by two separate models.

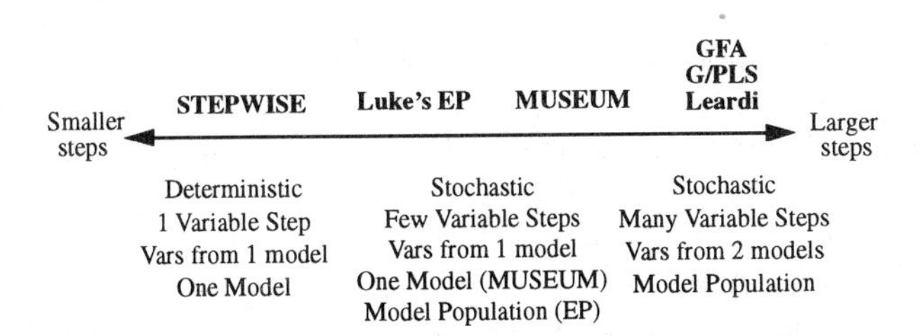

Figure 9 Different model construction techniques can be placed on a continuum which reflects the relative amount of change a current model may undergo while the method selects the variables for a new model.

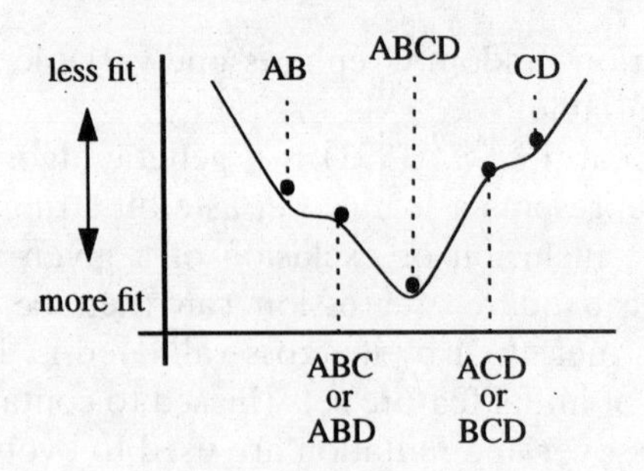

Figure 10 A smooth fitness landscape showing the relative fitnesses of six different models composed of the four variables A, B, C and D. In this case, a single-variable mutation algorithm such as stepwise regression can discover the minima ABCD starting at either AB or CD.

The differences between the methods are the factors the modeler should use in determining which of these methods should be used in a given situation. These differences will be discussed further in the next section.

Illustration of issues with a simplified fitness landscape

First, let us consider what the *fitness landscape* of the space of possible models looks like. Each possible model is a point in this landscape; the height or depth of the point is a measure of the quality of the model as rated by some fitness measure. For illustration, I will use a simplified 2-dimensional representation.

If there are few local minima (a 'smooth' landscape), a technique can reliably use local information in its search for a quality minima. Stepwise regression tries all single-variable steps, and always chooses the step that improves the model the most. Thus, it requires an ever-improving pathway through the fitness landscape to discover the better minima. Such a smooth fitness landscape is illustrated in a simple four-variable example shown in Figure 10.

In this example, the fitness landscape shows the relative fitnesses of six different models composed of the four variables A, B, C and D. The fittest model is the combination ABCD. Starting at either AB or CD, stepwise regression can discover the minima ABCD using only local information available from single-variable mutations.

If there are more local minima, techniques such as stepwise regression that rely on a smooth fitness landscape can be trapped in local minima and miss preferable better minima. A less-smooth fitness landscape is illustrated in Figure 11.

In this example, the steps away from AB and CD result in reduced-fitness models; this creates local minima at AB and CD. No one-variable alteration

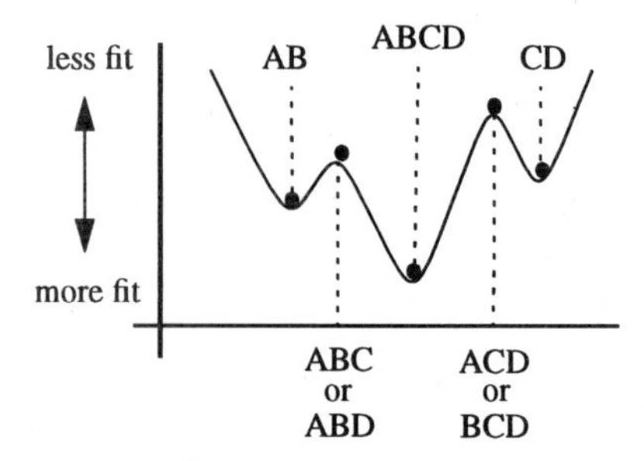

Figure 11 A complex fitness landscape showing the relative fitnesses of six different models composed of the four variables A, B, C and D. In this case, the landscape will not allow discovery of the minima ABCD with a single-variable mutation from either minima AB or CD.

step is sufficient to escape these minima; thus, stepwise regression is unable to discover the combination ABCD.

A model in Luke's EP method thus would not be able to escape the minima directly. However, the weaker child, if not eliminated from the larger population, would be able to complete the path to the global minima. Here we see a possible advantage of keeping a population of models: weaker models on the path to the global minima may be kept for some length of time.

MUSEUM tries both one-variable and few-variable mutations, and so has the possibility of making the leap AB → ABCD or CD → ABCD. However, even if it discovers AB and CD at different times, it is unable to combine these separate 'partial solutions'. Instead, it must start from one model or the other, and fortuitously discover the appropriate combination to make the leap to the better minima. This illustrates a situation in which a genetic algorithm has an advantage in discovery; it can recombine separately-discovered partial solutions in its search for better minima.

Of course, the effort that a genetic algorithm spends on recombination search is effort that could have been spent on mutation-based/evolutionary search. In the former, we take larger steps through function space, often ending up with poorly-performing models, but occasionally discovering a novel model from the combination of partial solutions contained in each parent. In the latter, we take smaller steps through function space, and so perform a more thorough search of function space in the region nearby the current 'best' model. Simply stated, I would claim that search algorithms utilizing crossover are less likely to miss possibly-useful local minima in the function space; the price they pay for this extra computation is a less-thorough exploration of function space nearby the 'best' model.

One possibility which would mediate these techniques is to combine them; begin with some components of both crossover and mutation, and use the fitness of the discovered models to estimate the smoothness of the fitness landscape. If smoother, perform more mutation and less crossover; if more complex, shift the bias towards more crossover. The continuum in Figure 9 was meant to suggest just this: that evolutionary and genetic techniques are not competitors, but two examples of stochastic search procedures that can be used either individually or combined. Indeed, the basic components of MUSEUM or Luke's EP method could be added to GFA as available mutation techniques, giving the combined algorithm much of the power of its components. Such a combination algorithm may be the most appropriate way to handle the disparate structure of QSAR data sets.

Kubinyi (1994a) states that '[p]reliminary studies showed that random mutations of steadily-improving models perform better than crossover strategies'. The example above suggests that this is more a comment on the structure of the specific data set studied than a general statement of the relative performance of the two classes of algorithms. However, he raises the intriguing possibility that perhaps many or most QSAR data sets share a fundamental similarity in having a relatively smooth fitness landscape. Techniques such as the rapid exhaustive search procedure he outlines in Kubinyi (1994b) could survey the fitness landscapes for a large number of QSAR data sets and provide us with such an answer. I would suspect, alas, that modern QSAR data sets, containing information, estimates, and measurements from disparate sources and procedures, are representatives of the class of data sets with complex fitness landscapes, since the interrelationships between descriptors may be quite complex, causing a lot of fine structure in the landscape. This would make it unlikely that any single strategy, however efficient, is suitable across all QSAR data sets. Wise modelers will explore their data sets with a variety of techniques, any one of which may be able to expose solutions not easily generated by others.

QSAR models with small numbers of variables give recombination-based variables little advantage, as the size of the partial solutions is restricted by the small size of the models. Thus, a recombination-based approach may have increasing advantage against mutation-based approaches as the number of variables in the models increases. However, a limit to the size of models generated by each of these techniques is their reliance on least-squares regression. This limitation can be overcome by replacing least-squares regression by another technique such as partial least-squares regression (PLS), in which case models which larger numbers of basis functions can more safely be constructed. A variant of GFA called *genetic partial least-squares* (G/PLS) is presented in Chapter 5 of this book.

CONCLUSIONS

This chapter outlines a number of issues and applications surrounding the GFA algorithm. As such, it was designed to supplement and expand upon theory and applications which were described in less detail in previously published work, such as Rogers (1992), Rogers and Hopfinger (1994) and Chapter 5 of this book. A general overview of GFA is presented in the section on pages 88–89.

Devising a metric for the comparison of models with different complexities is an area of much ongoing research (Wahba, 1990). When least-squares regression is used, as in GFA, the LOF measure of Friedman (1988c) is one candidate. The origin of this metric is explained on pages 89–92, both to justify its use in GFA, and to inspire readers to both consult the extensive source literature from spline modeling, and to consider development of other measures which may be superior in balancing the complexity of QSAR models with their fit to the training data. The development of such improved measures would have a significant impact in the ability of GFA and other evolutionary algorithms to search model space for quality QSARs.

The recent interest in neural network models derives in part from their ability to discover nonlinear relationships between descriptors and activity. Pages 92–98 describe how GFA can be used to build nonlinear QSARs with spline terms, and show the application of nonlinear modeling against a published QSAR example of β-hydroxylase inhibitors. GFA performs both variable selection and nonlinear basis function construction to generate a compact model of activity. Scatterplots of the nonlinear descriptors confirms that the nonlinearity discovered by GFA is indeed in the data, and has a logical interpretation.

In QSAR studies, PLS is often used in place of variable selection methods. The generated data set example in the section on pages 98–100 is designed to illustrate a type of data set in which variable section will perform superior to PLS. This type of data set is one in which the useful information is concentrated in a few descriptors. In such a case, GFA may provide more compact models which are also more predictive and interpretable than the model generated by PLS.

Finally, the recent interest in stochastic methods of model building is discussed, with a comparison of GFA to three other techniques: Kubinyi's MUSEUM method, Luke's EP, and Leardi's feature selection method. A framework is proposed which distinguishes these algorithms based upon the step-size between the current model(s) and a newly-generated model. A hypothetical fitness landscape for model space is described, and for each landscape a preferred type of algorithm is suggested. It is the author's hope that this preliminary discussion can be formalized, perhaps leading to a new class of algorithms which automatically adjust themselves based upon actual measurements of the shape of the model landscape.

Evolutionary and genetic algorithms in modeling are still quite new, and the goal of this chapter was not only to present what is known but to suggest possibly useful areas of research into what is not known. For ease of comparison and to encourage continued development of novel techniques, all data sets described in this work are available freely *via* FTP by contacting the author.

ACKNOWLEDGMENTS

The author wishes to thank Drs Kubinyi, Luke and Leardi for their discussion and sense of excitement at being fellow travellers in this novel algorithmic landscape. Drs Devillers and Domine deserve thanks for planning and executing this book, remaining patient but determined in the face of recalcitrant and difficult authors, tight deadlines, and the requirements of quality and peer-review in a still-evolving field. Molecular Simulations Incorporated allowed and supported my continued efforts in this area, encouraging publication rather than secrecy. My parents and my partner-in-crime Doug Brockman deserve mention for their pride and support in my work and papers, which they often try to read (though due to my still-developing skills as a writer, full understanding often eludes them!). Finally, thanks are due to the reader, for being willing to consider these theories, and for whom I hope my writing skills did not obscure the essence of what I believe is an exciting and valuable area of discovery and research.

REFERENCES

Berglund, A., Ränner, S., and Wold, S. (1995). Nonlinear QSAR problems. Possible preprocessings and nonlinear PLS algorithms. In, *Proceedings of the 10th European Symposium on Structure–Activity Relationships: QSAR and Molecular Modeling*. Prous Science Publishers, Barcelona, Spain.

Burke, B.J. and Hopfinger, A.J. (1990). 1-(Substituted-benzyl)imidazole-2(3H)-thione inhibitors of dopamine beta-hydroxylase. *J. Med. Chem.* **33**, 274–281.

Craven, P. and Wahba, G. (1979). Smoothing noisy data with spline functions. Estimating the correct degree of smoothing by the method of generalized cross-validation. *Numerische Mathematik* **31**, 317–403.

Frank, I.E. and Friedman, J.H. (1991). A statistical view of some chemometric regression tools using adaptive splines. *Technical Report No. 109*, Stanford University Department of Statistics.

Friedman, J.H. (1988a). Fitting functions to noisy data in high dimensions. In, *Computer Science and Statistics: Proceedings of the 20th Symposium of the American Statistical Association* (E. Wegman, D. Gantz, and J. Miller, Eds.). American Statistical Association, Washington DC, pp. 13–43.

Friedman, J.H. (1988b). Fitting functions to noisy data in high dimensions. *Technical Report No. 101*, Stanford University Department of Statistics.

Friedman, J.H. (1988c). Multivariate adaptive regression splines. *Technical Report No. 102*, Stanford University Department of Statistics.

Friedman, J.H. (1991). Adaptive spline networks. *Technical Report No. 107*, Stanford University Department of Statistics.

Friedman, J.H. and Silverman, B.W. (1989). Flexible parsimonious smoothing and additive modeling. *Technometrics* **31**, 3–39.

Hahn, M. and Rogers, D. (1995). Receptor surface models. 2. Application to quantitative structure–activity relationship studies. *J. Med. Chem.* **38**, 2091–2102.

Holland, J.H. (1975). *Adaption in Natural and Artificial Systems*. University of Michigan Press, Ann Arbor, MI.

Kruse, L.I., Kaiser, C., deWolf, W.E., Jr., Frazee, J.S., Ross, S.T., Wawro, J., Wise, M., Flaim, K.E., Sawyer, J.L., Erickson, R.W., Ezekiel, M., Ohlstein, E.H., and Berkowitz, B.A. (1987). *J. Med. Chem.* **30**, 486.

Kubinyi, H. (1994a). Variable selection in QSAR studies. I. An evolutionary algorithm. *Quant. Struct.-Act. Relat.* **13**, 285–294.

Kubinyi, H. (1994b). Variable selection in QSAR studies. II. A highly efficient combination of systematic search and evolution. *Quant. Struct.-Act. Relat.* **13**, 393–401.

Leardi, R., Boggia, R., and Terrile, M. (1992). Genetic algorithms as a strategy for feature selection. *J. Chemom.* **6**, 267–281.

Luke, B. (1994). Evolutionary programming applied to the development of quantitative structure–activity relationship, and quantitative structure–property relationships. *J. Chem. Inf. Comput. Sci.* **34**, 1279–1287.

Rogers, D. (1991). G/SPLINES: A hybrid of Friedman's multivariate adaptive regression splines (MARS) algorithm with Holland's genetic algorithm. In, *Proceedings of the Fourth International Conference on Genetic Algorithms*. (R.K. Belew and L.B. Booker, Eds.). Morgan Kaufmann Publishers, San Diego, CA, pp. 384–391.

Rogers, D. (1992). Data analysis using G/SPLINES. In, *Advances in Neural Processing Systems 4*. Morgan Kaufmann, San Mateo, CA.

Rogers, D. (1995). Genetic function approximation: A genetic approach to building quantitative structure–activity relationship models. In, *Proceedings of the 10th European Symposium on Structure–Activity Relationships: QSAR and Molecular Modeling*. Prous Science Publishers, Barcelona, Spain.

Rogers, D. and Hopfinger, A.J. (1994). Application of genetic function approximation to quantitative structure–activity relationships and quantitative structure–property relationships. *J. Chem. Inf. Comput. Sci.* **34**, 854–866.

Sekulic, S., Seasholtz, M.B., Wang, Z., Kowalski, B.R., Lee, S.E., and Holt, B.R. (1993). *Anal. Chem.* **65**, 835.

Wahba, G. (1990). *Spline Models for Observational Data*. CBMS-NSF regional conference series in applied mathematics. Society for Industrial and Applied Mathematics. Philadelphia, Pennsylvania.

5 Genetic Partial Least Squares in QSAR

W.J. DUNN[†*] and D. ROGERS[‡]
[†]*College of Pharmacy, University of Illinois at Chicago, 833 S. Wood Street, Chicago, IL 60612, USA*
[‡]*Molecular Simulations Incorporated, 9685 Scranton Road, San Diego, CA 92121, USA*

Partial least squares, (PLS), regression is a newly developed regression method that has found considerable use in many areas of chemistry. PLS is like ordinary least squares regression, (OLSR), in that it provides a least-squares estimate of the dependent variable(s), **Y**, *as a function of the independent variables,* **X**. *It differs from OLSR by extracting latent variables from* **X** *and* **Y** *that are: 1) approximately along the axes of greatest variation in* **X** *and* **Y**; *and 2) are optimally correlated. In doing this, PLS takes advantage of the collinearity that may exist in the* **X**-*data and provides more robust estimates of the* **Y**-*data. Unlike OLSR, it can be applied in cases where there are many more variables than samples or compounds. Genetic algorithms, (GAs), are another also newly-proposed method of problem-solving using aspects of the process of evolution to search for solutions. They recently have been applied to problems in quantitative structure–activity relationship (QSAR), with OLSR as the modeling medium. Here we combine the advantages of PLS with the model generating ability of GAs to create what we term genetic partial least squares (GPLS). The approach is discussed and applied to previously studied QSAR data.*

KEYWORDS: *PLS; genetic function approximation; genetic partial least squares; outlier limiting.*

INTRODUCTION

The now classic quantitative structure–activity relationship (QSAR) work of Hansch (1968) initiated the use of computer-based technology in the

* Author to whom all correspondence should be addressed.

In, *Genetic Algorithms in Molecular Modeling* (J. Devillers, Ed.)
Academic Press, London, 1996, pp. 109–130.
ISBN 0-12-213810-4

discovery, design and development of pharmaceutical agents. As with the introduction of new methodologies, this was not widely accepted at first, but presently the use of the so-called Hansch approach is routine. It is based on the assumption that the change in biological activity that is observed within a series of similar compounds is a function of the change in chemical structure within the series. This whole philosophy of biological structure–activity relationships evolved from the work of Hammett (1970) who studied the effect of substituents on the ionization constants of benzoic acids and defined the Hammett σ-constant as a quantitative measure of electronic effects on equilibria. It was extended to include the effect of substituents on reaction rates. Likewise, Hansch, in order to place molecular description on a numerical bases, defined the π-constant as a measure of a substituent's effect on processes sensitive to lipophilicity or hydrophobicity. He combined this with the Hammett σ-constant to give the Hansch model for QSAR.

$$\log BR = -k_1\pi^2 + k_2\pi + k_3\sigma + k_4 \qquad (1)$$

$$BR = \text{biological response}$$

If additional effects, such as steric effects, are thought to be important, they can be included by linearly expanding the model. Ordinary least squares regression (OLSR) is typically used as the modeling method. As such, this is a heuristic approach, and should be used only when OLSR is appropriate. The use of OLSR under inappropriate circumstances continues to be a common error in QSAR studies.

While partial least squares (PLS) (Wold *et al.*, 1984) was not designed specifically to be applied to QSAR studies, it is not surprising that its first definitive application was a study of substituent effects on classifying compounds as agonists and antagonists of the β-adrenergic receptor but also as predicting their level of response in a post-classification analysis. Partial least squares (PLS) is still not well-known outside of the chemometrics literature or well understood by statisticians but it is becoming the method of choice for many QSAR studies. This is partly due to the emergence of 3-dimensional QSAR (3D-QSAR), in which the geometry of the drug is considered in the derivation of descriptors. Two methods of 3D-QSAR are commonly cited: (1) Comparative Molecular Field Analysis (CoMFA) (Cramer *et al.*, 1994) and (2) Molecular Shape Analysis (MSA) (Burke and Hopfinger, 1994). Both rely on the generation of many descriptors for few compounds making the use of PLS necessary as the data analytic method.

BACKGROUND

The standard QSAR data set for correlation analysis is given in Figure 1. Here, the **Y**-data are the dependent variables and the **X**-data are the

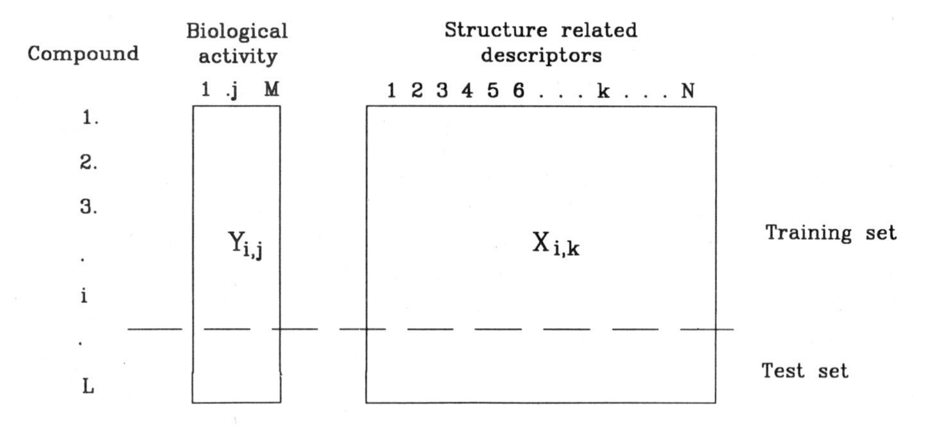

Figure 1 The standard QSAR data matrix.

independent variables. The index *i* is used to denote compounds, *j* to denote the biological tests and *k* to denote the descriptor. The primary objective of the QSAR study is to find a predictive relationship between the independent **X**-data and the dependent **Y**-data for the training set data. Both ordinary least squares regression, OLSR, and PLS can be used to generate QSAR models. The model can then be used to predict the activity of the test compounds which may be unknown or untested.

One advantage of PLS over OLSR is that it can be used in cases with more than one dependent variable. OLSR requires that only one dependent variable be analysed at a time.

PLS

The PLS model for the QSAR problem is given below in Eqs. (2)–(4). PLS extracts principal components-like latent variables from the **X**- and **Y**-data. The number of components is *A*. The PLS scores, *t* and *u*, respectively, are computed along the axes of greatest variation in **X** and **Y** and are optimally correlated. The latent variables are similar to principal component scores but are rotated to take into account the correlation between the two vectors. The loadings or weights for the *t*s and *u*s are *p* and *q*, respectively. The usual convention is to translate the analysis to the mean, x_k and y_j, of each of the columns of **X** and **Y**. The data can also be autoscaled.

$$y_{ij} = y_j + \sum_{a=1}^{A} t_{ia} p_{aj} + e_{ij} \tag{2}$$

$$x_{ik} = x_k + \sum_{a=1}^{A} u_{ai} q_{ak} + e_{ik} \tag{3}$$

$$\hat{u} = bt \tag{4}$$

The latent variables are correlated through Eq. (4), the inner relation. PLS is a least squares method in the sense that the residuals, e_{ij} and e_{ik}, are minimized, It can be shown that OLSR is a special case of PLS. When the number of PLS components is the same as the number of independent variables; OLSR and PLS results are equivalent.

The development of PLS models is not a straightforward process, especially if the data do not result from an experimental design. This is often the case in QSAR studies. One consideration is the number of components, A, to use. The accepted protocol is to use a cross-validation technique (Wold, 1978) with some stopping rule based on a computed statistic, which depends on the objective of the modeler. Typically, a predicted sum of squares of residuals, PRESS, is computed for a given model and this is compared with the PRESS for a new, more complex model. If there is a significant reduction in the new PRESS, the number of components is retained. Another method is to compute r^2 for the observed and predicted **Y**-data for the deleted compounds in the cross-validation. This strategy seems to give good results and a variation of it is used later in this report.

Another problem that must be addressed when analysing data sets with many independent variables is whether to reduce the set to isolate the significant independent variables, i.e. variable selection. The simplest thing to do is to run a PLS analysis with all independent variables included and delete variables which have low loadings. A new model is then derived with a smaller number of variables. This process is repeated until some deterioration of the model is observed. The processes of model complexity (number of components) and variable selection are obviously interrelated so several cycles may be necessary in order to converge on to an appropriate model.

PLS is optimized for prediction. Some researchers are reluctant to use it because the latent variables extracted by PLS become linear combinations of original variables. This leads to the original variables losing their intuitive meaning so that a mechanistic interpretation of the results becomes difficult. In order to overcome this problem with PLS, for this report, the results are rotated back into the original data space. The algorithm for this rotation has been published (Glen *et al.*, 1989).

GENETIC ALGORITHMS

Genetic algorithms (GAs) are a newly described problem solving method, invented and discussed by Holland (1975). They have recently been reviewed

(Forrest, 1993). The genetic function approximation algorithm (GFA) of Rogers (1991) has been applied to derive QSARs (Rogers and Hopfinger, 1994). Other QSAR applications (Kubinyi, 1994a,b) are beginning to appear in the literature. GAs have been used to parameterize potential energy surfaces and direct dynamics calculations for organic reactions in an interesting chemical application (Rossi and Truhlar, 1995). GAs use evolutionary operations to drive the process in computer-aided problem solving. The operators used are *random mutation* and *genetic recombination* (crossover) and their use leads to solutions optimized to predetermined selection criteria. In QSAR, the solution required is a functional relationship between biological activity and chemical structure that will predict the biological activities of new or untested compounds. Most QSAR studies are OLSR or PLS. Since PLS has a number of advantages over OLSR it is natural that GAs and PLS be combined to give a new algorithm related to the previously reported GFA algorithm which we term genetic partial least squares (GPLS). A preliminary account of this work has been reported (Rogers and Dunn, 1995, manuscript submitted).

Genetic function approximation

Genetic function approximation (GFA) is a GA derived from the G/SPLINES algorithm of Rogers (1991) and has been used to construct QSARs. A GA requires that an individual be represented as a linear string which is analogous to DNA in a living organism. In GFA, the string is the series of basis functions, as shown in Figure 2. Using the information in the string, it is possible to reconstruct the QSAR model by using OLSR to generate the regression coefficients for the basis functions. The initial models are generated by randomly selecting a number of features from the training data set, building basis functions from these features using the user-specified basis function type, and then constructing the models from random sequences of these basis functions. The models are rated using a modified form of Friedman's (Friedman, 1988) 'lack of fit' (LOF), function which is given in Eq. (5).

F_1:{ $\mathbf{var_1}$; $\mathbf{var_3}$; $\mathbf{var_{30}}$; $\mathbf{var_{45}}$; $\mathbf{var_{21}}$; $\mathbf{var_{70}}$; $\mathbf{var_{56}}$;}
F_2:{ $\mathbf{var_{10}}$; $\mathbf{var_{43}}$; $\mathbf{var_{35}}$; $\mathbf{var_{54}}$; $\mathbf{var_2}$; $\mathbf{var_{38}}$; $\mathbf{var_{61}}$;}
F_3:{ $\mathbf{var_{11}}$; $\mathbf{var_{39}}$; $\mathbf{var_{60}}$; $\mathbf{var_{71}}$; $\mathbf{var_{41}}$; $\mathbf{var_7}$; $\mathbf{var_6}$;}

F_k:{ $\mathbf{var_{13}}$; $\mathbf{var_9}$; $\mathbf{var_{50}}$; $\mathbf{var_{31}}$; $\mathbf{var_8}$; $\mathbf{var_{72}}$; $\mathbf{var_{26}}$;}

Figure 2 Examples of a population of k models represented for the GFA algorithm. Each model is represented as a linear string of basis functions. The activity models can be reconstructed using OLSR to regenerate the regression coefficients. In this case, each basis function is a single variable.

$$LOF = \frac{LSE}{\left[1 - \frac{((d + 1)c)}{M}\right]^2} \tag{5}$$

The smaller the LOF, the better the model. The terms in the function are defined as follows. LSE is the ordinary least-squares error, M is the number of compounds in the analysis, d is a smoothing parameter, and c is the number of basis functions (independent variables) in the model. The smoothing parameter is the only user-adjustable term in the equation. It can be seen that for a given value of d (the default is 1), as M becomes large LOF approaches LSE, the index frequently used for evaluating OLSR models. The same effect is expected for small values of c. Thus, LOF is sensitive to the case with a small number of compounds and a large number of variables. The effect of d on LOF can be seen from the following. For a given value of d, as new terms are added to a model, LOF decreases but as the number of variables increases, LOF passes through a minimum. The value of d determines the position of the minimum. The minimum shifts to smaller values of c for larger values of d. GFA is a stochastic optimization technique which is able to sample variable space and escape local minima in the LOF. It will, given sufficient time, converge on the global minimum of this selection function.

GFA generates a population of models which are rated by LOF. It then performs repeated genetic crossover operations. This consists of selecting two good models as 'parents'. The parents are chosen based on the inverse of their LOF scores. Each parent is then randomly cut into two sections and a new model is created by using a piece from each parent. This is shown in Figure 3. The coefficients of the new model are generated from OLSR. Next, mutation operators may be used to alter a newly created model. In this work, the mutation operator is the *new* operator which appends a new random basis function. The default is for the new operator to be applied 50% of the time to the newly created model.

The overall process is ended when the average LOF score of the models in the population no longer indicates improvement. For a population of 100 to 300 models, 3000 to 10 000 genetic operations are usually sufficient to achieve 'convergence'. For a typical data set, this process takes between 10 minutes on a workstation and 1 hour on a Macintosh-IIfx computer. Upon completion, one can simply select the model from the population with the lowest score, though it is usually preferable to inspect the different models and select one on the basis of the appropriateness of the features, the basis function, and the feature combinations. Selecting a single model may not be always desirable. Many strategies can be used to converge on a solution. A small subset of the population can be used to provide information on feature use. Also, models may be valid for certain domains so that models can be averaged to give more stable estimates for prediction purposes. The GFA

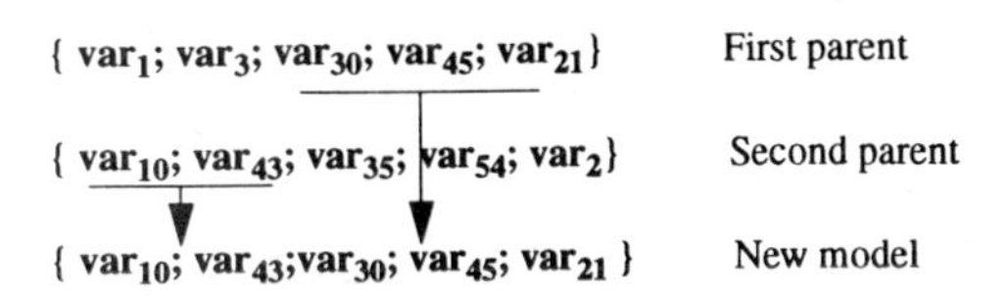

Figure 3 The crossover operation. Each parent is cut at a random point, and a piece from each parent is used to construct the new model which now uses some basis functions from each parent.

algorithm has a number of advantages over other techniques. Other techniques tend to build single models, while GFA generates several models.

GENETIC PARTIAL LEAST SQUARES

The GFA technique uses a GA to select appropriate basis functions to be used to model QSAR data. Once the basis functions are chosen, it is still necessary to use some fitting technique to weight their relative contributions in the final model. As originally developed, the GFA algorithm used OLSR to generate the coefficients for linear combinations of these basis functions. Previous studies by one of us (Rogers and Hopfinger, 1994) suggest that this is a satisfactory technique if the final model contains few basis functions.

With the use of 3D-QSAR methods which generate many more variables than compounds, this OLSR is of little use. Cruciani *et al.* (1993) among others, have described the dangers of using this technique as a fitting technique. Specifically, as the number of basis functions increases, the risk of overfitting also increases. The incorporation of OLSR into the core GFA algorithm makes it susceptible to this weakness. As a solution we have replaced OLSR in the GFA algorithm with PLS regression. We call this new algorithm Genetic Partial Least Squares (GPLS), since it incorporates many of the advantages of the GA as developed in the GFA algorithm, and many of the advantages of PLS regression.

There are some gains for incorporating PLS into the GFA algorithm. The objective of the GFA algorithm is to construct functional models of data. By compressing the feature data into latent variables (linear combinations of the original variables), PLS takes advantage of the collinearity which is inherent in feature data. By using PLS in the modeling phase, more robust (stable) models usually result. PLS does not fail to generate coefficients, although users seldom rotate the results back into the original feature space as we do here. Also, PLS with the UNIPALS© algorithm (Glen *et al.*, 1989) is more efficient in memory use and lower in computational time for data sets with a large number of variables and few compounds as is the case here.

One of the critical issues which arises with the use of PLS with large numbers of independent variables is variable selection. There have been some attempts to develop a strategy for variable selection in PLS. Most notable is the GOLPE (Cruciani *et al.*, 1993) approach. GOLPE is based on carrying out a PLS analysis with all independent variables included and then using D-optimal design methods to select those variables which are most heavily loaded for inclusion in the model. The danger of this approach is that variables which contribute to the model but are masked by those with high loadings will be deleted. Also, PLS takes advantage of the collinearity in the data and this can be partially removed by the GOLPE method.

In order to understand the role of variable selection with PLS, it is necessary to consider the idea of 'noise'. Because of the way PLS treats the residuals, noise has a broader meaning with PLS than with OLSR. PLS components are extracted from the **X**- and **Y**-data which are along the axes of greatest variation **and** optimally correlated. This means that systematic variation in the **X**-data which is not correlated with variation in the **Y**-data is noise. GFA, by using subsets of variables with cross-validated r^2, considers models with different variable combinations. This subselection process filters 'noise' and greatly improves the quality of PLS models.

OUTLIER LIMITING

Outliers (samples which have descriptor values that are very different from training descriptor values) are difficult for data modeling algorithms. Often, an algorithm may work excellently on generated or artificial data sets, but then fail on data sets derived from real-world applications. This is often due to the presence of such outliers in the real data. Variable reduction techniques, such as GFA or step-wise multiple regression, may reduce the effect of such variables, as they are often less predictive and hence usually eliminated. Since PLS models explicitly include all variables, at best, the offending variable may be given a small loading, and so leave the model somewhat resistant in the face of large variance caused by the variable. More commonly, these variables cause the full PLS model to make a wild prediction for the sample with the outlying variable(s). As many different descriptors may contain outliers, many different samples may also generate these highly unreasonable predictions. The approach we have taken to stabilize the models we call *outlier limiting*. It is motivated by a common-sense analysis of the situation; that is, how would a scientist deal with real data containing such variables? Suppose a sample contains such an outlier, and its presence is detected; what action would be appropriate? A number of possibilities exist:

- *Do nothing*. Accept the value as is, and calculate the response using the value. Since the variable is often linear in the model, this can cause large changes in the response.

- *Delete the sample.* Signal to the user that no prediction could be generated for that sample. A bit depressing, but certainly an improvement over a wildly inaccurate prediction.
- *Ignore the model.* This is not possible with standard PLS, since there is only one model. However, it would be a possibility for the GFA, which keeps a population of possible models. One scheme would have a model disqualify itself, and let the next-ranked model make the prediction for certain samples.
- *Limit the response.* At training time, the model records the mean of the response of the training set, and the distance to the furthest response from the mean. If a calculated response is outside of some accepted range about the mean (usually given as a multiple of the distance to the furthest response), it is limited to the 'furthest acceptable' value. This new response is returned in place of the calculated response.
- *Limit the variable.* At training time, the model records the mean of each variable, and the distance to the furthest value of the variables from the mean. If an input variable is outside of some accepted range about the mean (usually given as a multiple of the distance to the furthest value), it is returned to the 'furthest acceptable' value. This new value is used to calculate the response in the model.

The first option (do nothing) is the most common, but for the reasons outlined above, it is unacceptable. The second option (delete the sample) is better, but for data sets with many outliers, samples may be eliminated with alarming regularity. The third option (ignore the model) is interesting, and we plan to explore it further in future work. The fourth option (limit the response) and the fifth option (limit the variable) are the most interesting; we have implemented these techniques, and believe they can significantly improve the results obtainable from both PLS and GFA.

Outlier limiting is done as follows. The mean of the variable (or response) is calculated from the training samples. The distance to the furthest variable (or sample) is recorded; this distance is given an outlier value of 1.0. An allowed outlier distance is given by the user; for example, 1.25, which means that a variable (or response) cannot be further than 1.25 times the distance from the mean to the furthest outlier in the training set. This allows some measure of extrapolation for samples that are just beyond values provided by the training set; this is useful since often the range of most interest is just outside values found in the training set. Response limiting takes the value of the response, and sets values beyond the limit to the limiting value. Variable limiting takes the value of each input variable, and sets any values beyond the limit to that limiting value.

Table I *Variables used in the Selwood data set.*

LOGP: Partition coefficient
MPNT: Melting point
DIPMOM: Dipole moment
VDWVOL: van der Waals volume
SURFA: Surface area
MOLWT: Molecular weight
DIPVX, DIPVY, DIPVZ: Dipole vector components in X, Y, and Z
MOFIX, MOFIY, MOFIZ: Principal moments of inertia in X, Y, and Z
PEAXX, PEAXY, PEAXZ: Principal ellipsoid axes in X, Y, and Z
S81DX, S81DY, S81DZ: Substituent on atom 8 dimensions in X, Y, and Z
S81CX, S81CY, S91CZ: Substituent on atom 8 center in X, Y, and Z
ATCH1 to **ATCH10**: Partial atomic charges for atoms 1–10
ESDL1 to **ESDL10**: Electrophilic superdelocalizability for atoms 1–10
NSDL1 to **NSDL10**: Nucleophilic superdelocalizability for atoms 1–10
SUMF and **SUMR**: Sums of the F and R substituent constants

CASE STUDY

To illustrate GPLS a frequently analysed data set was chosen to demonstrate the application of the GFA algorithm. The Selwood (Selwood *et al.*, 1990) data set illustrates the application of GPLS to a data set of scientific interest and the resulting analysis illustrates the advantage of GPLS as a tool in QSAR.

The Selwood data set (Selwood *et al.*, 1990) illustrates a number of the issues that arise when analysing real-world data, and its treatment by GPLS shows the strengths of the approach. The data set contains 31 compounds and 53 features with corresponding antifilarial antimycin activities. In order to save space, the complete data set is not presented here. The original report should be consulted for details on the structure–activity data. The series of analogues are of the general form shown in Figure 4. This data set was of particular interest because it contains a large number of features relative to the number of compounds. The list of features is given in Table I.

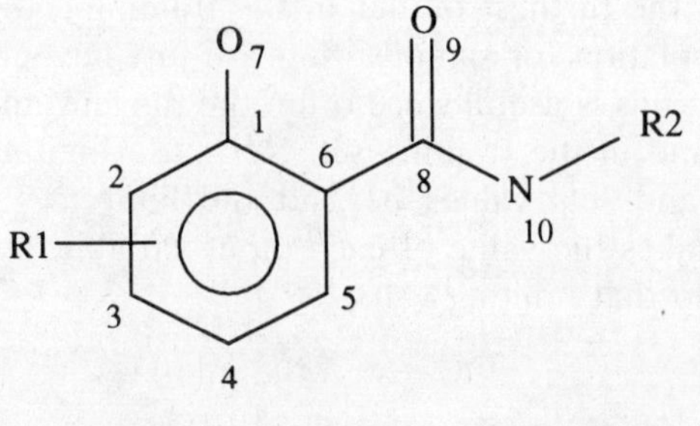

Figure 4 The generic structure of the antifilarial antibiotics in the Selwood data set.

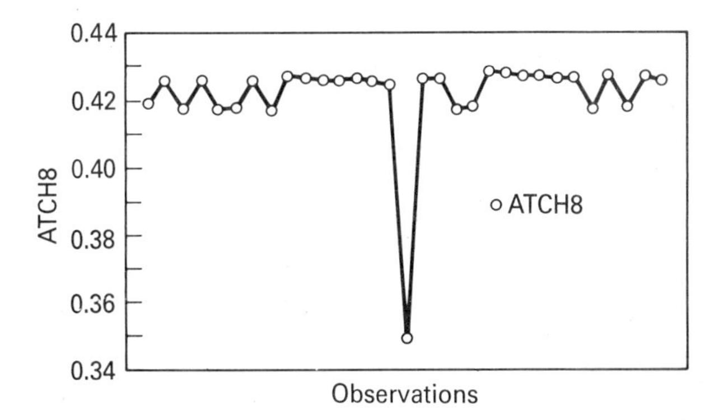

Figure 5 The value of ATCH8 plotted against the compound index. ATCH8 for all compounds falls in a tight range, except for compound 16, which has a value far outside of the range of the other compounds.

One reason the Selwood data set is interesting is that it has been well-studied, and illustrates the complexities faced by the modeler in many real-world data sets. One such issue is the presence of outliers in the feature variables. By outliers, we mean feature values which are far outside the range of values of those of the remaining samples. As shown in Figure 5 compound 16 (chemical 31 in Selwood *et al.*, 1990) contains such a variable, ATCH8. All values of ATCH8 fall in a tight range, except for compound 16, which has a value far from the range of the others. There are many possible reasons this may be so: there may be a typographic error in copying the number; a computational error in calculating the number; or perhaps the 'error' is due to compound 16 truly being from a molecule with significantly different electronic properties from the remaining samples. In any case, a useful modeling algorithm must deal with this effect efficiently if it is to be useful in the face of such an all-too-common data anomaly.

Selwood *et al.* (1990) used OLSR to develop a QSAR from these data. Later, this same data set was studied by Wikel and Dow (1993) who used a neural network to select features for their QSAR model. Other groups have also studied this data set (Livingstone *et al.*, 1991; Rose *et al.*, 1991a,b, 1992; McFarland and Gans, 1994; Luke, 1994; Kubinyi, 1994). This data set was studied using GFA with OLSR (Rogers and Hopfinger, 1994).

The methods employed by Selwood *et al.* (1990), Wikel and Dow (1993) and Rogers and Hopfinger (1994) all used *feature reduction* to reduce the number of features used in the final model(s). The full data set was used in selecting features; once the subset was chosen, the error introduced by regression on these features was measured by cross-validation. For example, Eqs. (6)–(8) show a model generated by Selwood *et al.* (1990), a model by Wikel and Dow (1993), and the top-rated model by Rogers and Hopfinger (1994), respectively:

$$-\log(IC_{50}) = -3.93 + 0.44 \text{ logP} + 0.01 \text{ MPNT} - 0.30 \text{ ESDL10} \quad (6)$$
$r^2 = 0.54$; regression-only CV $r^2 = 0.44$

$$-\log(IC_{50}) = -1.63 + 0.23 \text{ logP} + 4.42 \text{ ATCH4} + 0.01 \text{ MOFIX} \quad (7)$$
$r^2 = 0.60$; regression-only CV $r^2 = 0.46$

$$-\log(IC_{50}) = -2.50 + 0.58 \text{ logP} + 1.51 \text{ SUMF} - 0.00008 \text{ MOFIY} \quad (8)$$
$r^2 = 0.72$; regression-only CV $r^2 = 0.65$

In Eqs. (6)–(8), the correlation coefficient, r^2, and regression-only CV r^2 for models using the features of Selwood *et al.* (1990), Wikel and Dow (1993) and the top-rated models discovered by the GFA algorithm using $d = 2.0$ for the smoothing parameter are given. Cross-validation was conducted only after the features were selected using the full data set, and do not measure the effects of GFA on variable selection. For comparison purposes, the cross-validation step was conducted only after the features were selected using the full data set. To perform a full cross-validation, the features should have been reselected for each cross-validation group, resulting in different models for each group. This was not possible as the feature selection techniques of the original authors were not available.

GPLS analysis of the Selwood data set

PLS does not attempt to reduce the feature set; instead, it builds a model over the full data set. Since the features in the data set have different metrics, the data were autoscaled to unit variance before model building. The autoscaling is done implicitly, and is reflected in the coefficients, rather than by user-transformation of the data set. In this way, new samples can be predicted from the model without any preprocessing. The full, 4 component PLS model of the Selwood data set is shown in Eq. (9). The model has been rotated back into original feature space. The terms are sorted in ascending order by their PLS loadings, not to be confused with the regression coefficients. In this form it appears imposing. On closer examination, as discussed later, it can be simplified.

$-\log(IC_{50})_{PLS-4}$ = 0.836219	+ 0.000892 * **M_PNT**	– 0.046056 * **NSDL6**
+ 3.626831 * **ATCH4**	+ 0.000059 * **MOFI_X**	– 0.002927 * **S8_1CX**
+ 0.180160 * **LOGP**	– 0.025472 * **DIPV_X**	+ 0.045206 * **NSDL3**
– 0.061843 * **DIPMOM**	– 0.043469 * **S8_1CZ**	+ 0.082696 * **NSDL10**
+ 0.036801 * **DIPV_Y**	+ 1.214713 * **ATCH1**	– 1.333264 * **ATCH8**
+ 0.612836 * **SUM_F**	– 2.449233 * **ATCH6**	– 0.002359 * **ESDL2**
– 0.191473 * **PEAX_Y**	– 0.000004 * **MOFI_Y**	+ 0.028981 * **ESDL10**
+ 0.254936 * **ESDL3**	– 0.000004 * **MOFI_Z**	+ 0.002265 * **ESDL4**
+ 4.972921 * **ATCH7**	– 0.005132 * **S8_1CY**	– 0.000206 * **MOL_WT**
+ 0.115087 * **NSDL8**	– 0.025014 * **NSDL2**	– 0.001689 * **PEAX_X**
+ 0.061980 * **ESDL6**	+ 0.784907 * **ATCH3**	+ 0.355106 * **ATCH9**

$$
\begin{aligned}
&+ 0.085727 * \mathbf{S8\_1DZ} && - 0.015156 * \mathbf{S8\_1DX} && + 0.005343 * \mathbf{PEAX\_Z}\\
&- 0.061373 * \mathbf{DIPV\_Z} && - 0.009006 * \mathbf{NSDL5} && - 0.006737 * \mathbf{ESDL9}\\
&- 0.001913 * \mathbf{VDWVOL} && - 0.009221 * \mathbf{NSDL4} && - 0.001768 * \mathbf{ESDL8}\\
&+ 0.211676 * \mathbf{NSDL9} && - 0.000509 * \mathbf{SURF\_A} && - 0.005998 * \mathbf{ATCH2}\\
&- 0.023682 * \mathbf{NSDL1} && - 0.025742 * \mathbf{ESDL5} && + 0.000049 * \mathbf{ESDL1}\\
&+ 0.016955 * \mathbf{S8\_1DY} && + 0.140923 * \mathbf{SUM\_R} && - 0.000214 * \mathbf{ESDL7}\\
&- 0.162420 * \mathbf{NSDL7} && + 0.633422 * \mathbf{ATCH5} &&\\
&- 1.057671 * \mathbf{ATCH10} && && \qquad (9)
\end{aligned}
$$

$r^2 = 0.83$, cross-validated $r^2 = 0.20$

The cross-validated r^2 value from PLS is less than those from the models derived from other techniques. This is due to two factors:

- PLS is sensitive to outlier values of the biological activity and feature data; the Selwood data set, like many other real-world data sets, contains features that have extreme values in only one or two samples. These outliers cause extreme predictions to be made during cross-validation. Feature reduction techniques often eliminate such variables.
- The cross-validation scores of the three models (Eqs. (6)–(8)) were calculated after feature selection, so any bias introduced during that step was not measured. PLS does not perform feature selection. Thus, these cross-validated r^2 scores are not directly comparable to the cross-validated r^2 value for the PLS model as given in Eqs. (6)–(8).

We discuss each of these issues in turn. First, we will use outlier limiting to improve the predictiveness of PLS (and other) models in the face of outliers. Next, we will perform experiments that cross-validate both the feature selection and regression steps, and so give us a cross-validated r^2 comparable to that of full PLS.

Outlier limiting with PLS

Figure 6 shows the effect of outlier limiting on the performance of PLS. In this case, the performance of standard PLS is poor on the scaled data set, with a cross-validated r^2 of 0.20 for four components, and with a maximum cross-validated r^2 of 0.28 with 5 components; limiting the response appears to correct the worst cases, but does not improve predictivity overall. In this sense 'fixing' the response after it has already been damaged by outliers does not succeed in improving predictivity, but only in keeping the predictivity from breaking down in the worst cases.

However, when we limit the value of the input variables, predictivity is significantly and consistently improved, with a cross-validated r^2 of 0.357 using 4 PLS components, and an improvement in cross-validated r^2 is seen for models built with any number of components.

For this case, outlier limiting improves the ability of a PLS model to predict, as measured by cross-validated r^2. While not demonstrated here,

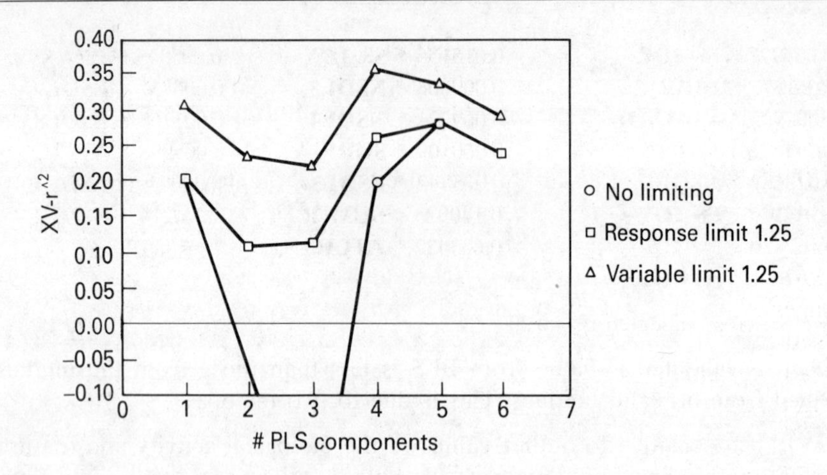

Figure 6 Cross-validated r^2 versus the number of PLS components for standard PLS, PLS with a response limit of 1.25, and PLS with an input variable limit of 1.25.

outlier limiting has a similar, though smaller advantage, in improving the ability of GFA models to predict.

Cross-validated feature selection

At first, it would appear that there is little to be gained by extending the cross-validation process through the feature selection step. It can significantly increase computational overhead; it is complex, necessitating the construction and manipulation of as many models as there are cross-validation groups; and it may not be needed, since it would at first appear that leaving out a single sample (or a few samples) could hardly change the features selected. Indeed, if the feature selection process produces the same set of features for each cross-validation group, then the cross-validated r^2 is equal to the cross-validated r^2 calculated for the regression step, and the extra work would add nothing.

Of the three points given above, the first and second are true, but the third is quite false. Removal of a single sample can greatly affect the features chosen, and there can be a significant degradation of cross-validated r^2 associated with bias introduced in the feature selection phase of model construction. This bias must be measured if a cross-validated r^2 is to be obtained that is both more predictive and can be compared to the cross-validated r^2 done with PLS.

The Selwood data set offers an excellent test case for studying whether different features are chosen for different cross-validation groups. Eqns (10)–(40) show the models derived for each of the cross-validation groups in a leave-one-out process. The default settings for GFA using OLSR were used for this example. The first result is that the cross-validated r^2 is greatly decreased when we cross-validate through the feature selection phase: 0.32 versus 0.65 when we only cross-validate the top-rated model in the regression step. This means that the bias introduced during feature selection can be significant, and needs to be measured if we are to properly estimate predictiveness. The source of the bias is clearer when we inspect the models.

What is most surprising is the great variety of models generated; the removal of a single sample from the data set can apparently cause large changes in the features which are selected. Some of the differences may be caused by the inherent randomness of the genetic algorithm; however, repeated runs with the same data set usually result in the same top model being discovered. The reason for this is not that model scores are usually greatly altered by the removal of a single sample; rather, many models have very similar scores, and it only takes a minor change in the training set to reorder these similarly-scoring models. In effect, these multiple models are similarly-rated hypotheses about the structure of the data set, and the choice of the 'best' hypothesis is rather arbitrary.

It is somewhat nonintuitive to be faced with this collection of models; what indeed are we cross-validating? If someone were to give you a test sample and ask you to predict its activity, which of these models would you use?

$$-\log(IC50)\ \mathrm{CV1} = -2.53 + 0.59\ \mathrm{logP} + 1.52\ \mathrm{SUMF} + 0.00008\ \mathrm{MOFIY} \tag{10}$$

Calculated – Observed = –0.07

$$-\log(IC50)\ \mathrm{CV2} = 0.02 + 6.45\ \mathrm{ATCH4} + 18.83\ \mathrm{ATCH5} - 0.13\ \mathrm{DIPVX} \tag{11}$$

Calculated – Observed = 1.66

$$-\log(IC50)\ \mathrm{CV3} = -1.73 + 0.48\ \mathrm{logP} + 3.44\ \mathrm{ATCH4} + 9.99\ \mathrm{ATCH5} + 0.00005\ \mathrm{MOFIY} \tag{12}$$

Calculated – Observed = 0.13

$$-\log(IC50)\ \mathrm{CV4} = -2.41 + 0.55\ \mathrm{logP} - 0.00007\ \mathrm{MOFIZ} + 1.45\ \mathrm{SUMF} \tag{13}$$

Calculated – Observed = 1.03

$$-\log(IC50)\ \mathrm{CV5} = 1.72 + 0.50\ \mathrm{logP} + 2.81\ \mathrm{ATCH4} - 0.19\ \mathrm{PEAXX} + 0.82\ \mathrm{ESDL} \tag{14}$$

Calculated – Observed = 0.19

$$-\log(IC50)\ \mathrm{CV6} = -1.47 + 0.42\ \mathrm{logP} + 4.73\ \mathrm{ATCH4} + 13.35\ \mathrm{ATCH5} - 0.11\ \mathrm{DIPVX} - 0.00005\ \mathrm{MOFIY} \tag{15}$$

Calculated – Observed = 0.60

$$-\log(IC50)\ \mathrm{CV7} = -1.11 + 0.19\ \mathrm{logP} + 5.65\ \mathrm{ATCH4} + 16.22\ \mathrm{ATCH5} - \mathrm{DIPVX} \tag{16}$$

Calculated – Observed = 1.40

$$-\log(IC50)\ \mathrm{CV8} = 2.06 + 0.52\ \mathrm{logP} + 4.33\ \mathrm{ATCH4} - 0.21\ \mathrm{PEAXX} + 0.85\ \mathrm{ESDL3} \tag{17}$$

Calculated – Observed = –1.05

$$- \log (IC50)\ CV9 = -1.18 + 0.38\ \mathrm{logP} + 4.98\ \mathrm{ATCH4} + 12.54\ \mathrm{ATCH5} - 0.12\ \mathrm{DIPVX} - 0.00005\ \mathrm{MOFIZ} \quad (18)$$

Calculated – Observed = 0.37

$$- \log (IC50)\ CV10 = 2.00 + 0.50\ \mathrm{logP} + 2.91\ \mathrm{ATCH4} - 0.21\ \mathrm{PEAXX} + 0.87\ \mathrm{ESDL3} \quad (19)$$

Calculated – Observed = 0.64

$$- \log (IC50)\ CV11 = -1.82 + 0.50\ \mathrm{logP} + 3.51\ \mathrm{ATCH4} + 9.44\ \mathrm{ATCH5} - 0.00006\ \mathrm{MOFIZ} \quad (20)$$

Calculated – Observed = –0.34

$$- \log (IC50)\ CV12 = 1.65 + 0.53\ \mathrm{logP} + 2.60\ \mathrm{ATCH4} - 0.21\ \mathrm{PEAXX} + 0.77\ \mathrm{ESDL3} \quad (21)$$

Calculated – Observed = 0.51

$$- \log (IC50)\ CV13 = -2.50 + 0.60\ \mathrm{logP} - 0.00008\ \mathrm{MOFIY} + 1.42\ \mathrm{SUMF} \quad (22)$$

Calculated – Observed = 0.21

$$- \log (IC50)\ CV14 = 2.47 + 0.44\ \mathrm{logP} + 3.39\ \mathrm{ATCH4} + 0.71\ \mathrm{ESDL3} - 0.01\ \mathrm{VDWVOL} + 0.0005\ \mathrm{MOFIX} \quad (23)$$

Calculated – Observed = 0.78

$$- \log (IC50)\ CV15 = 0.08 + 0.52\ \mathrm{logP} + 2.42\ \mathrm{ATCH4} + 1.05\ \mathrm{ESDL3} - 0.00007\ \mathrm{MOFIZ} \quad (24)$$

Calculated – Observed = –0.73

$$- \log (IC50)\ CV16 = 2.95 + 0.56\ \mathrm{logP} - 0.01\ \mathrm{SURFA} + 0.88\ \mathrm{ESDL3} \quad (25)$$

Calculated – Observed = –0.79

$$- \log (IC50)\ CV17 = -3.75 + 0.47\ \mathrm{logP} + 2.24\ \mathrm{ATCH4} - 0.00006\ \mathrm{MOFIZ} + 1.40\ \mathrm{NSDL8} \quad (26)$$

Calculated – Observed = 0.99

$$- \log (IC50)\ CV18 = -1.73 + 0.48\ \mathrm{logP} + 3.58\ \mathrm{ATCH4} + 9.64\ \mathrm{ATCH5} - 0.00006\ \mathrm{MOFIZ} \quad (27)$$

Calculated – Observed = –0.23

$$- \log (IC50)\ CV19 = -1.91 + 0.52\ \mathrm{logP} + 3.57\ \mathrm{ATCH4} + \mathrm{ATCH5} - 0.00006\ \mathrm{MOFIZ} \quad (28)$$

Calculated – Observed = 0.72

$$- \log (IC50)\ CV20 = -1.26 + 0.40\ \mathrm{logP} + 4.88\ \mathrm{ATCH4} + 12.17\ \mathrm{ATCH5} - 0.00005\ \mathrm{MOFIZ} - 0.12\ \mathrm{DIPVX} \quad (29)$$

Calculated – Observed = 0.08

$$- \log (IC50)\ CV21 = 1.76 + 0.50\ \mathrm{logP} + 2.79\ \mathrm{ATCH4} - 0.20\ \mathrm{PEAXX} + 0.84\ \mathrm{ESDL3} \quad (30)$$

Calculated – Observed = –0.20

$$- \log (IC50)\ CV22 = 1.74 + 0.50\ \mathrm{logP} + 2.78\ \mathrm{ATCH4} - 0.20\ \mathrm{PEAXX} + 0.82\ \mathrm{ESDL3} \quad (31)$$

Calculated – Observed = –0.28

$$- \log (IC50)\ CV23 = 1.81 + 0.50\ \mathrm{logP} + 2.83\ \mathrm{ATCH4} - 0.20\ \mathrm{PEAXX} + 0.85\ \mathrm{ESDL3} \quad (32)$$

Calculated – Observed = 0.17

$$-\log(IC50)\ CV24 = -1.78 + 0.49\ \text{logP} + 3.51\ \text{ATCH4} + 9.39\ \text{ATCH5} - 0.00006\ \text{MOFIZ} \quad (33)$$

Calculated – Observed = –0.40

$$-\log(IC50)\ CV25 = 2.58 + 0.44\ \text{logP} + 3.20\ \text{ATCH4} + 0.0004\ \text{MOFIX} + 0.77\ \text{ESDL3} \quad (34)$$

Calculated – Observed = 0.99

$$-\log(IC50)\ CV26 = -1.24 + 0.40\ \text{logP} + 4.66\ \text{ATCH4} + 12.29\ \text{ATCH5} - 0.00005\ \text{MOFIY} - 0.12\ \text{DIPVX} \quad (35)$$

Calculated – Observed = –0.14

$$-\log(IC50)\ CV27 = -1.74 + 0.48\ \text{logP} + 3.57\ \text{ATCH4} + 9.65\ \text{ATCH5} - 0.00005\ \text{MOFIZ} \quad (36)$$

Calculated – Observed = –0.24

$$-\log(IC50)\ CV28 = -3.38 + 0.49\ \text{logP} + 2.35\ \text{ATCH4} - 0.00005\ \text{MOFIZ} + 3.29\ \text{NSDL8} - 6.78\ \text{NSDL9} \quad (37)$$

Calculated – Observed = –1.08

$$-\log(IC50)\ CV29 = -1.70 + 0.48\ \text{logP} + 3.53\ \text{ATCH4} + 9.41\ \text{ATCH5} - 0.00006\ \text{MOFIZ} \quad (38)$$

Calculated – Observed = –0.51

$$-\log(IC50)\ CV30 = -1.35 + 0.39\ \text{logP} + 4.47\ \text{ATCH4} + 11.99\ \text{ATCH5} - 0.00005\ \text{MOFIZ} - 0.10\ \text{DIPVX} \quad (39)$$

Calculated – Observed = –0.61

$$-\log(IC50)\ CV31 = -1.11 + 0.19\ \text{logP} - 5.65\ \text{ATCH4} + 16.22\ \text{ATCH5} + 0.14\ \text{DIPVX} \quad (40)$$

Calculated – Observed = 1.40

The answer is that we are cross-validating *the process* which builds the models; that is, what is our best estimate of the predictiveness of the top-rated model generated over the full data set? Eq. (41) shows that model; it was generated over the full data set using the same parameters that were used for the cross-validation study.

First, we note that the model generated (Eq. (41)) over the full Selwood data set contains different features than any of the top models during cross-validation. The regression-only cross-validation in which the terms of the model are fixed, and only the coefficients are recalculated for each cross-validation group gives an estimate of predictivity that is too high. A better estimate is the fully cross-validated r^2, in which a new population of models is evolving for each cross-validation group. The large difference in the two cross-validation scores demonstrates that, for the Selwood data set, variable selection is a much larger source of bias than the regression step. Clearly, the removal of any one sample is enough to change the 'best' combination of features. Another important point is the wide difference between the regression-only cross-validated r^2 and the fully cross-validated r^2. This gap shows that feature selection is a larger source of bias than the regression step. This result suggests that great caution must be used when reporting statistics on any process that removes

$$-\log (IC_{50}) = -1.82 + 0.50 \log P + 0.50 ATCH4 + 2.05 SUMF - 0.000007 MOFIY - 0.13 DIPVX \quad (41)$$

$$r^2 = 0.81; \text{ regression-only CV } r^2 = 0.67; \text{ fully CV } r^2 = 0.32$$

features. All too often, authors publish only the final selected features from a study and the model and statistics derived from those features; our study with the Selwood data set cautions against this approach. Instead, all features studied should be made available, and the validation must encompass the feature selection step, in order for the statistics to be a useful measure of a model's predictiveness.

GPLS results on the Selwood data set

With the techniques explained, we are in a position to show the results when GPLS is applied to the Selwood data set. For fairness of comparison, we compare against a version of PLS which uses outlier limiting to resist errors. In each experiment, the GPLS models were constructed to contain a given number of variables; crossover was restricted to only allow the creation of children with the same number of variables. This was necessary because the LOF function could only be used to estimate model length when relatively few variables were used. A 4-component PLS regression was performed on the selected variables. The evolution was conducted for 5000 steps. For cross-validation, leave-one-out was the standard method, and the best model (rated by LSE) was used to predict the left-out sample.

Standard PLS optimizes the number of components using the cross-validation score; in this experiment, we hold the number of components constant and optimize over the number of variables. The results are shown

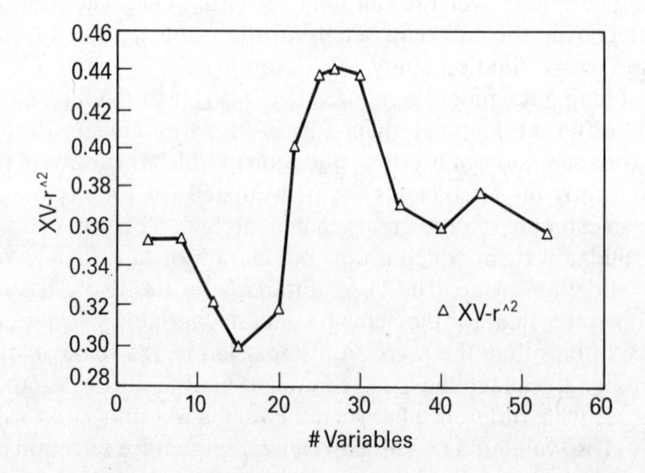

Figure 7 Cross-validated r^2 versus the number of variables holding the number of components constant.

in Figure 7. First, we note that the GPLS procedure, coupled with outlier limiting, is superior at any number of variables to standard PLS, which had a cross-validated r^2 of 0.20. In fact, standard GFA outperformed PLS, with a cross-validated r^2 of 0.32. Second, outlier limiting improves the results of both modeling processes on the Selwood data set. This is due to the large number of variable outliers in the data values in this set. Third, GPLS successfully modeled the data for all numbers of variables we tried, and outperformed the PLS with outlier removal when 25–30 variables were used. Fourth, while there is some variation in cross-validated r^2 associated with changes in the number of variables in the models, fixed-length models allow the user to successfully model without using the LOF measure to estimate model size. In fact, as the size of the models increases, the LOF measure becomes increasingly poor at making appropriate estimates, as it was originally derived for models built with LSR. Last, it is tempting to conclude that the GPLS models of length 25–30 are definitely superior to the models generated at other lengths or from PLS with outlier removal; however, as we will show in the next section, the amount of variation in cross-validated r^2 scores is such that this conclusion can only be provisional. However, it demonstrates the ability of GPLS to discover compact models which are superior to the PLS model of the full data set as measured by the standard cross-validated r^2 score.

Sidenote: comparisons with few cross-validation groups

The most common method for cross-validating is *N*-fold (also called 'leave-one-out') cross-validation. While this technique has disadvantages, as discussed by Cruciani and coworkers (Cruciani *et al.*, 1993), it does have two advantages versus using a fewer number of cross-validation groups. First, it is well-defined, and so the work is easily reproduced; when more samples are contained in each cross-validation group, reproduction can only be accomplished if the sample memberships are given along with the number of cross-validation groups, information that is often not given. Second, assuming that the samples were selected using good experimental design principles, important information is likely to be concentrated in few samples; the larger number of samples in a cross-validation group make it more likely that such information is not available when the cross-validation groups is removed, with the result of poorer estimates on the removed samples. It is from issues such as these that we believe Fisher's randomization test is a superior validation technique. However, given the established nature of cross-validation testing, it will likely remain a commonly-used tool, and so it is important to be aware of limitations such as those above.

We were curious about the magnitude of the variation in cross-validated r^2 from the selection of samples for membership in cross-validation groups. For example, in the original CoMFA study of 21 steroids 4 cross-validation

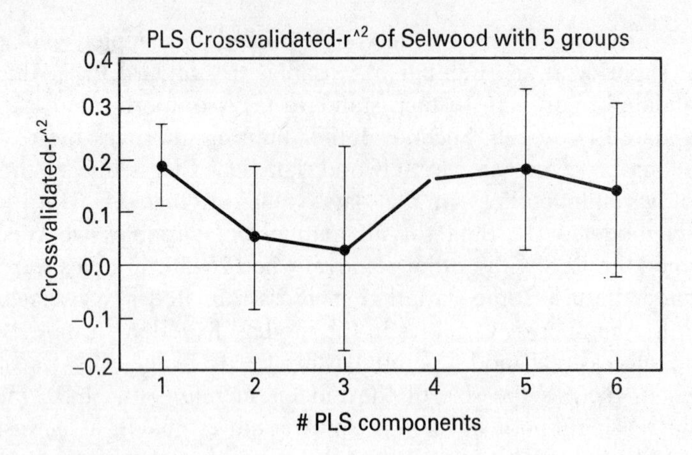

Figure 8 Cross-validated r^2 versus the number of PLS components for standard PLS with 5 cross-validation groups. The Selwood data set was mean centered and scaled for this example.

groups were used. We asked: is the variation in cross-validated r^2 significantly changed by different sample assignments? Figure 8 shows such variance for the Selwood data set. The large one standard-derivation error bars imply that comparison of different regression techniques using a single cross-validation run is dangerous. Cruciani and coworkers (Cruciani *et al.*, 1993) are correct in insisting on averaging over multiple runs to get more stable estimates. We would also recommend that workers publish more extensive details about the sample membership of the cross-validation groups for each run; this would allow the most safety when comparing methods. Given the complexity of this process, workers may wish to simply continue using *N*-fold cross-validation scores to compare different techniques on the same data set, or abandon cross-validation entirely and use randomization testing.

CONCLUSION

In this paper, we described genetic partial least-squares (GPLS), a new algorithm for constructing functional models of data. It is derived from two previously-published techniques: the genetic function approximation (GFA) algorithm, which uses a genetic algorithm to evolve populations of models; and the partial least-squares (PLS) algorithm, increasingly being used as a regression technique in QSAR. Both GFA and PLS have shown themselves to be valuable analysis tools in the case where the data set has more descriptors than samples; GPLS is shown to retain this value, combining strengths from each of its parent algorithms.

REFERENCES

Burke, B.J. and Hopfinger, A.J. (1994). Advances in molecular shape analysis. In, *3D-QSAR in Drug Design: Theory, Methods and Applications* (H. Kubinyi, Ed.). ESCOM, Leiden, pp. 276–306.

Cramer III, R.D., DePriest, S.A., Patterson, D.A., and Hecht, P. (1994). The developing practice of comparative molecular field analysis. In, *3D-QSAR in Drug Design: Theory, Methods and Applications* (H. Kubinyi, Ed.). ESCOM, Leiden, pp. 443–485.

Cruciani, G., Clementi, S., and Baroni, M. (1993). Variable selection in PLS analysis. In, *3D-QSAR in Drug Design: Theory, Methods and Applications* (H. Kubinyi, Ed.). ESCOM, Leiden, pp. 551–566.

Forrest, S. (1993). Genetic algorithms: Principles of natural selection applied to computation. *Science* **261**, 872–878.

Friedman, J. (1988). Multivariate adaptive regression splines. *Technical Report No. 102*, Laboratory for Computational Statistics, Department of Statistics, Stanford University, Palo Alto, CA (revised 1990).

Glen, W.G., Dunn III, W.J., and Scott, D.R. (1989). Principal components analysis and partial least squares regression. *Tetrahedron Comput. Technol.* **2**, 349–376.

Hammett, L. (1970). *Physical Organic Chemistry* 2nd edition. McGraw-Hill, New York, Chapter 11.

Hansch, C. (1968). A quantitative approach to biochemical structure–activity relationships. *Accts. Chem. Res.* **2**, 232–239.

Holland, J.H. (1975). *Adaptation in Natural and Artificial Systems*. University of Michigan Press, Ann Arbor, MI.

Kubinyi, H. (1994a). Variable selection in QSAR studies. I. An evolutionary algorithm. *Quant. Struct.-Act. Relat.* **13**, 285–294.

Kubinyi, H. (1994b). Variable selection in QSAR studies. II. A highly efficient combination of systematic search and evolution. *Quant. Struct.-Act. Relat.* **13**, 393–401.

Livingstone, D.J., Hesketh, G., and Clayworth, D. (1991). Novel method for the display of multivariate data using neural networks. *J. Mol. Graphics* **9**, 115–118.

Luke, B.T. (1994). Evolutionary programming applied to the development of quantitative structure–activity relationship and quantitative structure–property relationships. *J. Chem. Inf. Comp. Sci.* **34**, 1279–1287.

McFarland, J.W. and Gans, D.J. (1994). On identifying likely determinants of biological activity in high-dimensional QSAR problems. *Quant. Struct.-Act. Relat.* **13**, 11–17.

Rogers, D. (1991). G/SPLINES: A hybrid of Friedman's multivariate adaptive regression splines (MARS) algorithm with Holland's genetic algorithm. In, *Proceedings of the Fourth International Conference on Genetic Algorithms* (R.K. Belew and L.B. Booker, Eds.). Morgan Kaufmann Publishers, San Diego, California, pp. 384–391.

Rogers, D. and Hopfinger, A.J. (1994). Application of genetic function approximation (GFA) to Quantitative Structure–Activity Relationships (QSAR) and Quantitative Structure–Property Relationships (QSPR). *J. Chem. Inf. Comp. Sci.* **34**, 854–866.

Rose, V.S., Croall, I.F., and MacFie, H.J.H. (1991a). An application of unsupervised neural network methodology (Kohonen topology-preserving mapping) to QSAR analysis. *Quant. Struct.-Act. Relat.* **10**, 6–15.

Rose, V.S., Wood, J., and MacFie, H.J.H. (1991b). Single class discrimination using principal component analysis (SCD-PCA). *Quant. Struct.-Act. Relat.* **10**, 359–368.

Rose, V.S., Wood, J., and MacFie, H.J.H. (1992). Generalized single class discrimination (GSCD). A new method for the analysis of embedded structure–activity relationships. *Quant. Struct.-Act. Relat.* **11**, 492–504.

Rossi, I. and Truhlar, D.G. (1995). Parameterization of NDDO wavefunctions using genetic algorithms. An evolutionary approach to parameterizing potential energy surfaces and direct dynamics calculations for organic reactions. *Chem. Phys. Lett.* **233**, 231–236.

Selwood, D.L., Livingstone, D.J., Comley, J.C., O'Dowd, A.B., Hudson, A.T., Jackson, P., Jandu, K.S., Rose, V.S., and Staples, J.N. (1990). Structure–activity relationships of antifilarial antimycin analogues: A multivariate pattern recognition study. *J. Med. Chem.* **33**, 136–146.

Wikel, J. and Dow, E. (1993). The use of neural-networks for variable selection in QSAR. *Bioorg. Med. Chem. Lett.* **3**, 645–651.

Wold, S. (1978). Cross-validatory estimation of the number of components in factor and principal components models. *Technometrics* **20**, 397–405.

Wold, S., Ruhe, A., Wold, H., and Dunn III, W.J. (1984). The collinearity problem in linear regression. The partial least squares (PLS) approach to generalized inverses. *SIAM J. Sci. Stat. Comput.* **5**, 735–743.

6 Application of Genetic Algorithms to the General QSAR Problem and to Guiding Molecular Diversity Experiments

A.J. HOPFINGER* and H.C. PATEL

Laboratory of Molecular Modeling and Design, M/C 781, The University of Illinois at Chicago, College of Pharmacy, 833 S. Wood Street, Chicago, IL 60612–7231, USA

Genetic algorithms (GAs) and, in particular, the genetic function approximation (GFA) have been applied to two major problems in computer-assisted molecular design. In one application we have explored how GFA can be used to establish reliable quantitative structure–activity relationships (QSARs) when multiple conformations, shape references and observed biological activities are given for each compound in the training data base. The second application involves applying QSAR analysis to a molecular diversity experiment. The underlying idea in this application is that the accumulating structure–activity data during a molecular diversity experiment can be harnessed by QSAR techniques to guide compound selection in the on-going diversity generation. The successful application of GA (GFA) to each of these problems requires the development of algorithms to structure and restructure the training data base over the course of the GFA analysis. These algorithms are viewed in terms of the generalization of the mutation operator currently employed in GAs.

KEYWORDS: *genetic function approximation; genetic algorithm; conformation; molecular diversity.*

* Author to whom all correspondence should be addressed.

In, *Genetic Algorithms in Molecular Modeling* (J. Devillers, Ed.)
Academic Press, London, 1996, pp. 131–157.
ISBN 0-12-213810-4

INTRODUCTION AND BACKGROUND

Since the late 1970s our laboratory has been working to develop accurate, reliable and robust formalisms, and corresponding methodologies, to construct 3-dimensional quantitative structure–activity relationships, 3D-QSARs (Kubinyi, 1993). The larger part of this work has been focused upon the analysis of the structure–activity relationships (SARs) of ligand analogs where the geometry of the receptor is not known (Burke and Hopfinger, 1993). We term a 3D-QSAR developed for a ligand analog series in which their activities are known, but not the receptor geometry, an *intramolecular* 3D-QSAR. The idea of constructing a QSAR equation to complement molecular binding models for ligand analogs bound to a known receptor geometry, which we term an *intermolecular* 3D-QSAR was pioneered in our laboratory (Hopfinger *et al.*, 1981). The latest state of evolution of our intermolecular 3D-QSAR research is the free energy force field FEFF, 3D-QSAR formalism (Hopfinger, 1995, manuscript submitted). The use of genetic algorithms, GAs (Holland, 1975), coupled with taboo searching (Cvijovic and Klinowski, 1995) and an expert rule based mutation operator in the GA, are key to developing FEFFs. However, intermolecular 3D-QSARs are not the subject of this chapter.

The subject of this chapter is to describe our efforts to advance the construction of intramolecular 3D-QSARs. Our intramolecular 3D-QSAR formalism is called molecular shape analysis (MSA). The molecular modeling operations are defined in Figure 1 (Hopfinger and Burke, 1990). Unrestricted successful application of these seven MSA operations would constitute a general solution to the intramolecular 3D-QSAR problem when each member of the SAR training set has a single biological activity measure which is reasonably accurate. Unfortunately, not all of the MSA operations can be done in a way which leads to an unambiguous QSAR model. In particular, the assignments of an active conformation, shape reference, and molecular superposition (alignment) are usually problematic owing to the large number of possibilities (solutions) in each category. GA can be an effective tool for searching out the optimum MSA 3D-QSAR model when an independent variable (molecular descriptor/property) is degenerate in value for each observation (compound).

METHODS

Shape reference

In order to map out space occupancy by the ligands with regard to a single common comparison scale, a shape reference is needed. The shape reference can be any molecular structure. In the large majority of previous MSA

Molecular Shape Analysis

MSA

Basic Operations to Construct an MSA 3D-QSAR From a Structure-Activity Dataset

1. Conformational Analysis
2. Hypothesize an "Active" Conformation
3. Select a Candidate Shape Reference Compound
4. Perform Pair-Wise Molecular Superpositions
5. Measure Molecular Shape
6. Determine Other Molecular Features
7. Construct a Trial QSAR

Use the Optimized QSAR Ligand Design

Figure 1 The operations of molecular shape analysis to develop intramolecular 3D-QSARs.

3D-QSAR studies each ligand in the training set was evaluated as the shape reference. The statistical significance of fit, usually measured by the correlation coefficient, r, of the QSAR was the basis for selecting a particular shape reference. It is generally found that the shape reference is one of the most active, and/or largest analogs in the training set. This observation has been translated into a simplifying assumption in the selection of trial shape references.

One of our recent intramolecular MSA 3D-QSAR applications involved a series of 3-(acylamino)-5-phenyl-2H-1,4-benzodiazepine cholecystokinin-A (CCK-A) antagonists (Tokarski and Hopfinger, 1994). The common overlap volume between a reference shape and each test analog was used as a molecular shape descriptor (Hopfinger and Burke, 1990). The corresponding non-overlap volume (Hopfinger and Burke, 1990) of the test analog was also considered in the construction of 3D-QSAR models. It was found that the best MSA 3D-QSARs required the reference shape to be the composite (union) shape of three specific analogs. The shape of any single analog, including the most active analog cannot provide a reference molecular shape which captures the overall shape-dependent component of the SAR. Moreover, no arbitrary collection of analogs could be used to construct a reference shape that would optimize the MSA 3D-QSAR. A specific set of three analogs had to be selected in order to generate the optimal reference shape. We call reference shapes constructed from two, or more, molecular entities *mutants*.

While there are different ways of selecting trial shape references, we must also recognize the multiplicity of independent variable selection this MSA operation introduces into the derivation of the corresponding optimum MSA 3D-QSAR equation.

Active conformation

The search for the active, or biologically relevant, conformation of a ligand is a general goal in most computer-assisted drug design applications. We have evolved a strategy of searching for the active conformation that first seeks to identify stable low-energy intramolecular conformer states of active ligands that are high intramolecular energy states for one or more, inactive ligand. The next level of trying to identify the active conformation is to use all intramolecular minimum energy states within some cutoff energy of the apparent global minimum as candidate active conformations. The final level of searching for the active conformation is to enlarge the intramolecular minima sample set to include a uniform sampling of low-energy conformer states, usually constrained to be in different molecular shape classes.

Seeking an active conformation in MSA 3D-QSAR analysis leads to multiple representations in corresponding molecular shape measures.

Molecular alignment

There are usually multiple ways of superimposing a test ligand on a shape reference. In an analog series the most likely superposition (common mode of receptor interaction) is easily identified. As the ligands become increasingly structurally diverse from one another, and/or from the shape reference, the number of test molecular superpositions increases. Multiple molecular superpositions may even need to be explored in highly congeneric analogs. In these cases the various molecular superpositions would be minor perturbations of one another reflecting small differences in ligand binding to a single target receptor due to the structural differences among the ligands.

Measures of biological activity

The discussion above has focused upon the role of shape reference alignment and conformation upon the multiplicity of molecular shape descriptor representations. Nothing has been said regarding the dependent variables in the training set – the biological activities. However, questions regarding the accuracy and reliability of the biological activity (dependent variable) measure of the training set are usually of interest to an investigator constructing a QSAR. In particular, the extraction of information, in terms of comparative rankings of the significance of the independent variables from SAR databases having, at best, semi-quantitative dependent-variable measures remains problematic.

The sequential crossover operation in GA may serve to extract information from semi-quantitative dependent-variable data sets. Comparative plots of independent variable usage over the crossover evolution of a GA-QSAR represents a comparison of relative significance of independent variables in the evolving QSAR models, as well as the consistency and stability of this evolving comparative ranking.

Another relatively unexplored class of QSAR problems, with respect to dependent variables, is the analysis of two, or more, dependent variable measures in a single training set. The most common example of this type of problem is a SAR dataset containing a useful therapeutic activity measure and a toxicity measure. The tertiary endpoint in this type of QSAR study is the derivation of a therapeutic index. A variant of this good–bad activity pair problem is that in which an *in vitro* and an *in vivo* activity are given for each compound in the training set. In this type of QSAR problem the goal is to discover which molecular properties responsible for *in vitro* activity do, and do not, hinder *in vivo* activity, as well as to identify molecular properties not related to *in vitro* activity, but which correlate with *in vivo* potency.

Another type of multiple dependent variables problem is that in which the goal is to globally optimize activity over two, or more, nearly parallel activities. An example of this type of problem is the search for herbicides that are active against a variety of target plant species (and selectively inactive against crop plants).

Partial least-squares, PLS (Glen *et al.*, 1989) offers the opportunity to optimize the fitting of the variances of both multiple dependent and independent variables in establishing QSAR models. Thus, PLS permits a probing to see if better QSARs can be generated for multiple activity (dependent variable) datasets by simultaneously considering the different activities, or if developing a set of independent variable-single activity QSAR models is the preferred approach.

Nevertheless, the reality of the situation is that in most cases common molecular properties are responsible (correlate) for each of the dependent variables (biological activities). Thus, the design goals in these problems are to: (a) establish the extent of functional similarity of molecular properties found to be common correlates to multiple activities (the global optimization goal) and/or, (b) the identification of molecular properties which distinctly correlate with specific activity measures (the specificity goal).

GA appears to be well-suited to drive multivariate linear regression (MLR) and/or PLS fittings to identify both global optimization and specificity QSAR models of multiple dependent variable datasets. The efficient exploration of combinatorial independent variable sets, subject to optimizing the correlation function, is the key factor to make GA solutions (QSAR models) reliable and significant.

Formulation of the general MSA 3D-QSAR problem

The set of physicochemical features available to explore the construction of a significant molecular shape analysis 3-dimensional quantitative structure–activity relationship (MSA 3D-QSAR) can be functionally partitioned into three classes:

1. Intrinsic molecular shape (IMS) features which are usually highly dependent upon conformation. The IMS features provide information on molecular shape within the steric contact surface of the molecule.
2. Molecular field (MF) features which also are highly dependent upon conformation. The MF features provide information on molecular shape beyond the steric contact surface.
3. The remaining set of physicochemical features which are computed such as lipophilicity, aqueous solubility, conformational entropy, etc. These features may, or may not, exhibit a dependence on conformation.
   The set of experimental physicochemical features that have been measured for the compounds of interest. These features may, or may not, exhibit conformational dependence. Moreover, any conformational dependence may be realized only as a Boltzmann average for the feature owing to the nature of the experimental measurement. It is also important to note that one or more of these measured properties may, in fact, be used as the dependent variable end-points in the construction of the QSAR.

Table I *Definitions of molecular features and entities used to construct the general MSA 3D-QSAR formalism.*

u	any compound in the training set, $\{M_u\}$
v	a reference compound
α	the set of conformations for the training set
β	the alignments for the training set
s	the set of intrinsic molecular shape features
p	the set of field probes
$r_{i,j,k}$	the spatial positions at which the molecular field is evaluated
h_p	the set of non-(molecular shape) and non-(molecular field) features
e_p	the set of experimental measures
f	field-related molecular features not derived from the p

Table I contains a set of definitions to facilitate the formulation of the MSA 3D-QSAR problem for a set of molecules, $\{M_u\}$. Four distinct molecular feature tensors can be constructed from the definitions in Table I to incorporate the information associated with each of the four classes of physicochemical features. The four tensors, in turn, can be linearly combined to yield the most general representation of the MSA 3D-QSAR problem (Hopfinger *et al.*, 1994).

$$\mathbf{P}_{u,v} = \mathbf{T}_{u,v} \otimes [\mathbf{V}_{u,v}(s,\alpha,\beta), \mathbf{F}_{u,v}(p,\mathbf{r}_{i,j,k},f,\alpha,\beta), \mathbf{H}_{u,v}(h_p,\alpha,\beta), \mathbf{E}_{u,v}(e_p,\alpha,\beta] \quad (1)$$

where $\mathbf{P}_{u,v}$ is the property matrix, $\mathbf{V}_{u,v}$ is the IMS tensor, $\mathbf{F}_{u,v}$ is the MF tensor, and $\mathbf{H}_{u,v}$ the tensor includes the physicochemical features which do not exhibit a dependence on conformation. $\mathbf{E}_{u,v}$ is the experimental physicochemical features tensor. The remainder of the terms in Eq. (1) are defined in Table I with the exception of $\mathbf{T}_{u,v}$. $\mathbf{T}_{u,v}$ is the transformation tensor which 'solves' the problem by yielding the optimum relationship between $\mathbf{P}_{u,v}$ and the physicochemical features. A graphical representation of Eq. (1), less the **E** tensor, is given in Figure 2.

A variety of approaches to determining $\mathbf{T}_{u,v}$ are being explored. At this time GA appears to be an effective approach to determining $\mathbf{T}_{u,v}$ for some classes of MSA 3D-QSAR problems embedded in Eq. (1). An example of the use of GA in solving a major type of MSA 3D-QSAR problem is given in this paper.

SAR training set architecture

A number of complicating contributions making intramolecular 3D-QSAR analysis either exhaustively difficult in regard to searching through possible models, or incomplete, have been presented above. Each of these complications can also be viewed from the perspective of the type of corresponding data set architecture. An analysis of the architecture of a data set can be instructive in defining the strategy used to uncover interrelationships in the data set. Each of the two studies reported in the next section, consequently, are described not only in terms of their SAR, but also by the characteristic statistical features of the data comprising the SAR.

Custom mutation operator

The successful application of GA to new types of QSAR problems may require custom algorithms functioning during the sequential crossover operations which deal with,

- selection of specific independent variables;
- selection of independent variable sub-datasets;
- introduction and/or deletion of observations (compounds);

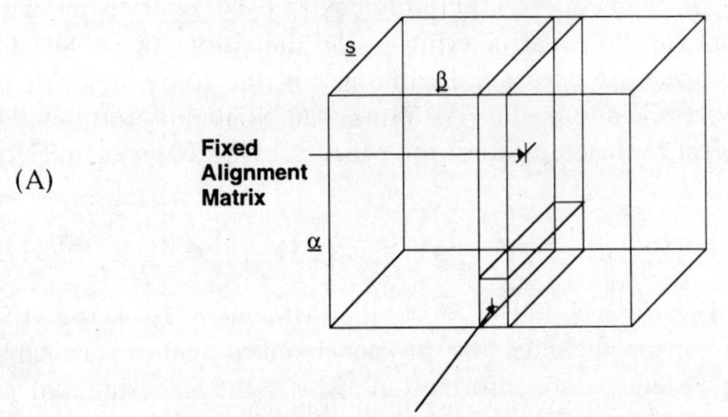

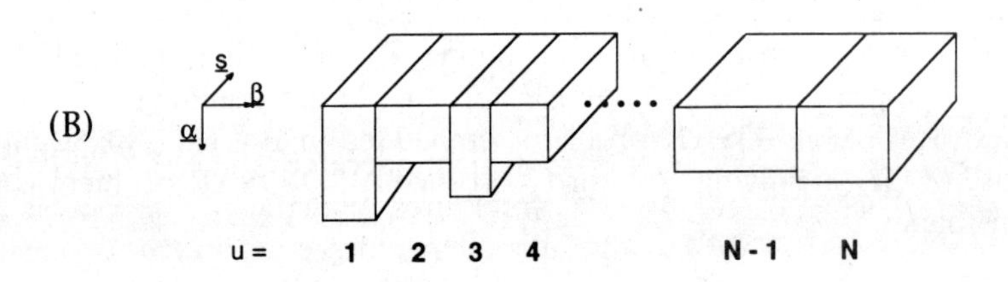

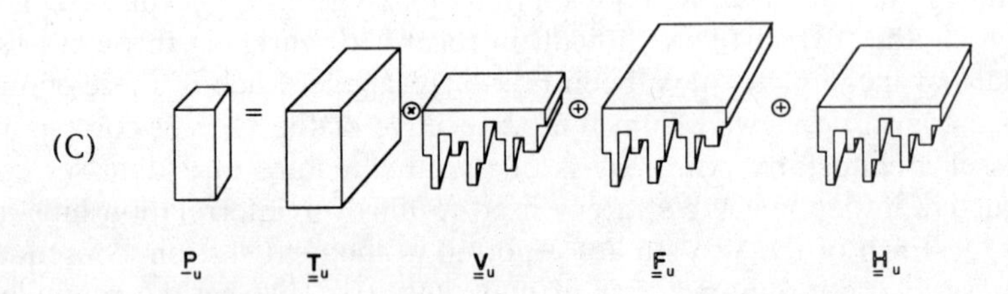

Figure 2 Schematic representation of Eq. (1) and its components.
(A) The IMS tensor for a single compound.
(B) The complete IMS tensor for a set of N compounds.
(C) The complete form of Eq. (1) less the $\mathbf{E}_{u,v}$ tensor.

- change and/or addition of the basis function representation of the independent variables;
- the adoption of the capabilities listed above according to constraints that may depend on the statistical merits of the evolving models.

These types of custom algorithms are most readily identified with the general concept of a mutation operator in the GA paradigm. Hence, we are advocating the generation of custom mutation algorithms to facilitate the widest use of GA to specific and distinct applications. Custom GA mutation operators had to be developed for each of the two applications reported below.

RESULTS

HIV protease inhibitor SAR database

Biological activity

The SAR is for a set of HIV protease inhibitors where the biological activities are the relative nanomolar concentrations of inhibitor required for 50% inhibition of the protease *in vitro*, IC_{50}. Inhibition of two HIV proteases, the parent and a cys-mutant, have been measured and expressed as $-\log(IC_{50})$ and $-\log(cysIC_{50})$, respectively, in the training set.

Architecture of the database

This dataset can be first characterized by the number of compounds (observations), 40, per type of activity as being moderately small in comparison to the number of possible property measures (independent variables), 206, arising from 180 possible molecular shape descriptors and 26 electronic descriptors. The second factor in the classification of this dataset is that many molecular properties are inter-dependent and/or highly colinear to one another. However, the 'regions' of non-colinearity among molecular properties measures can contain important information, unique to each molecular property, that is essential to establishing a correlation relationship.

Molecular properties

Common overlap steric volume, $\mathbf{V}_{u,v}$ (Hopfinger and Burke, 1990) was used as a molecular shape descriptor for multiple active conformation candidates and shape references for each compound. All free-space intramolecular minimum energy conformations within 6 kcal/mol of the global minimum of each ligand analog, were included in the training set. A single alignment was assigned in making the relative molecular shape measurements, $\mathbf{V}_{u,v}$. The number of distinct conformations and shape references varied from four to twelve over the set of ligand analogs. Twenty-six different electronic properties were also considered, and include dipole moments (both magnitude and direction), HOMO and LUMO energies and superdelocalization coefficients.

Representation of MSA 3D-QSAR

This QSAR problem, within the formalism encompassed by Eq. (1), can be defined as

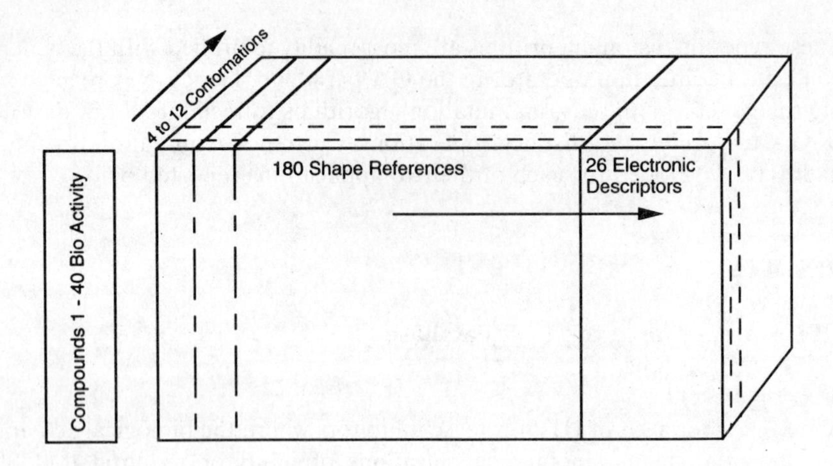

Figure 3 A schematic representation of the HIV protease inhibitor database. The entries in the *Conformation-Shape-Reference* subdata block are the $\mathbf{V}_{u,v}$.

$$\mathbf{P}_{\mathbf{u},\mathrm{v}} = \mathbf{T}_{\mathbf{u},\mathrm{v}} \otimes [\mathbf{V}_{\mathrm{u},\mathrm{v}}(s^*,\alpha,\beta^*), \mathbf{H}_{\mathbf{u},\mathrm{v}}(h_{\mathrm{p}},\alpha,\beta^*)] \qquad (2)$$

where the * indicates only a single measure of this variable has been considered. One schematic representation of the database associated with Eq. (2) is given in Figure 3. The activity matrix, $\mathbf{P}_{\mathbf{u},\mathrm{v}}$ can be either a (1,40) vector of single enzyme activities, or a (2,40) matrix of activities for both enzyme measures.

Analysis and construction of MSA 3D-QSARs

The multiple conformation problem makes this database difficult to study. What is needed is to identify a single conformer state, but possibly multiple overlap volume measures arising from multiple shape references for each compound, in the construction of trial MSA 3D-QSAR models. There is currently no technique which permits a sampling of the independent variables in model construction that is consistent with the required conformer state selection process. A GA formalism can be tailored to enforce this conformer state selection constraint by construction and adoption of a particular mutation operator. More will be said about the development of this mutation operator as part of a larger discussion of mutation operators as vehicles to customize GA to specific problem applications.

The analysis of this HIV protease SAR dataset began with an analysis of sub-populations of the independent variables to assess how important the inclusion of multiple conformations is for successful analysis of this SAR dataset.

Level 1 analysis: The $\mathbf{V}_{u,v}$ for each conformation of each compound measured relative to a single *common* shape reference, along with the 26 electronic descriptors, were used to construct independent variable sub-datasets. Thus, for any of the 180 possible model QSAR construction experiments (there are 180 possible shape references available from the set of conformers) each compound was allowed four to twelve active conformations as measured by $\mathbf{V}_{u,v}$. All 180 shape reference model QSAR experiments using the genetic function approximation (GFA) (Rogers, 1991, 1992; Rogers and Hopfinger, 1994) form of GA with both multivariate regression, MLR, and PLS data-fitting were performed. In addition, power splines (PS), (Friedman, 1988; Rogers, 1991) were combined with MLR-fitting in a third set of experiments. Only –log (IC_{50}) was considered as the dependent variable in this Level 1 analysis because the poor results found using this activity were also anticipated for –log (cysIC_{50}).

In every experiment 300 randomly generated QSARs were generated to seed the GFA run followed by 10 000 crossover operations. In every experiment convergence (optimization) of model QSAR generation appears to have been realized as measured by reaching a constant lack-of-fit (LOF) (Friedman, 1988) measure, and a constant usage of a particular set of molecular properties as independent variables.

The results of the Level 1 experiments are summarized in Table II. Succinctly, no Level 1 QSAR is significant, as can be seen in the range of maximum r^2 values (0.18–0.48), and the only common variable among the top five models is LUMO which is largely independent of conformation and does not reflect molecular shape.

Taken at face value, the Level 1 results suggest that conformation and/or molecular shape is not important in establishing a QSAR, or this form of data representation is not adequate for QSAR model development. This prompted the following set of Level 2 experiments.

Level 2 analysis: Sub-databases were constructed by using the global minimum energy conformation (*conformation 1*) and the second-lowest

Table II *Summary of results of the Level 1 GFA QSAR models. LUMO is the lowest unoccupied molecular orbital energy for each compound, and r^2 is the square of the correlation coefficient.*

Evaluation measure	r^2	Common independent variables among the five best models
MLR	0.18–0.48	LUMO
PLS	0.18–0.48	LUMO
MLR-<PS>	0.18–0.48	LUMO

energy relative minimum (*conformation 2*) in combination with all 180 possible shape references (for each of these two conformations) to generate two sets of $\mathbf{V}_{u,v}$. The $\mathbf{V}_{u,v}$ were then combined with the 26 electronic descriptors to yield two sub-datasets for GFA analysis. Only the –log (IC_{50}) values were used as the dependent variables in each data set because, as in the Level 1 analysis, the poor results found for the –log (IC_{50}) measures were also anticipated for –log ($cysIC_{50}$).

The two GFA experiments were performed in an identical fashion to those of the Level 1 analysis. The results are summarized in Table III. No significant MSA 3D-QSARs could be generated, as reflected by the low r^2 values (0.48 maximum) for these sub-datasets. The $\mathbf{V}_{ij}$ generated from the shape references, defined by the third lowest-energy minimum of compound 15, C153, and the global minimum of compound 35, C351, appear as common independent variables among the five best QSAR models for *conformation 1* and *conformation 2*, respectively. However, the low r^2 values negate any real meaning to this observation.

Level 3 analysis: The Level 1 and 2 analyses suggest that it is necessary to simultaneously consider all candidate active conformations of all compounds and all shape references in the construction of the SAR dataset. In other words, it appears that the entire independent variable data block, schematically illustrated in Figure 3, must be used in the construction of a MSA 3D-QSAR. Unfortunately, different compounds have different numbers of conformations contributing to the $\mathbf{V}_{u,v}$. This difference in numbers of conformations for the compounds leads to an ambiguity in the mapping of the $\mathbf{V}_{u,v}$ to the biological activities.

Table III *Summary of results of the Level 2 GFA QSAR models. CXXY refers to shape reference number (XX, Y) in the sub-dataset and corresponds to compound XX in its Y-lowest-energy conformer. LUMOBE refers to LUMO of atoms B and E (based upon an arbitrary user coding).*

Evaluation measure	r^2	Common independent variables among the five best models
• *Conformation 1*		
MLR	0.36–0.48	C153, LUMO, LUMOBE
PLS	0.36–0.48	C153, LUMO, LUMOBE
MLR-<PS>	0.33–0.48	C153, LUMO
• *Conformation 2*		
MLR	0.36–0.48	C351, HOMO
PLS	0.36–0.48	C351, HOMO
MLR-<PS>	0.17–0.31	HOMO

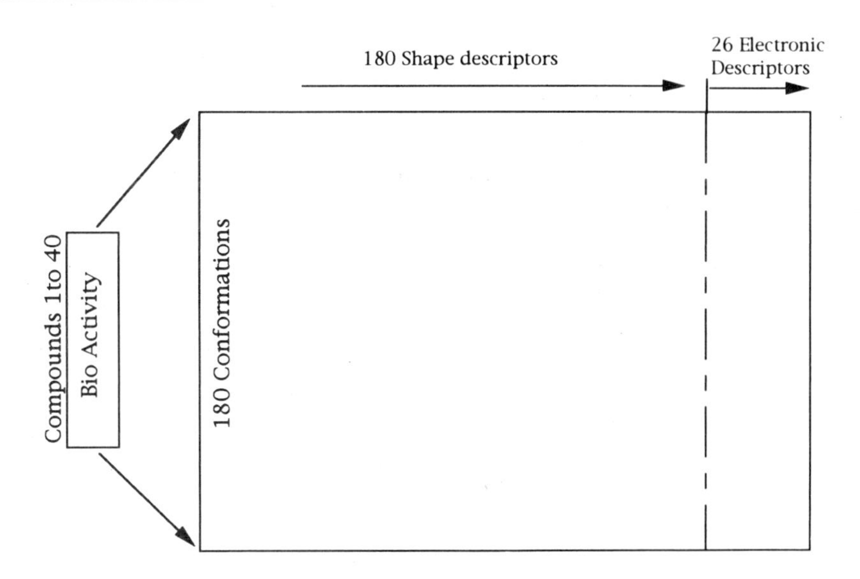

Figure 4 A schematic representation of the Level 3 HIV protease inhibitor database. The compound column is expanded to 180 'observations' reflecting the degeneracy in activities introduced due to multiple conformers for each compound.

One way this ambiguity in the independent variables ($\mathbf{V}_{u,v}$) has been partially overcome has been to 'transfer' some ambiguity to the dependent variable representation (the biological activities). In this particular case each $\mathbf{V}_{u,v}$ of a compound arising from multiple conformer states is mapped to its biological activity. This creates a degeneracy in the SAR dataset. Different measures of a molecular shape property, $\mathbf{V}_{u,v}$ are mapped to the same biological activity for a given compound. The set of 40 biological activities described in Figure 3 becomes expanded to 180 in a manner reflecting the number of conformations associated with each compound. The artificial introduction of degeneracy into the biological activity vector to create a one-to-one mapping between $\mathbf{V}_{u,v}$ and biological activity measures constitutes Level 3 analysis. The degeneracy created in the biological activity measures is also included in the representation of the 26 electronic descriptors in the data set. Figure 4 is a schematic representation of the database structure used in the Level 3 analysis. Obviously, both degenerate representations would disappear if the $\mathbf{V}_{u,v}$ were not included in the data set.

A summary of the results of performing GFA analyses on the degenerate data set in a manner identical to that done in Levels 1 and 2 is given in Table IV. Both –log (IC_{50}) and –log ($cysIC_{50}$) have been considered as dependent variables in separate studies. Values of r^2 reported for the best models in Table IV indicate that some significant MSA 3D-QSAR models have been

Table IV *Summary of results of the Level 3 GFA QSAR models. LUMOA refers to LUMO of atom A (based upon an arbitrary user coding).*

Evaluation measure	r^2	Common independent variables among the five best models
• *–log* (IC_{50})		
MLR	0.81–0.85	LUMO, C151
PLS	0.75–0.79	C351, C151, C341, LUMO, LUMOBE
MLR-<PS>	0.76–0.80	C351, C151, and splines of LUMO, C154 and C181
• *–log* $(cysIC_{50})$		
MLR	0.69–0.73	C101, C151, C401, LUMOA and LUMOBE
PLS	0.67–0.72	C21, C151, C292, C405, LUMOA and LUMOBE
MLR-<PS>	0.72–0.77	C21, C151, HOMO, LUMOA, LUMOBE and splines of C292

generated. The best correlation equations by GFA-MLR fitting for –log (IC_{50}) are:

$$
\begin{aligned}
-\log(IC_{50}) = {} & 7.385737 \\
& + 0.002862(\pm 0.0004) \times C151 \\
& + 1.347681(\pm 0.1101) \times LUMO \\
& - 0.003452(\pm 0.0005) \times C371 \\
& - 0.004112(\pm 0.0003) \times C351 \\
& + 0.003046(\pm 0.0002) \times C181 \\
& + 0.000240(\pm 0.0004) \times C366 \\
& - 0.001289(\pm 0.0004) \times C13 \\
& - 0.002461(\pm 0.0004) \times C341 \\
& - 0.000940(\pm 0.0004) \times C404 \\
& + 0.010358(\pm 0.0015) \times C132 \\
& - 0.000033(\pm 0.0003) \times C202 \\
& + 0.001137(\pm 0.0004) \times C363 \\
& - 0.002824(\pm 0.0003) \times C381 \\
& - 0.001183(\pm 0.0004) \times C406
\end{aligned}
\tag{3}
$$

with

$N = 180$; LOF $= 0.055$; $r^2 = 0.85$; SE $= 0.20$

The corresponding cross-validation measures, using the leave-one-out evaluation scheme, are:

$N = 180$; $r^2 = 0.73$; SE $= 0.22$

For the –log ($cysIC_{50}$) training set, the optimum MSA 3D-QSAR correlation equation is found to be:

$$
\begin{aligned}
-\log(cysIC_{50}) = {} & 11.3793 \\
& + 0.003310(\pm 0.0005) \times C251 \\
& + 0.249734(\pm 0.0745) \times IR \\
& + 3.093678(\pm 0.6550) \times LUMOBE \\
& + 0.002462(\pm 0.0007) \times C21 \\
& - 0.001748(\pm 0.0004) \times C403 \\
& + 0.004020(\pm 0.0005) \times C151 \\
& + 0.006751(\pm 0.0015) \times C292 \\
& - 8.281934(\pm 1.1169) \times LUMOA \\
& + 0.972395(\pm 0.1390) \times HOMO \\
& + 0.001018(\pm 0.0003) \times C185 \qquad (4)
\end{aligned}
$$

with

$$N = 180;\ LOF = 0.067;\ r^2 = 0.78;\ SE = 0.20$$

The corresponding cross-validation statistics for Eq. (4) are:

$$N = 180;\ r^2 = 0.73;\ SE = 0.22$$

In Eq. (3) two compounds, 36 and 40, each have two $\mathbf{V}_{u,v}$ terms due to two conformations [36 (*conformations 3 and 6*), 40 (*conformations 4 and 6*)]. The regression coefficient for $\mathbf{V}_{u,v}$ of C363 is about five times larger than C366 suggesting that C363 is a more significant correlation descriptor. C404 and C406 both are negative and of about the same magnitudes, see Table III for definition of the CXXY. Figure 5 contains plots of the usage of the independent variables as a function of the number of crossover operations performed, and the common independent variables among the best five models are given as a part of Table IV. An inspection of Table IV and Figure 5 suggests that the $\mathbf{V}_{u,v}$ of compounds 36 and 40 are not global correlates in Eq. (3).

Eq. (4) does not contain any $\mathbf{V}_{u,v}$ values based upon multiple conformations from a single compound.

Both Eqs. (3) and (4) have significant cross-validation r^2 values (0.73 in each case). This is important to note in regard to model stability since each of these MSA 3D-QSARs contains many independent variable terms (14–Eq. (3) and 10–Eq. (4)) which are permitted only as a result of 180 observations from the training set of 40 compounds through an activity-conformation degeneracy mapping. The high significance of the QSAR models developed in the Level 3 analysis, as compared to the Levels 1 and 2 analyses, strongly suggests that all conformations and shape references must be *simultaneously* included in GFA model development.

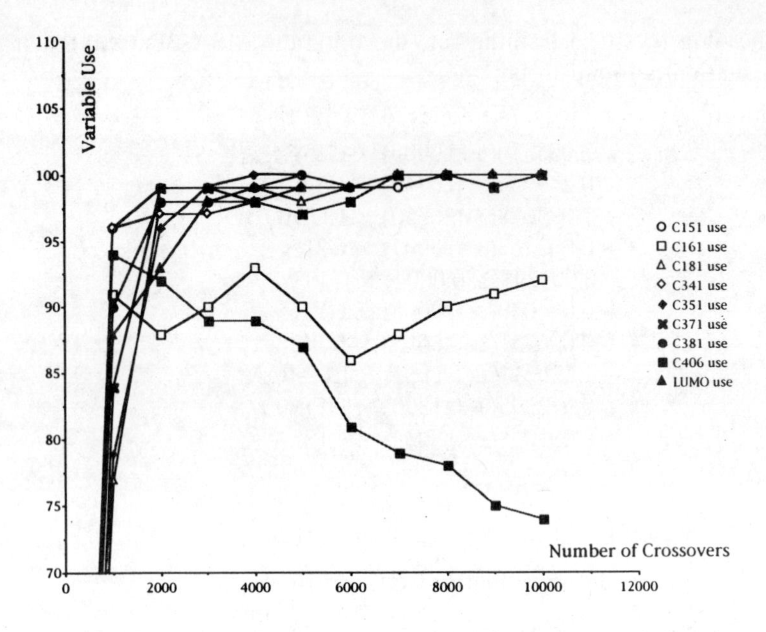

Figure 5 Usage of independent variables as a function of number of executed crossover operations in the GFA experiment.

Custom mutation operator

Deletion of the activity–conformation degeneracy in the Level 3 analysis would lend greater credence to the resulting MSA 3D-QSARs and to the overall treatment of multiple conformations and shape references, that is, determination of the transformation tensor, $\mathbf{T}_{u,v}$, in Eq. (1).

One way to avoid the degeneracy is to first randomly select a single conformation for each compound in datasets used to construct the initial GFA population. As the crossover operation is carried forward from this initial population a check is made to see if independent variables derived from two, or more, conformations of a single compound are found in a child. If so, an on the spot evaluation is made to see if any of the multiple-conformation based independent variables can be eliminated from the child without marked loss in statistical significance. If the child can be mutated to a model having only single-conformation based independent variables, the mutation replaces the child in the population. If, on the other hand, the child needs to be retained because its mutations do not meet statistical significance requirements, then the child is flagged as not being interested in mating. The child is not used in subsequent crossovers until all single-conformation independent variable based QSAR models have been explored and found to be inferior to the child.

Renin inhibitors

Biochemical classification

A set of sixteen peptide-mimetics derived from angiotensinogen-II, the natural substrate of the enzyme renin, have been made and tested as inhibitors of the enzyme (Epps *et al.*, 1990). The general form of these inhibitors is,

$$X_1 - X_2 - X_3 - C_1 - C_2 - A_1 - A_2 - A_3 - A_4 - A_5$$

X_i – fixed, or not present
C_i – fixed, (Phe–His)
A_i – variable with any residue possible, or none

where the variable residue region, the A_i, can consist of two to five amino acid residues and/or mimetics. The binding constants, K_i, have been measured, among other ligand-receptor properties, for these analog inhibitors.

Architecture of the database

Seven of the sixteen inhibitor analogs, two tight binders, two weak binders, and three medium binders were used to construct a forced-fit QSAR between the $-\log K_d$ observed and a set of computed side chain residue physico-chemical features of the variable residue sequences of the inhibitors.

$$-\log K_d = -1.43\ \pi(1) + 5.37\ \pi(2) + 1.93\ \chi_1(3) + 1.23\ Q_+(4) + 1.97\ Q_I(5) - 4.56 \quad (5)$$

$$N = 7; \quad r^2 = 0.98; \quad SD = 0.04$$

The Forced Fit QSAR

In Eq. (5) $\pi(i)$ refers to the lipophilicity (Hansch and Leo, 1979) of the side chain of residue i, $\chi_i(3)$ is the path-1 connectivity index (Kier and Hall, 1976) of the side chains of residue 3, $Q_+(4)$ is the largest positive charge density on an atom of the side chain of residue 4, and $Q_I(5)$ is the residual charge on the ionizable group of the side chain of residue 5. If residue 5 is not ionizable, $Q_I(5) = 0$. The charge densities were computed using the Del Re method (Del Re *et al.*, 1963). This forced-fit QSAR, which is highly significant in statistical fit to the seven observations, nevertheless may reflect little connection to reality, was used to generate the 'observed' $-\log K_d$ for 15 600 possible peptide inhibitor analogs in which the allowed substitutions at A_1 through A_5 are defined in Table V. These 15 600 possible analogs, and their corresponding computed ('observed') $-\log K_d$ values, were used to perform a computer simulated molecular diversity experiment. Thus, the use of Eq. (5) introduces an artificial reality into the molecular diversity simulations, but, by so doing, permits a detailed monitoring of how well each simulation evolves towards optimum behavior.

The architecture of this database is one in which the number of observations (compounds) is changing (increasing) over the course of the series of

Table V *The set of residues, by position A_i, used to generate the renin inhibitors in the simulated molecular diversity experiments.*

Position	Residue choices
A_1(10)	Gly, Leu, Ile, Val, Ser, Phe, Tyr, Thr, Asn, Cys
A_2(10)	Gly, Leu, Ile, Val, Ser, Phe, Tyr, Thr, Asn, Cys
A_3(05)	Gly, Ala, Ile, Ser, Thr
A_4(05)	Gly, Ala, His, Phe, Asn
A_5(05)	Lys, Glu, Gly, Ser, Cys
	Total number of analogs 15 600

GFA crossover operations, and, depending upon the success of evolving GFA QSAR models during the simulation, the average value of $-\log K_d$ is increasing. The number of observations is very large compared to the number of independent variables. All of the independent variables (molecular features) are non-(3D-QSAR) descriptors.

Molecular diversity

The ability to make large numbers of compounds (thousands to millions) and rapidly screen (seconds to minutes per compound) the compounds is referred to as generating molecular diversity for drug candidate discovery (Gordon *et al.*, 1994; Moos *et al.*, 1993). The compounds made and tested in most current molecular diversity experiments are randomly generated. The large number of compounds made and tested is expected to compensate for, and overcome, any drawbacks to not using rational SAR guidelines in the molecular diversity experiment. Nevertheless, a vast SAR database is evolving over the course of a molecular diversity experiment, and it would be advantageous to use this information to guide the later stages of the molecular diversity experiment. The simulated molecular diversity experiment, incorporating a QSAR guide for compound selection, reported here represents an attempt to harness and deploy the evolving SAR information.

The principal chemical means of generating molecular diversity is by combinatorial chemistry in which a set of chemical groups are put together in all possible ways to create a large population of distinct compounds. Linear polymer syntheses are ideal chemical pathways to generate molecular diversity using combinatorial chemistry building blocks. In particular, the solid-state coupling of monomer units called peptoids by robotics, and the subsequent screening of oligomers generated, is a molecular diversity strategy being successfully exploited by Chiron Corporation and others (Gordon *et al.*, 1994; Martin *et al.*, 1995; Moos *et al.*, 1993). The simulated molecular diversity experiment reported here corresponds to a robot-driven generation of oligomeric peptides (a subset of peptoids) as defined in Table V.

Molecular properties

A set of approximately 90 non-(3D-QSAR) descriptors, including those of Eq. (5), were considered for the 15 600 peptides defined in Table V. In addition to the descriptor types used in Eq. (5), the χ_2 and χ_3 connectivity indices, molecular weight, M, all side chain atoms charge densities, Q_I (Del Re *et al.*, 1963), molecular volume, V, sum of atomic polarizabilities for a molecule, $\Sigma\alpha$ (Hopfinger, 1973), and the HOMO, LUMO and their difference, ΔE, computed by EHT (Hoffman, 1963) were considered as QSAR descriptors in the molecular diversity experiment.

The reason for considering all of these molecular properties as QSAR descriptors was to see if the best QSAR models being evolved, and used to guide the generation of compounds, were 'homing in' on Eq. (5). In other words, the realization of Eq. (5) over the course of the molecular diversity simulation would be an indicator validating this modeling approach.

The simulated QSAR-directed molecular diversity experiments

The application of QSAR analysis to guide compound selection in a molecular diversity experiment will be referred to as *dynamic QSAR analysis*. The combined (robot-based) synthesis-screening instrumentation, along with the QSAR hardware and software components, gives rise to a dynamic QSAR analysis engine. One configuration of such an engine is given in Figure 6. The synthesis and screening is carried out in the two boxes labeled 'Synthesis' and 'Biological Screens' and can be viewed as simply an input/output device with respect to the QSAR analysis components. The box labeled 'Random Initialization' refers to the random generation of N_1 compounds to produce an initial SAR for the construction of trial QSAR models. This unit is subsequently employed only if the evolving QSAR models cannot provide adequate numbers of compounds for synthesis over the course of the molecular diversity experiment.

The boxes labeled 'Model QSAR Generator', 'Descriptor Corrector' and 'Model QSAR Evaluator', in composite, provide the evolving QSARs used for the selection of new compounds. We will not give an in-depth description of what happens 'in' these three boxes. Rather, we will overview how GA is used, and is particularly useful, for this part of the dynamic QSAR analysis.

Once the N_1 compounds and their screening activities are available the GFA routine is used (in conjunction with other software) to define the first set of 'best' QSAR models. These models are evaluated, first by cross-validation (Cramer *et al.*, 1988), and then by using some of the new ΔN compounds being made and screened in parallel with the construction of the QSARs. If the QSARs are not validated as being predictive, the QSARs are further refined/re-defined by GFA. New compounds, from the ΔN set, are added to the SAR training set and inactive compounds, which do not fit the QSARs, are deleted from the training set. Furthermore, additional

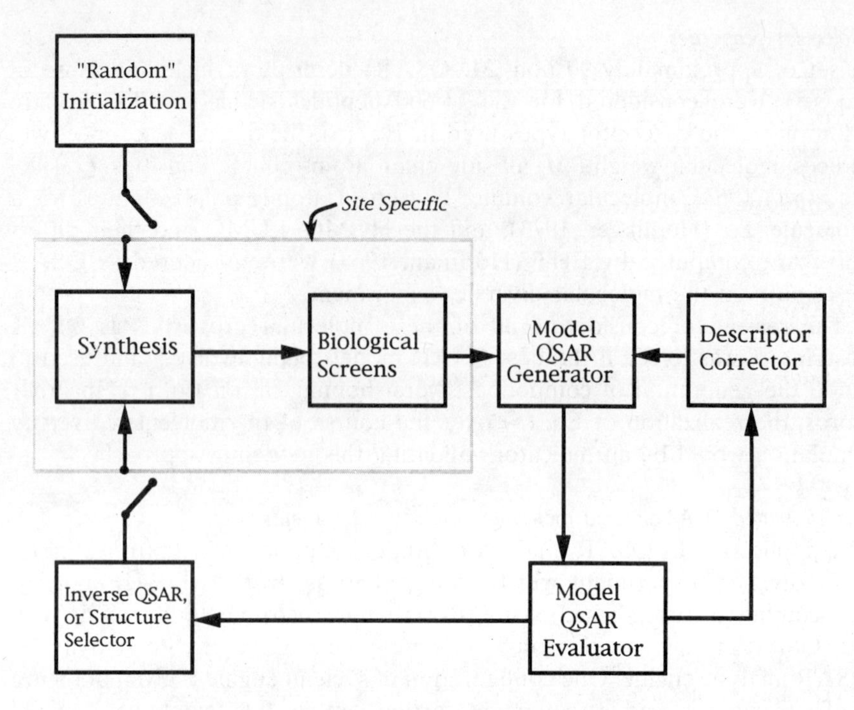

Figure 6 A possible dynamic QSAR analysis engine.

descriptors (independent variables) are considered in the GFA application. The new set of best QSARs are again evaluated for predictive validation. Until a validated QSAR is found, the molecular diversity experiment is executed by the random generation of compounds.

The steps in the construction and validation of QSARs in the molecular diversity experiment can be viewed, and treated, as a unique GA mutation operator for dynamic QSAR analysis. Thus, this application of GA to a QSAR problem leads to a corresponding distinct mutation operator which is needed to make GA effective in the application.

If a validated QSAR is established at any point over the course of the molecular diversity experiment, it is used to select new compounds for synthesis. We emphasize that the development of a general approach to define chemical structures with desired dependent-variable property measure from a QSAR is a 'world-class' problem. This task is referred to as the *inverse QSAR operation*, the *reverse engineering process*, and/or the *structure–selector operation*. We have *not* solved this problem. The inverse QSAR operation, shown in Figure 6 and as applied in the 'contrived reality' molecular diversity simulation of the renin analogs, consists simply of selecting residues side chains for the A_i, and seeing how well they 'work' by evaluating

Table VI *Contrived reality. Model for the mechanism of action.*

1. K_d varies only as a function of the sidechains of the A_i.
2. Each residue backbone considered as an amide or imide.
3. –log K_d is assumed to be a linear function of specific sidechain properties of the A_i.
4. Only non-3D descriptors are used to construct QSARs.
5. If A_i is not present in the peptide, its descriptors' measures are all zero.

Table VII
Contrived reality. The conditions used in the molecular diversity experiments.

1. Introduce error into the –log K_d using a gaussian distribution error generator where one standard deviation corresponds to $\pm X\%$ of error from the actual value.
2. Randomly 'synthesize' and 'screen' N_1 analogs to initiate the simulated molecular diversity experiment.
3. Only substituent (sidechain) analogs are permitted in the inverse QSAR step.
4. Only an analog predicted to have a –log K_d no lower than 20% (8.80) of the most analog in the initial training set passes the inverse QSAR operation.
5. A genetic algorithm is used to construct and evolve the QSAR models.
6. The 'control' for the experiment is the random selection of all compounds.
7. Duplicate compounds are not allowed.

the current validated QSAR models. Analogs of predicted high activity (within 20% of the currently most active renin inhibitor) are targeted for synthesis. Less active selections are put on hold, or forgotten.

Table VI gives a summary overview of the 'contrived reality' mechanism of action of the renin inhibitors in the molecular diversity simulation. 'Rule 2' deletes mimetics from the simulation data set and rule 5 provides a simple means of handling peptides of different lengths. The simulation conditions used in the molecular diversity experiments are given in Table VII. Condition 1 of Table VII permits the monitoring of the significance of dynamic QSAR analysis upon the molecular diversity experiment as a function of the extent of error in determining biological activities. The errors are assigned to the –log K_d computed from Eq. (5). The reference point, or 'control', for the dynamic QSAR analysis molecular diversity simulation experiment is the comparison to the results achieved based upon random compound selection (condition 6).

Numerous molecular diversity simulation experiments have been done. The results of the simulation experiment given in Table VIII are quite representative of our overall findings regarding the impact of dynamic QSAR analysis on the optimization of activity. The simulation control parameters

Table VIII *Summary of the results of a dynamic QSAR analysis molecular diversity experiment.*

Dynamic QSAR						Control (random)					
I	n_s	n_h	t	$\Delta A\%$	r^2	I	n_s	n_h	t	$\Delta A\%$	r^2
0	–	8	1.8	3.5	0.62	0	–	8	0.5	3.5	–
1	113	27	2.7	8.9	0.71	1	–	9	0.8	–1.9	–
2	137	21	2.9	10.2	0.76	2	–	7	1.1	0.6	–
3	98	22	3.8	13.9	0.82	3	–	8	0.9	0.9	–
4	77	16	5.0	16.7	0.84	4	–	4	0.7	0.7	–
5	62	11	6.7	18.3	0.83	5	–	1	0.9	0.9	–
6	47	6	9.8	18.9	0.85	6	–	3	1.0	1.0	–
7	44	7	14.3	19.1	0.86	7	–	0	0.7	0.7	–
8	45	0	20.9	19.3	0.83	8	–	0	0.8	0.8	–
9	39	3	22.1	19.4	0.84	9	–	1	0.6	0.6	–
10	37	0	21.7	19.4	0.84	10	–	0	0.9	0.9	–

[$N_1 = 500$, $X\% = 6\%$, $\Delta N = 500$, $n_m = 500$, $I = 1500$]

I is the simulation iteration number,
n_s is the number of compounds passing the inverse QSAR test,
n_h is the number of compounds at a given iteration step that are more active than the most active compound found in the previous iteration step,
t is the average processing time per compound at a given iteration,
ΔA is the average increase, in percent, per compound in activity for a given iteration, and
r is the correlation coefficient of the best QSAR.

and boundary values are defined in Part A of Table IX, while the measures used to monitor the results of the simulation at uniform steps through diversity generation are defined in Part B of Table IX. The maximum number of inverse QSAR test compounds evaluated in an update, n_m, refers to the upper limit in the number compounds tested by the inverse QSAR operator at each interval step (the generation of ΔN new compounds) in the simulation, and n_s is the number of successful compounds out of the n_m at each step. $\Delta A\%$ refers to the increase, compared to the initial set of analogs, in screened activity averaged over the number of compounds made and screened to that point in the simulation. The time, t, to process a compound refers to the CPU time used (per analog) at each interval step along the molecular diversity simulation pathway. The r^2 of the 'best' QSAR is for the most significant GFA-derived QSAR realized to that point in the simulation.

The two most significant columns to respectively compare in Table VIII are the n_h and $\Delta A\%$ of the 'Dynamic QSAR' and 'Control' (random) simulations. The dynamic QSAR simulation identifies a significantly greater number of increasingly active analogs than found in the random simulation. Moreover, the average activity of the evolving SAR data set, $\Delta A\%$, increases almost 20% in the dynamic QSAR analysis molecular diversity simulation

Table IX *Measures to monitor a molecular diversity simulation experiment.*

Part A:	*Control parameters and boundary values*
N_1	initial set of randomly synthesized compounds
$X\%$	'average' error in the activity measures
ΔN	number of compounds considered in each QSAR update
n_m	maximum number of inverse QSAR test compounds evaluated in an update
I	QSAR update step number for Nu processed compounds
Part B:	*Indices to monitor results*
n_s	number of successful inverse QSAR compounds
n_h	number of compounds more active than most active found to date
t	'time' to process a compound
$\Delta A\%$	percent increase in average activity relative to the initial set of compounds
r^2	correlation coefficient of the 'best' QSAR

while no increase in $\Delta A\%$ is realized in the random simulation. An even better 'bottom line' comparison defining the benefits of applying dynamic QSAR analysis in molecular diversity experiments can be gleaned from Table X. The activities of the five most active compounds (based upon the contrived reality expressed in Table VI and Eq. (5)) using dynamic QSAR analysis, and by random compound selection, respectively, are given in Table X. No specific compounds are given in order to avoid any possible use of this artificial data in actual research applications. The most active compound found using dynamic QSAR analysis is nearly 8× (log scale) more active than the most active compound identified in the random screening molecular diversity simulation. Moreover, all five of the most active analogs determined by dynamic QSAR analysis are more active than the most active analog found from random screening.

The n_h, t, $\Delta A\%$ and r^2 data in Table VIII suggest that the molecular diversity experiment using dynamic QSAR analysis has converged in the $I = 8$ to 10 range. The highest activity reported in Table X is the highest possible activity in the data set defined by the Eq. (5) and the compounds of Table V. We have kept track of the processing time, t, over the course of the simulation to see if dynamic QSAR analysis is practical to apply in real molecular diversity experiments. A processing time of about 30 CPU seconds per compound per update cycle on a small workstation suggests there should be no problems in applying dynamic QSAR analysis. Of course, the complexity of the inverse QSAR operator and/or the level of molecular modeling used to develop the QSARs could adversely impact the practical implementation of a particular form of dynamic QSAR analysis.

A final measure of the validation and convergence of the simulated dynamic QSAR analysis molecular diversity experiment is to compare the 'best' QSAR being used to guide compound selection at the end of the simulation to

Table X *Five highest activities found, –log K_d, at the end of the simulation.*

Dynamic QSAR	Control (random)
11.37	10.59
11.10	10.51
10.73	10.40
10.71	10.38
10.62	10.33

* Highest –log K_d in the original, randomly selected data set (N_1 = 500) is 9.43

Eq. (5) employed to construct 'reality'. The optimum QSAR at the end of the simulation is,

$$-\log K_d = -1.26\ \pi(1) + 6.09\ \pi(2) + 0.21\ M(3) + 0.91\ Q_+(4) + 0.06\ \Delta E(5) - 3.61 \qquad (6)$$

$$N = 316; \quad r^2 = 0.84; \quad SD = 0.13$$

The value of N in Eq. (6) is for the set of compounds in the final iteration of the molecular diversity simulation. It must be remembered, however, that this set of compounds has been 'filtered' by the application of GFA and the constraints given in Table VII over the course of the simulation. A term-by-term comparison of Eqs. (5) and (6) indicate that terms 3 and 5 of the dependent variables are different, while the terms 1, 2 and 4 are the same. $\chi_1(3)$ and M(3) of term 3 in Eqs. (5) and (6), respectively, have an r^2 value of 0.89, and r^2 for $Q_1(5)$ and $\Delta E(5)$ is 0.73 for term 5 using the compounds having the highest (most active) 40% –log K_d values. Only the most active compounds are considered in the comparison analysis of Eqs. (5) and (6) because the most active compounds are selectively used in developing Eq. (6) over the course of the simulation.

CONCLUDING REMARKS

Two SAR training sets of markedly different architecture have been investigated using GA (GFA) with the goal of developing QSARs. A set of flexible HIV protease inhibitors gives rise to a data set where there is a conformational degeneracy in the assignment of molecular shape feature to an observed activity. That is, there is an ambiguity in the assignment of active conformations. Independent analyses of different subsets of the entire SAR training database establish that the entire SAR training database must be considered in a single analysis in order to construct a significant and robust MSA 3D-QSAR.

The removal of the conformational degeneracy without introducing degeneracy anywhere else in the training database requires a special processing of the database for usage in the GFA. In essence, single conformation shape descriptors must be selected for each 'person' (model) considered in the GFA at both the level of random population generation, and at each crossover operation. This special processing of the data which is degenerate in conformational representation can be viewed as a customized mutation operator.

It is important to keep in mind that the ability to unambiguously identify the set of distinct molecular shape descriptors as a function of conformation, alignment, shape reference, and representation, does *not* tell us how to use this information in a molecular design mode. In essence, the molecular shape descriptors and the rest of the MSA 3D-QSAR must be mapped into a common space which defines (displays) the composite 3D requirements for the activity. The formalism to accomplish this MSA 3D-QSAR to space mapping is being developed in our laboratory.

The application of GA (GFA) to guiding molecular diversity generation, which we term dynamic QSAR analysis, also requires its own custom mutation operator. In this application we take advantage of the evolving nature of GA, and update the evolution process by introducing additional observations (new SAR generated in the on-going synthesis of molecular diversity) at regular intervals in the sequential crossover operations. There is another feature that can be a useful property of this dynamic QSAR analysis mutation operator. The 'unhealthy' (low activity) compounds in the evolving diversity population can be deleted from consideration in the construction of the evolving QSAR models. Hence, only *hi-activity* QSARs emerge over the course of the molecular diversity experiment. These hi-activity QSARs are then used to select at least some of the new compounds to be made in the coming step of the on-going molecular diversity experiment.

Overall, the maximum usage of GA in QSAR problems may require the development of specific algorithms to prepare and/or process the training sets of data for analysis by GA. We view these specific algorithms within the context of a mutation operator in the conventional GA formalism. Custom mutation operators may be the tools which make GA the most effective approach to handling the diverse database architectures that arise in QSAR problems.

ACKNOWLEDGMENTS

We very much appreciate the very helpful discussions with Dr Mario Cardozo of Boehringer Ingelheim Pharmaceuticals Inc. and Dr David Rogers of Molecular Simulations Inc., over the course of work reported here. Dr. Dan Pernich of Dow Elanco provided valuable support and insight into the treatment of molecular diversity. Resources of the Laboratory of Molecular Modeling and Design were used in performing the research. Boehringer Ingelheim Pharmaceuticals Inc., The Procter and Gamble Company, and Dow Elanco are acknowledged for their generous financial support of this research.

REFERENCES

Burke, B.J. and Hopfinger A.J. (1993). Advances in molecular shape analysis. In, *3D-QSAR in Drug Design. Theory, Methods and Applications* (H. Kubinyi, Ed.). ESCOM Science Publishers B.V., Leiden, pp. 276–292.

Cramer III, R.D., Bunce, J.D., Patterson, D.E., and Frank, I.E. (1988). Crossvalidation, bootstrapping, and partial least squares compared with multiple regression in conventional QSAR studies. *Quant. Struct.-Act. Relat.* **7**, 18–25.

Cvijovic, D. and Klinowski, J. (1995). Taboo search: An approach to the multiple minima problem. *Science* **267**, 664–666.

Del Re, G., Pullman, B., and Yonezawa, T. (1963). Electronic structure of the α-amino acids of proteins, I. Charge distributions and proton chemical shifts. *Biochim. Biophys. Acta* **75**, 153–158.

Epps, D.E., Cheney, J., Schostarez, H., Sawyer, T.K., Praire, M., Kreuger, W.C., and Mandel, F. (1990). Thermodynamics of the interaction of inhibitors with the binding site of recombinant human renin. *J. Med. Chem.* **33**, 2080–2086.

Friedman, J. (1988, revised 1990). Multivariate adaptive regression splines. *Technical Report No. 102*, Stanford University, Stanford, California.

Glen, W.G., Dunn III, W.J., and Scott, D.R. (1989). Principal components analysis and partial least squares regression. *Tetrahedron Comp. Methodol.* **2**, 349–376.

Gordon, E.M., Barrett, R.W., Dower, W.J., Fodor, S.P.A., and Gallop, M.A. (1994). Applications of combinatorial technologies to drug discovery. 2. Combinatorial organic synthesis, library screening strategies, and future directions. *J. Med. Chem.* **37**, 1385–1401.

Hansch, C. and Leo, A. (1979). *Substituent Constants for Correlation Analysis in Chemistry and Biology.* Wiley-Interscience, New York.

Hoffman, R. (1963). An extended Hueckel theory. I. Hydrocarbons. *J. Chem. Phys.* **39**, 1397–1403.

Holland, J. (1975). *Adaptation in Artificial and Natural Systems*. University of Michigan Press, Ann Arbor, Michigan.

Hopfinger, A.J. (1973). Calculations of residual charges. In, *Conformational Properties of Macromolecules*. Academic Press, New York, pp. 50–52.

Hopfinger, A.J. and Burke, B.J. (1990). *Concepts and Applications of Molecular Similarity* (M.A. Johnson and G. Maggiora, Eds.). Wiley, New York, pp. 173–210.

Hopfinger, A.J., Nakata, Y., and Max, N. (1981). Quantitative structure activity relationships of anthracycline antitumor activity and cardiac toxicity based upon intercalation calculations. In, *Intermolecular Forces* (B. Pullman, Ed.). Reidel-Dordrecht, pp. 431–444.

Hopfinger, A.J., Burke, B.J., and Dunn III, W.J. (1994). A generalized formalism of three-dimensional quantitative structure-property relationship analysis for flexible molecules using tensor representation. *J. Med. Chem.* **37**, 3768–3774.

Kier, L.B. and Hall, L.H. (1976). *Molecular Connectivity in Chemistry and Drug Research*. Academic Press, New York.

Kubinyi, H. (1993). *3D-QSAR in Drug Design. Theory, Methods and Applications*. ESCOM Science Publishers B.V., Leiden.

Martin, E.J., Blaney, M.A., Siani, D.C., Spellmeyer, D.C., Wong, A.K., and Moos, W.H. (1995). Measuring diversity: Experimental design of combinatorial libraries for drug discovery. *J. Med. Chem.* **38**, 1431–1436.

Moos, W.H., Green, G.D., and Pavia, M.R. (1993). Recent advances in the generation of molecular diversity. In, *Ann. Rep. Med. Chem.* (J.A. Bristol, Ed.). Academic Press, New York, Vol. 28, pp. 315–322.

Rogers, D. (1991). G/SPLINES: A hybrid of Friedman's multivariate adaptive regression splines (MARS) algorithm with Holland's genetic algorithm. In, *The Proceedings of the Fourth International Conference on Genetic Algorithms* (R.K. Belew and L.B. Booker, Eds.). Morgan Kaufmann Publishers, San Diego, California, pp. 384–391.

Rogers, D. (1992). Data analysis using G/SPLINES. In, *Advances in Neural Processing Systems 4*. Kaufmann Publishers, San Mateo, California, pp. 103–109.

Rogers, D. and Hopfinger, A.J. (1994). Application of genetic function approximation to quantitative structure-activity relationships and quantitative structure-property relationships. *J. Chem. Inf. Comput. Sci.* **34**, 854–866.

Tokarski, J.S. and Hopfinger, A.J. (1994). Three-dimensional molecular shape analysis-quantitative structure-activity relationship of a series of cholecystokinin-A receptor antagonists. *J. Med. Chem.* **37**, 3639–3654.

7 Prediction of the Progesterone Receptor Binding of Steroids using a Combination of Genetic Algorithms and Neural Networks

S.P. VAN HELDEN*, H. HAMERSMA, and V.J. VAN GEERESTEIN
Department of Computational Medicinal Chemistry, NV Organon, P.O. Box 20, 5340 BH Oss, The Netherlands

In this paper a combination of a genetic function approximation (GFA) and neural networks (NNs) is used to predict the progesterone receptor binding affinity (RBA) of a set of 56 steroids. In total 52 quantum chemical and steric properties were calculated for each of the relevant substituents in these steroids. A training set of 43 steroids is used to derive the models of which the predictive power is assessed by using a test set of the remaining 13 compounds. In previous work we tested a number of different traditional QSAR methods on this data set. In this paper, a GFA is applied to select those properties that are important for describing the RBA of the steroids. In the evolutionary process, quadratic and/or spline mutations of the original properties are allowed to account for nonlinear relationships. This procedure results in models comparable to those obtained by techniques like stepwise regression and PLS ($r^2_{training\ set}$ = 0.64, $r^2_{test\ set}$ = 0.49).

The 10 most frequently used properties give a good description of the range of substituents present in the data set and are subsequently used as input for a feed-forward backpropagation NN. Leave-one-out cross-validation shows that no more than 500 training cycles should be performed on this data set to avoid overtraining and that the best model is obtained with 5 hidden neurons ($r^2_{training\ set}$ = 0.88, $r^2_{test\ set}$ = 0.57). In order to test the reproducibility of the NN

* Author to whom all correspondence should be addressed.

In, *Genetic Algorithms in Molecular Modeling* (J. Devillers, Ed.)
Academic Press, London, 1996, pp. 159–192.
ISBN 0-12-213810-4

each of the 43 networks created by cross-validation is used to predict the compounds in the test set. Typically, the standard deviation of the predictions is less than 10% of the mean predicted value. These results show that a combination of a GFA for selection of variables and a NN for model building leads to superior results compared with conventional statistical techniques.

KEYWORDS: *steroids; receptor binding affinity; neural network; genetic algorithm.*

INTRODUCTION

Modern drug research is primarily directed towards the discovery of structurally specific drugs aimed at specific targets. Drug companies are normally interested in developing drugs that have high affinity and selectivity for a particular receptor. The understanding of the forces governing the activity of a particular compound or class of compounds may help in designing new drugs for specific purposes. We now know that the specific biological properties of a drug molecule are dependent upon its physicochemical properties and 3-dimensional structure which in turn are determined by its chemical structure. Obviously, the study of quantitative structure–activity relationships (QSAR) is of both great scientific interest as well as of commercial importance.

Classical QSAR approaches try to describe biological activity in terms of simple physicochemical properties of the whole molecule or of certain substituents in the drug molecule. An example of this approach is the basic Hansch equation:

$$\log (1/C) = K_1 \log P + K_2\sigma + K_3 \tag{1}$$

where C is the concentration of the drug required to produce a specific effect; log P is the partition coefficient of the drug distributed between water and (typically) *n*-octanol; σ is a constant characteristic of the electronic effect of a substituent, and the Ks are constants. The K constants are normally determined by means of a least-squares procedure on data of a congeneric series. At the time of its invention, about 30 years ago, the Hansch approach was an exciting progression in understanding structure–activity relationships (Hansch and Fujita, 1964). At the same time Free and Wilson developed a model of additive group contributions (Free and Wilson, 1964). However, it soon became clear that more sophisticated methods were needed.

A recent development in QSAR includes the 3D structural properties of the drug molecules. In particular the so-called comparative molecular field analysis (CoMFA) has become very popular and is now a standard technique (Cramer *et al.*, 1988). These new QSAR techniques became possible after mathematical techniques were introduced that were new to the field of

QSAR and also because the advent of computer hardware technology made it practically feasible to use these calculation methods.

In this paper we will show that the application of modern mathematical methods to more or less traditional parameters also gives very satisfactory results. One of the problems of the traditional QSAR approaches (even CoMFA) is that the models are inherently additive and thus do not allow for nonlinearity of structure–activity relationships. A possible solution of this problem can be obtained by using artificial neural networks (NNs). In this paper we will present an example of the power and the pitfalls of these methods. Moreover, in contemporary QSAR, we are typically confronted with many more parameters than observed values because it is quite often relatively easy to automatically calculate a large series of steric and electronic properties based on substituents or the whole molecule. For that reason, the selection of relevant parameters to derive a meaningful model has become a problem in itself. A very recent approach is to perform variable selection by genetic algorithms (GAs) which can also be used to transform mathematically the variables and thus derive nonlinear models. In this paper we show an example of the application of GAs to this problem. NNs and GAs will be applied to a set of steroids for which the binding affinity to the progesterone receptor is known.

Since the introduction of the first progestagens as oral contraceptives in the late fifties a large number of progestagenic steroids has been prepared. As compounds of this class all share basically the same steroid skeleton, but show a wide variety of substituents with a generally well-defined orientation, this class of compounds lends itself particularly well for QSAR analysis. Indeed, despite occasional scepticism (Rozenbaum, 1982), a large number of analyses have been performed relating the progestagenic activity to the structure, preferentially the one determined by X-ray crystallography (Duax *et al.*, 1984), using steric (Teutsch *et al.*, 1973) or combinations of steric and electronic (Coburn and Solo, 1976) parameters. Other variables, such as NMR data (Hopper and Hammann, 1987), have also been used. The methods employed included Hansch-type analysis (van den Broek *et al.*, 1977), multiple regression (Lee *et al.*, 1977), cluster analysis (Belaisch, 1985), correspondence analysis (Doré *et al.*, 1986; Ojasoo *et al.*, 1988), pattern recognition using adaptive least squares (Moriguchi *et al.*, 1981), and CoMFA (Loughney and Schwender, 1992).

We have tested (van Helden and Hamersma, 1995) a number of different QSAR methods to derive a relationship between the structure of a number of progestagenic structures and their relative binding affinities (RBA) to the progesterone receptor, comparing, on the one hand, the performance of the various methods in deriving a structure–activity relationship between the members of a set of steroids (the 'training set') and, on the other hand, the ability to predict the activities of a different set of compounds (the 'test set'), chemically related to a varying degree to the training set. From this analysis, NNs and CoMFA appeared to give superior results compared with Partial Least Squares (PLS) or Principal Component Analysis (PCA) methods. In this paper, we want to discuss the performance, both in deriving a relationship for the training set and in predicting the test set, of a combination of a genetic algorithm for selecting the input variables and a neural network for model building. We would like to emphasize that in this paper the use of a training and test set is intended to simulate the situation where a small set of known compounds (the training set) is available for deriving and validating a model and where new compounds (the test set) have to be predicted at a later stage. For that reason, the test set will not be used at the stage of model building.

EXPERIMENTAL

Construction of the data set

The compounds used for our data set (see Figure 1 for structures and Tables I and II for a complete list including RBA) were selected from literature and in-house sources with two thoughts in mind. Firstly, the RBA for the progesterone receptor should cover a wide range, preferably several orders of magnitude. RBAs were determined using standard procedures (Bergink *et al.*, 1983) and the extremes of affinity in our data set were found to differ by a factor of 440. Secondly, the structures should offer a wide variation of substituents and, to a lesser degree, molecular skeletons. Particular care was taken to include pairs of compounds that differ only in a single feature. Thus, **1/44**, **45/47** and **5/9** differ solely in that they possess a Δ^4 or a $\Delta^{5(10)}$ double bond, respectively; **53** and **54** have *E*- and *Z*-fluoromethylene substituents; **17/35** and **15/55** have a hydrogen or a methyl group in position 7α; **37/42** and **18/34** are virtually isoelectronic, an ethynyl group having been replaced by a nitrile; **13/51** and **27/48** differ by a Δ^{15} double bond, whereas **26/51** and **45/48** differ by the presence of a 11-methylene substituent. Finally, **2**, **38**, and **45** have a single, double and triple 20–21 bond, respectively.

The distribution between training and test sets was made primarily in such a way that each of the sets would contain both high- and low-RBA compounds on the one hand, and at least one example of each basic molecular skeleton (Δ^4, $\Delta^{5(10)}$, $\Delta^{4,9}$, Δ^{15}, 17β-pregnan-20-one, etc.) on the other hand. In principle, it would be possible to search for combinations of molecules differing only in two features, placing three of the possible combinations in the training set, and the fourth one in the test set. This strategy was not actively persued, as this might direct the prediction too much; thus, the situation described above sometimes occurs (e.g., **26**, **27**, and **28** are placed in the training set, whereas **51** is placed in the test set), but examples of the

(a)

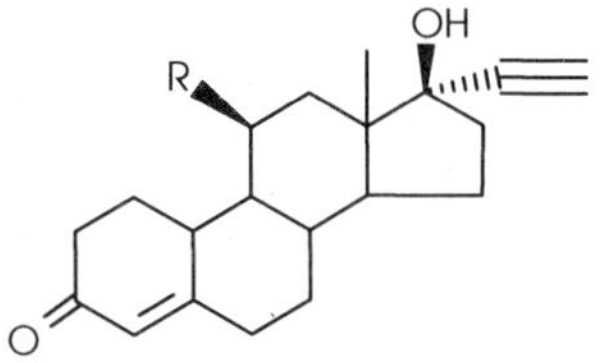

10 R = CH_3
12 R = OCH_3
14 R = CH_2Cl
15 R = C_2H_5
16 R = CH_2CH_2Cl
17 R = CH=CH_2
18 R = C≡CH
19 R = CH_2OCH_3
34 R = CN
39 R = CHF_2
45 R = H

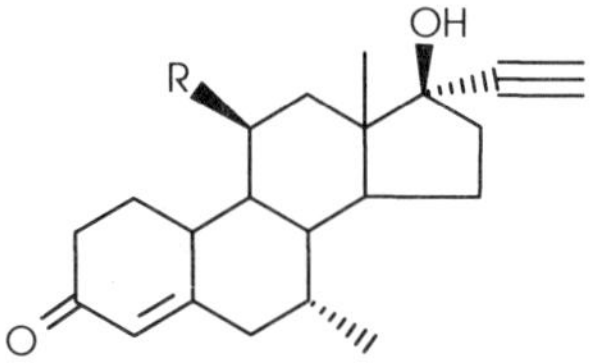

23 R = OCH_3
35 R = CH=CH_2
36 R = CH_2OCH_3
37 R = CN
40 R = CH_3
42 R = C≡CH
44 R = H
55 R = C_2H_5

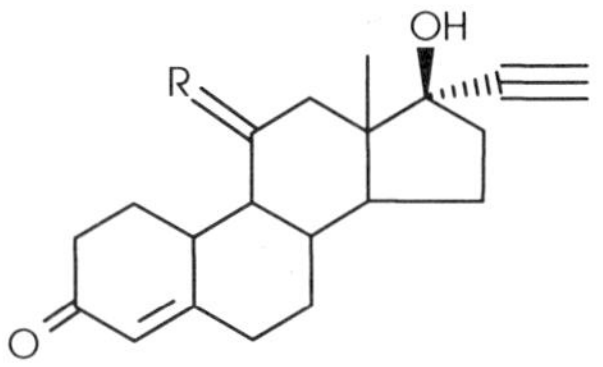

43 R = (E)$CHCH_3$
48 R = CH2
53 R = (E)CHF
54 R = (Z)CHF

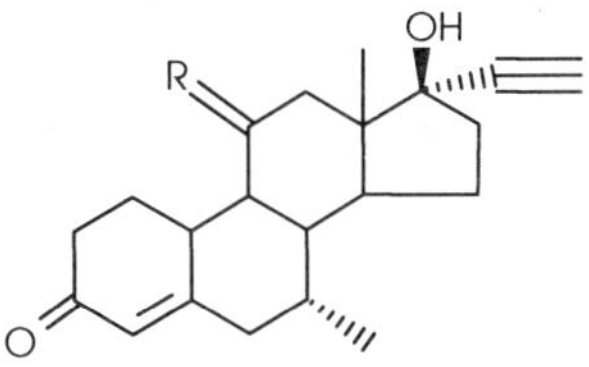

22 R = CH2
41 R = (E)$CH(CH_2)_3CH_3$

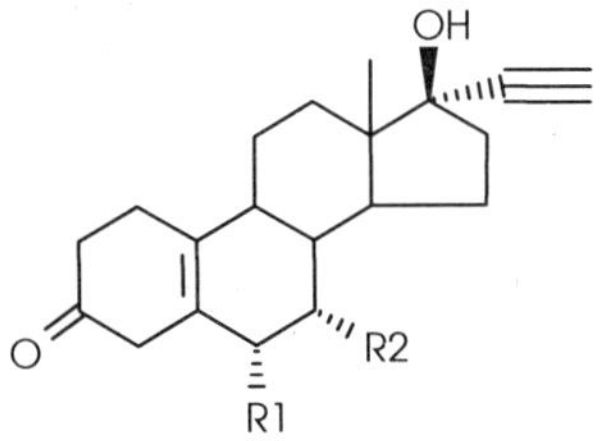

1 $R^1 = H, R^2 = CH_3$
9 $R^1 = CH_3, R^2 = H$
47 $R^1 = R^2 = H$

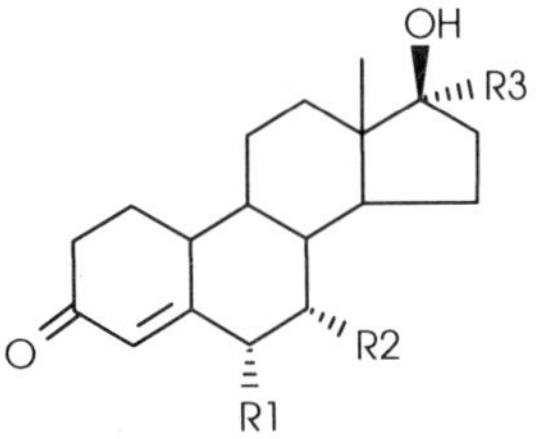

2 $R^1 = R^2 = H, R^3 = C_2H_5$
5 $R^1 = CH_3, R^2 = H$, R^3 = C≡CH
25 $R^1 = R^2 = H, R^3 = CH_2CH(CH_3)_2$
38 $R^1 = R^2 = H$, R^3 = CH=CH_2
50 $R^1 = H, R^2 = CH_3$, R^3 = CH_2CH=CH_2

Figure 1 Structure diagrams of the compounds used in this study.

(b)

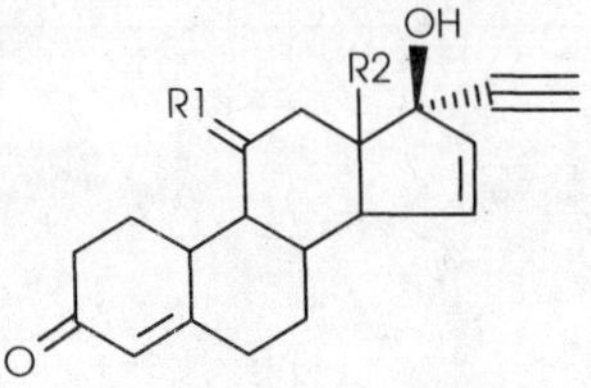

26 $R^1 = H_2$, $R^2 = C_2H_5$
27 $R^1 = CH_2$, $R^2 = CH_3$
28 $R^1 = H_2$, $R^2 = CH_3$
51 $R^1 = CH_2$, $R^2 = C_2H_5$

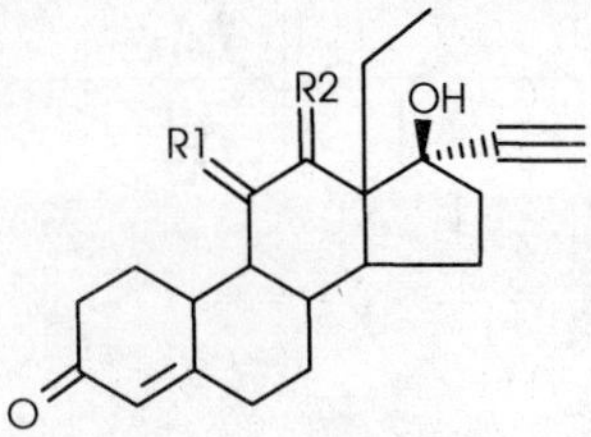

11 $R^1 = R^2 = H_2$
13 $R^1 = CH_2$, $R^2 = H_2$
30 $R^1 = H_2$, $R^2 = CH_2$
31 $R^1 = H_2$, $R^2 = (\alpha H, \beta CH_3)$

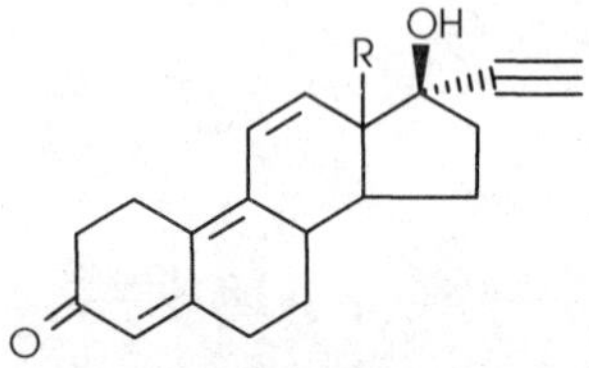

20 $R = C_2H_5$
24 $R = CH_3$

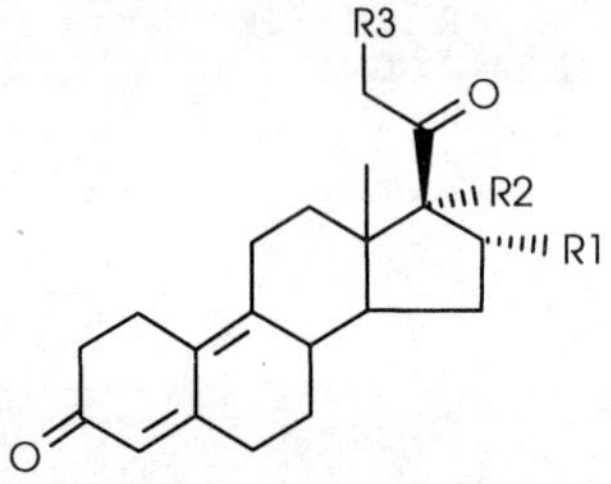

21 $R^1 = C_2H_5$, $R^2 = H$, $R^3 = OH$
49 $R^1 = R^3 = H$, $R^2 = CH_3$

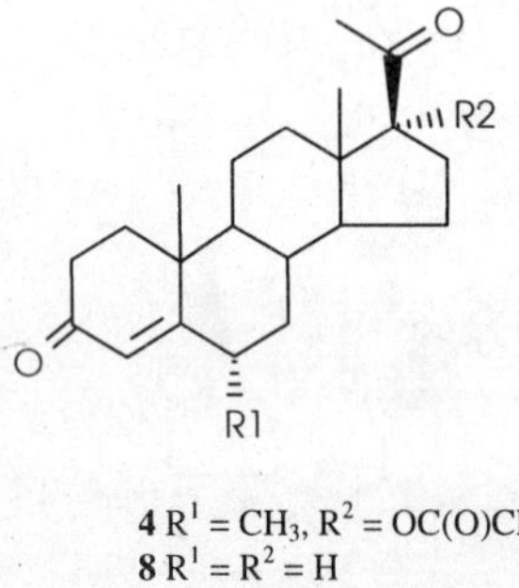

4 $R^1 = CH_3$, $R^2 = OC(O)CH_3$
8 $R^1 = R^2 = H$

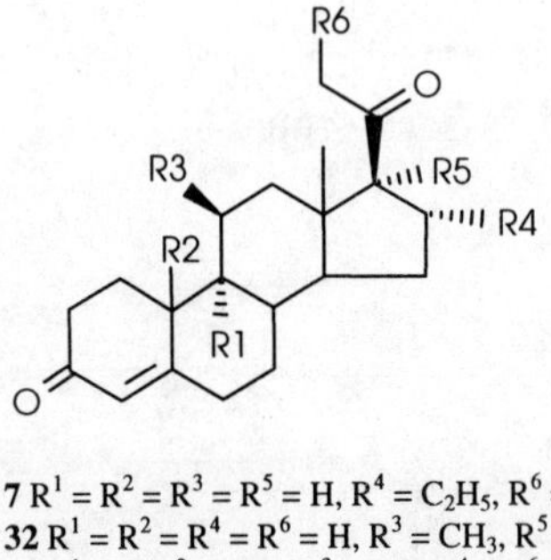

7 $R^1 = R^2 = R^3 = R^5 = H$, $R^4 = C_2H_5$, $R^6 = OH$
32 $R^1 = R^2 = R^4 = R^6 = H$, $R^3 = CH_3$, $R^5 = OC(O)CH_3$
33 $R^1 = F$, $R^2 = CH_3$, $R^3 = OH$, $R^4 = R^6 = H$, $R^5 = OC(O)CH_3$

Figure 1 *Continued.*

(c)

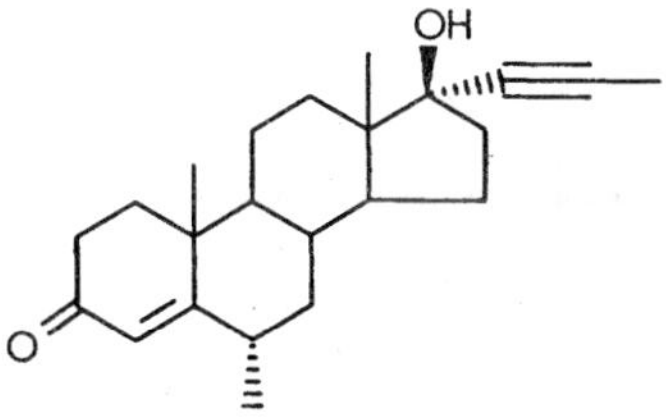

Figure 1 *Continued.*

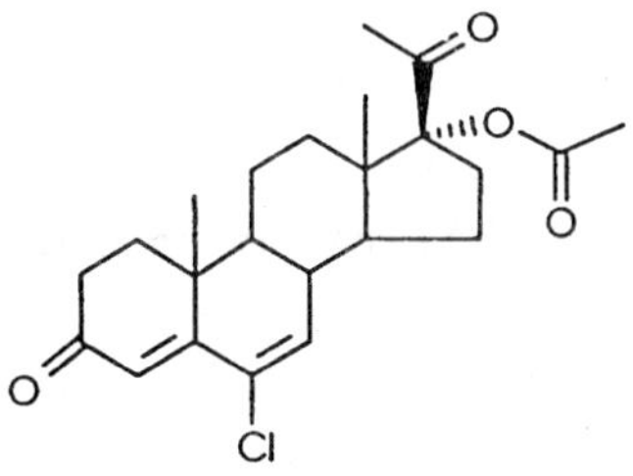

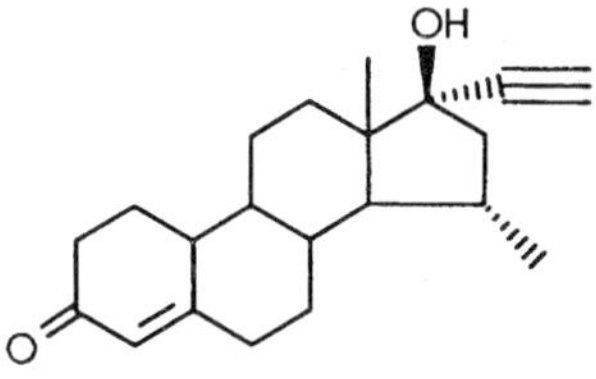

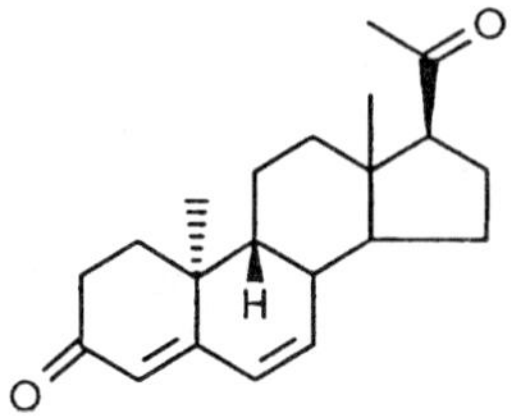

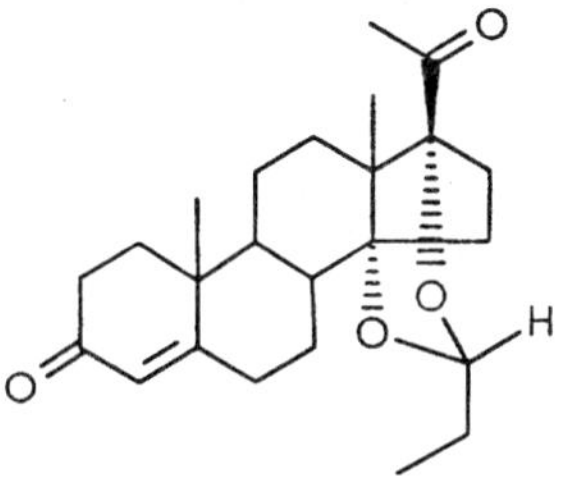

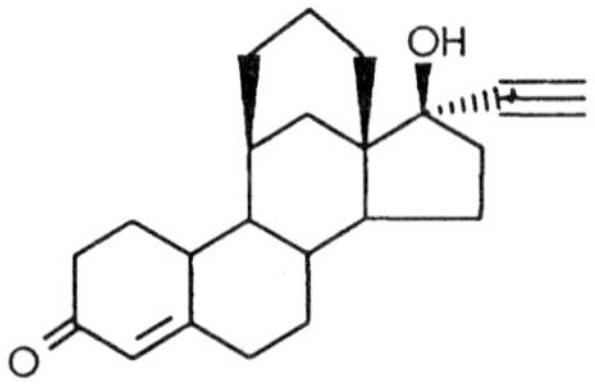

Table I *Scientific names of compounds used in the training set and their RBA (**7** = 100%).*

No.	Compound	RBA
1	(7α,17α)-17-Hydroxy-7-methyl-19-norpregn-5(10)-en-20-yn-3-one	3
2	(17α)-17-Hydroxy-19-norpregn-4-en-3-one	32
3	(6α,17β)-17-Hydroxy-6-methyl-17-(1-propynyl)androst-4-en-3-one	6
4	(6α)-17-(Acetyloxy)-6-methylpregn-4-ene-3,20-dione	55
5	(6α,17α)-17-Hydroxy-6-methyl-19-norpregn-4-en-20-yn-3-one	16
6	17-(Acetyloxy)-6-chloropregna-4,6-diene-3,20-dione	39
7	(16α)-16-Ethyl-21-hydroxy-19-norpregn-4-ene-3,20-dione	100
8	Pregn-4-ene-3,20-dione	14
9	(6α,17α)-17-Hydroxy-6-methyl-19-norpregn-5(10)-en-20-yn-3-one	0.5
10	(11β,17α)-17-Hydroxy-11-methyl-19-norpregn-4-en-20-yn-3-one	120
11	(17α)-13-Ethyl-17-hydroxy-18,19-dinorpregn-4-en-20-yn-3-one	80
12	(11β,17α)-17-Hydroxy-11-methoxy-19-norpregn-4-en-20-yn-3-one	17
13	(17α)-13-Ethyl-17-hydroxy-11-methylene-18,19-dinorpregn-4-en-20-yn-3-one	191
14	(11β,17α)-11-(Chloromethyl)-17-hydroxy-19-norpregn-4-en-20-yn-3-one	22
15	(11β,17α)-11-Ethyl-17-hydroxy-19-norpregn-4-en-20-yn-3-one	71
16	(11β,17α)-11-(2-Chloroethyl)-17-hydroxy-19-norpregn-4-en-20-yn-3-one	37
17	(11β,17α)-11-Ethenyl-17-hydroxy-19-norpregn-4-en-20-yn-3-one	40
18	(11β,17α)-11-Ethynyl-17-hydroxy-19-norpregn-4-en-20-yn-3-one	79
19	(11β,17α)-17-Hydroxy-11-(methoxymethyl)-19-norpregn-4-en-20-yn-3-one	1
20	(17α)-13-Ethyl-17-hydroxy-18,19-dinorpregna-4,9,11-trien-20-yn-3-one	10
21	(16α)-16-Ethyl-21-hydroxy-19-norpregna-4,9-diene-3,20-dione	129
22	(7α,17α)-17-Hydroxy-7-methyl-11-methylene-19-norpregn-4-en-20-yn-3-one	30
23	(7α,11β,17α)-17-Hydroxy-11-methoxy-7-methyl-19-norpregn-4-en-20-yn-3-one	16
24	(17α)-17-Hydroxy-19-norpregna-4,9,11-trien-20-yn-3-one	14
25	(17β)-17-Hydroxy-17-(2-methylpropyl)estr-4-en-3-one	12
26	(17α)-13-Ethyl-17-hydroxy-18,19-dinorpregna-4,15-dien-20-yn-3-one	182
27	(17α)-17-Hydroxy-11-methylene-19-norpregna-4,15-dien-20-yn-3-one	220
28	(17α)-17-Hydroxy-19-norpregna-4,15-dien-20-yn-3-one	49
29	(15α,17α)-17-Hydroxy-15-methyl-19-norpregn-4-en-20-yn-3-one	32
30	(17α)-13-Ethyl-17-hydroxy-12-methylene-18,19-dinorpregn-4-en-20-yn-3-one	2*
31	(12β,17α)-13-Ethyl-17-hydroxy-12-methyl-18,19-dinorpregn-4-en-20-yn-3-one	3*
32	(11β)-17-(Acetyloxy)-11-methyl-19-norpregn-4-ene-3,20-dione	182
33	(11β)-17-(Acetyloxy)-9-fluoro-11-hydroxypregn-4-ene-3,20-dione	112
34	(11β,17α)-17-Hydroxy-3-oxo-19-norpregn-4-en-20-yne-11-carbonitrile	21
35	(7α,11β,17α)-11-Ethenyl-17-hydroxy-7-methyl-19-norpregn-4-en-20-yn-3-one	8
36	(7α,11β,17α)-17-Hydroxy-11-(methoxymethyl)-7-methyl-19-norpregn-4-en-20-yn-3-one	0.3
37	(7α,11β,17α)-17-Hydroxy-7-methyl-3-oxo-19-norpregn-4-en-20-yne-11-carbonitrile	5
38	(17α)-17-Hydroxy-19-norpregna-4,20-dien-3-one	16
39	(11β,17α)-11-(Difluoromethyl)-17-hydroxy-19-norpregn-4-en-20-yn-3-one	13
40	(7α,11β,17α)-17-Hydroxy-7,11-dimethyl-19-norpregn-4-en-20-yn-3-one	41
41	(7α,11E,17α)-17-Hydroxy-7-methyl-11-pentylidene-19-norpregn-4-en-20-yn-3-one	6
42	(7α,11β,17α)-11-Ethynyl-17-hydroxy-7-methyl-19-norpregn-4-en-20-yn-3-one	25
43	(11E,17α)-11-Ethylidene-17-hydroxy-19-norpregn-4-en-20-yn-3-one	98

* Recalculated for the active (D) isomer from Broess *et al.* (1992).

Table II *Scientific names of compounds used in the test set and their RBA (**7** = 100%).*

No.	Compound	RBA
44	(7α,17α)-17-Hydroxy-7-methyl-19-norpregn-4-en-20-yn-3-one	15
45	(17α)-17-Hydroxy-19-norpregn-4-en-20-yn-3-one	25
46	(9β,10α)-Pregna-4,6-diene-3,20-dione	4.5
47	(17α)-17-Hydroxy-19-norpregn-5(10)-en-20-yn-3-one	3
48	(17α)-17-Hydroxy-11-methylene-19-norpregn-4-en-20-yn-3-one	107
49	17-Methyl-19-norpregna-4,9-diene-3,20-dione	36
50	(7α,17β)-17-Hydroxy-7-methyl-17-(2-propenyl)estr-4-en-3-one	27
51	(17α)-13-Ethyl-17-hydroxy-11-methylene-18,19-dinorpregna-4,15-dien-20-yn-3-one	200
52	(R)-14,17-[Propylidenebis(oxy)]pregn-4-ene-3,20-dione	41
53	(11E,17α)-11-(Fluoromethylene)-17-hydroxy-19-norpregn-4-en-20-yn-3-one	34
54	(11Z,17α)-11-(Fluoromethylene)-17-hydroxy-19-norpregn-4-en-20-yn-3-one	190
55	(7α,11β,17α)-11-Ethyl-17-hydroxy-7-methyl-19-norpregn-4-en-20-yn-3-one	19
56	(11α,13α,17α)-11,13-Cyclo-C-dihomo-19-norpregn-4-en-20-yn-3-one	50*

* Recalculated for the active (D) isomer from Pitt *et al.* (1979).

reverse situation can also be found (e.g., **1** in the training set, **44**, **45**, and **47** in the test set). Although not all possible combinations of such quadruplets in a data set of this size can be accounted for the distribution was made essentially randomly, causing some types of substituent (e.g. the fluoromethylene group in **53** and **54**) to be unique to the test set. In order to enable us to assess the performance of the various methods when extrapolating beyond the scope of the training set, a number of structures containing unique molecular skeletons were also included in the test set; examples are Dydrogesterone (**46**), which has the unusual retro (9β,10α) configuration; Proligestone (**52**), having a unique 14α,17α-bridge; and **56**, the 11β,13β-bridge of which at first sight would seem to emulate a combination of 13β- and 11β-ethyl groups (such as in **15** and **55**); however, in the latter compounds, the ethyl groups are known to occupy an anti conformation not possible in **56**; moreover, steric repulsion in this case causes the steroid to adopt a characteristic convex shape (Rohrer *et al.*, 1976; van Geerestein, 1988).

Modelling and data generation

3D starting structures were obtained by converting 2D diagrams into 3D coordinates using CORINA (Gasteiger *et al.*, 1990). These 3D structures were loaded into a Chem-X database to perform all modelling calculations (Chem-X, 1994). Atomic charges were calculated using MOPAC AM1 version 6.0 (Stewart, 1990) and all structures were optimized using the Chem-X MME force-field. Unless stated otherwise, Chem-X default values were used for all modelling tasks.

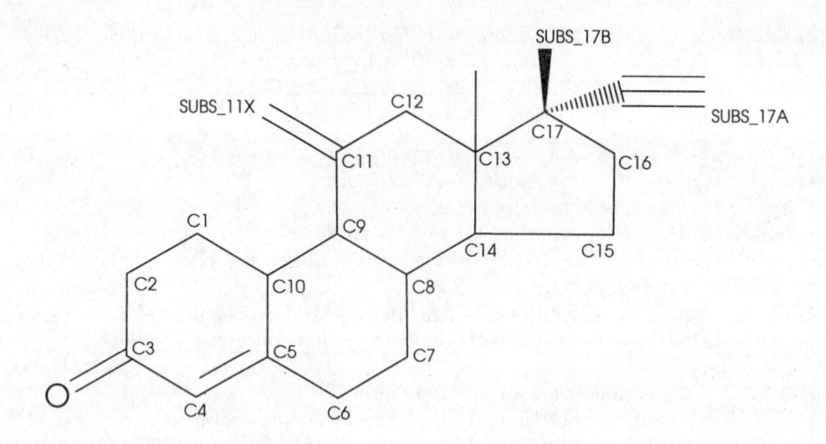

Figure 2 Example of atom naming scheme. In each substituent name a letter reflects the stereochemistry of the substituent (A for α, B for β and X for substituents connected to sp^2 carbon atoms).

Chem-X provides powerful tools to calculate a wide range of properties of (parts of) a molecule. In order to automate these calculations the atoms of all structures need to be numbered in a consistent way. For our purposes, it was necessary to use a combination of atom names, group numbers and residue names to distinguish between positions of substituents. Figure 2 shows an example of the numbering scheme. The carbon atoms in the steroid ring system are numbered C1 through C17 by convention and substituents connected to each of these carbon atoms were given a corresponding group number. Stereochemistry in a steroid is normally indicated by alpha (a substituent below the plane of the steroid) or beta (a substituent above the plane of the steroid). In Chem-X, these substituents are distinguished by adding a residue name ('a' or 'b') to the names of the substituents. Substituents connected to a sp^2 carbon atom in the ring are treated separately in our numbering scheme: as they are neither above nor below the plane of the steroid these substituents are labelled with residue name 'x'.

The QSAR module of Chem-X was used to calculate a total of 52 properties for each of the compounds. These properties include nine whole-molecule properties and 43 properties for individual atoms or substituents, and were chosen in such a way as to reflect the structural diversity in the data set. The resulting data set with 52 properties and RBA for each molecule as given in Table III was used for further analysis with genetic algorithms and neural networks. Throughout this paper log (RBA) will be used as the activity to be predicted.

Table III *Data set of 56 compounds with biological activity (ACT) and 52 properties as calculated by Chem-X.*

Compound	ACT	MME	CH_SKEL	CH_3	CH_7A	CH_11TOT	CH_11B	CH_11X	CH_13B	CH_17A
1	0.48	77.65	–1.39	–0.28	0.02	0.16	0.08	0.00	0.04	–0.18
2	1.51	85.98	–1.61	–0.29	0.09	0.16	0.08	0.00	0.04	0.01
3	0.78	57.03	–1.40	–0.29	0.09	0.17	0.09	0.00	0.04	–0.18
4	1.74	41.15	–1.51	–0.29	0.09	0.16	0.08	0.00	0.05	–0.17
5	1.20	73.80	–1.44	–0.29	0.09	0.16	0.08	0.00	0.05	–0.18
6	1.59	30.47	–1.41	–0.28	0.00	0.17	0.08	0.00	0.05	–0.17
7	2.00	107.04	–1.77	–0.29	0.09	0.16	0.08	0.00	0.04	0.11
8	1.15	75.16	–1.75	–0.29	0.09	0.16	0.08	0.00	0.04	0.10
9	–0.30	78.12	–1.40	–0.29	0.09	0.16	0.08	0.00	0.04	–0.18
10	2.08	76.12	–1.42	–0.29	0.09	0.09	0.02	0.00	0.05	–0.18
11	1.90	80.44	–1.49	–0.29	0.09	0.16	0.08	0.00	0.04	–0.18
12	1.23	64.85	–1.32	–0.29	0.09	–0.06	–0.13	0.00	0.06	–0.18
13	2.28	127.63	–1.39	–0.29	0.09	–0.02	0.00	–0.02	0.05	–0.18
14	1.34	76.16	–1.44	–0.28	0.09	0.07	–0.02	0.00	0.05	–0.18
15	1.85	83.82	–1.42	–0.29	0.09	0.09	0.01	0.00	0.05	–0.18
16	1.57	81.62	–1.42	–0.29	0.09	0.07	–0.01	0.00	0.05	–0.18
17	1.60	75.61	–1.39	–0.29	0.09	0.05	–0.04	0.00	0.05	–0.18
18	1.90	67.20	–1.32	–0.29	0.09	–0.06	–0.16	0.00	0.04	–0.17
19	0.00	80.99	–1.43	–0.29	0.09	0.07	–0.02	0.00	0.05	–0.18
20	1.00	72.05	–1.28	–0.29	0.09	0.13	0.00	0.13	0.05	–0.18
21	2.11	95.95	–1.63	–0.29	0.09	0.18	0.09	0.00	0.05	0.11
22	1.48	67.91	–1.32	–0.29	0.03	0.14	0.00	0.14	0.05	–0.18
23	1.20	58.14	–1.27	–0.29	0.02	–0.05	–0.13	0.00	0.06	–0.18
24	1.15	63.23	–1.30	–0.29	0.09	0.13	0.00	0.13	0.06	–0.18
25	1.08	74.30	–1.61	–0.29	0.09	0.16	0.08	0.00	0.04	0.00
26	2.26	83.43	–1.44	–0.29	0.09	0.16	0.08	0.00	0.04	–0.17
27	2.15	76.64	–1.32	–0.29	0.09	0.14	0.00	0.14	0.05	–0.17
28	1.69	77.56	–1.44	–0.29	0.09	0.16	0.08	0.00	0.05	–0.17
29	1.51	70.02	–1.45	–0.29	0.09	0.16	0.08	0.00	0.04	–0.17
30	0.30	80.48	–1.37	–0.29	0.09	0.18	0.09	0.00	0.05	–0.18
31	0.48	80.17	–1.43	–0.29	0.09	0.16	0.08	0.00	0.04	–0.18
32	2.26	50.05	–1.58	–0.29	0.09	0.10	0.02	0.00	0.05	–0.18
33	2.05	23.19	–1.22	–0.29	0.10	–0.04	–0.12	0.00	0.06	–0.17
34	1.32	65.60	–1.35	–0.28	0.09	–0.09	–0.19	0.00	0.05	–0.17
35	0.90	67.06	–1.33	–0.28	0.08	0.03	–0.06	0.00	0.03	–0.18
36	–0.52	68.35	–1.37	–0.29	0.08	0.07	–0.02	0.00	0.04	–0.17
37	0.70	57.20	–1.27	–0.28	0.03	–0.09	–0.20	0.00	0.06	–0.18
38	1.20	83.31	–1.58	–0.29	0.09	0.16	0.08	0.00	0.04	–0.04
39	1.11	83.86	–1.48	–0.29	0.09	0.07	–0.03	0.00	0.04	–0.17
40	1.61	74.83	–1.36	–0.29	0.07	0.10	0.02	0.00	0.05	–0.18
41	0.79	74.35	–1.32	–0.29	0.08	0.00	0.00	0.00	0.05	–0.18
42	1.40	66.69	–1.28	–0.29	0.08	–0.07	–0.18	0.00	0.05	–0.18
43	1.99	75.68	–1.38	–0.29	0.09	0.00	0.00	0.00	0.05	–0.18
44	1.18	72.94	–1.43	–0.29	0.07	0.16	0.08	0.00	0.05	–0.18
45	1.40	76.39	–1.51	–0.29	0.09	0.16	0.08	0.00	0.04	–0.17
46	0.65	77.15	–1.66	–0.28	0.00	0.16	0.08	0.00	0.04	0.10
47	0.49	81.15	–1.45	–0.28	0.09	0.16	0.08	0.00	0.04	–0.18

Table III *continued.*

Compound	ACT	MME	CH_SKEL	CH_3	CH_7A	CH_11 TOT	CH_11B	CH_11X	CH_13B	CH_17A
48	2.03	70.19	–1.37	–0.29	0.09	0.00	0.00	0.00	0.05	–0.18
49	1.56	73.08	–1.63	–0.29	0.09	0.18	0.09	0.00	0.04	0.04
50	1.43	79.45	–1.55	–0.29	0.02	0.16	0.08	0.00	0.04	0.00
51	2.30	114.54	–1.34	–0.28	0.09	–0.02	0.00	–0.02	0.06	–0.17
52	1.61	59.09	–1.48	–0.29	0.10	0.16	0.08	0.00	0.06	–0.25
53	1.53	76.94	–1.46	–0.28	0.09	0.07	0.00	0.07	0.05	–0.18
54	2.28	72.40	–1.45	–0.29	0.09	0.07	0.00	0.07	0.05	–0.18
55	1.29	74.02	–1.36	–0.29	0.08	0.09	0.01	0.00	0.05	–0.18
56	1.70	80.84	–1.42	–0.29	0.09	0.10	0.01	0.00	0.04	–0.19

Table III *continued.*

Compound	CH_17B	CHC1	CHC2	CHC3	CHC4	CHC5	CHC6	CHC7	CHC8	CHC9	CHC10
1	–0.10	–0.11	–0.21	0.24	–0.18	–0.12	–0.12	–0.10	–0.08	–0.06	–0.10
2	–0.12	–0.15	–0.21	0.26	–0.24	–0.04	–0.13	–0.15	–0.09	–0.09	–0.07
3	–0.11	–0.15	–0.21	0.26	–0.25	–0.02	–0.08	–0.15	–0.10	–0.09	–0.03
4	0.01	–0.15	–0.21	0.26	–0.24	–0.03	–0.08	–0.15	–0.09	–0.09	–0.02
5	–0.10	–0.15	–0.21	0.26	–0.24	–0.03	–0.08	–0.15	–0.09	–0.09	–0.07
6	0.02	–0.15	–0.21	0.26	–0.21	–0.02	–0.07	–0.13	–0.06	–0.09	–0.01
7	–0.03	–0.15	–0.21	0.26	–0.24	–0.04	–0.13	–0.15	–0.09	–0.09	–0.07
8	–0.02	–0.15	–0.21	0.26	–0.24	–0.04	–0.13	–0.15	–0.09	–0.09	–0.02
9	–0.10	–0.11	–0.21	0.24	–0.18	–0.11	–0.07	–0.14	–0.09	–0.06	–0.10
10	–0.10	–0.15	–0.21	0.26	–0.24	–0.04	–0.13	–0.15	–0.09	–0.09	–0.07
11	–0.10	–0.15	–0.21	0.26	–0.24	–0.04	–0.13	–0.15	–0.09	–0.09	–0.07
12	–0.10	–0.15	–0.20	0.26	–0.24	–0.04	–0.13	–0.15	–0.09	–0.08	–0.07
13	–0.08	–0.16	–0.21	0.26	–0.24	–0.04	–0.13	–0.15	–0.10	–0.06	–0.06
14	–0.10	–0.15	–0.21	0.26	–0.24	–0.04	–0.13	–0.15	–0.09	–0.09	–0.08
15	–0.10	–0.15	–0.21	0.26	–0.24	–0.04	–0.13	–0.15	–0.09	–0.08	–0.07
16	–0.10	–0.15	–0.21	0.26	–0.24	–0.04	–0.13	–0.15	–0.09	–0.08	–0.07
17	–0.10	–0.15	–0.21	0.26	–0.24	–0.04	–0.13	–0.15	–0.09	–0.08	–0.08
18	–0.10	–0.15	–0.21	0.26	–0.24	–0.04	–0.13	–0.15	–0.10	–0.09	–0.08
19	–0.10	–0.15	–0.21	0.26	–0.24	–0.04	–0.13	–0.15	–0.09	–0.08	–0.08
20	–0.09	–0.11	–0.21	0.26	–0.23	–0.01	–0.13	–0.15	–0.06	–0.05	–0.09
21	–0.03	–0.11	–0.20	0.26	–0.23	–0.01	–0.13	–0.15	–0.06	–0.07	–0.11
22	–0.10	–0.16	–0.21	0.26	–0.24	–0.04	–0.13	–0.10	–0.08	–0.05	–0.07
23	–0.10	–0.15	–0.21	0.26	–0.25	–0.04	–0.13	–0.10	–0.08	–0.08	–0.07
24	–0.09	–0.11	–0.21	0.26	–0.23	–0.01	–0.13	–0.15	–0.06	–0.05	–0.09
25	–0.12	–0.15	–0.21	0.26	–0.24	–0.04	–0.13	–0.15	–0.09	–0.09	–0.07
26	–0.09	–0.15	–0.21	0.26	–0.24	–0.04	–0.13	–0.15	–0.09	–0.09	–0.07
27	–0.09	–0.16	–0.21	0.26	–0.24	–0.04	–0.13	–0.15	–0.09	–0.06	–0.07
28	–0.09	–0.15	–0.21	0.26	–0.24	–0.04	–0.13	–0.15	–0.09	–0.09	–0.07
29	–0.10	–0.15	–0.21	0.26	–0.24	–0.04	–0.13	–0.16	–0.09	–0.09	–0.07
30	–0.09	–0.15	–0.21	0.26	–0.24	–0.04	–0.13	–0.15	–0.10	–0.09	–0.08
31	–0.10	–0.15	–0.21	0.26	–0.24	–0.04	–0.13	–0.15	–0.10	–0.09	–0.07
32	0.01	–0.15	–0.21	0.26	–0.24	–0.04	–0.13	–0.15	–0.09	–0.09	–0.07
33	0.01	–0.15	–0.20	0.26	–0.24	–0.04	–0.13	–0.15	–0.11	0.15	–0.03
34	–0.10	–0.15	–0.21	0.26	–0.23	–0.05	–0.13	–0.15	–0.10	–0.08	–0.08

Table III *continued.*

Compound	CH_17B	CHC1	CHC2	CHC3	CHC4	CHC5	CHC6	CHC7	CHC8	CHC9	CHC10
35	–0.09	–0.15	–0.21	0.26	–0.24	–0.04	–0.13	–0.10	–0.09	–0.09	–0.08
36	–0.10	–0.15	–0.21	0.26	–0.24	–0.04	–0.13	–0.10	–0.08	–0.09	–0.08
37	–0.10	–0.15	–0.21	0.26	–0.24	–0.04	–0.13	–0.10	–0.08	–0.08	–0.08
38	–0.12	–0.15	–0.21	0.26	–0.24	–0.04	–0.13	–0.15	–0.09	–0.09	–0.07
39	–0.10	–0.15	–0.21	0.26	–0.24	–0.05	–0.13	–0.15	–0.10	–0.08	–0.09
40	–0.10	–0.15	–0.21	0.26	–0.25	–0.04	–0.13	–0.10	–0.08	–0.09	–0.07
41	–0.10	–0.16	–0.21	0.26	–0.24	–0.04	–0.13	–0.10	–0.09	–0.05	–0.07
42	–0.11	–0.15	–0.21	0.26	–0.25	–0.03	–0.13	–0.10	–0.09	–0.10	–0.08
43	–0.10	–0.16	–0.21	0.26	–0.24	–0.04	–0.13	–0.15	–0.09	–0.05	–0.07
44	–0.10	–0.15	–0.21	0.26	–0.24	–0.04	–0.13	–0.10	–0.08	–0.09	–0.07
45	–0.10	–0.15	–0.21	0.26	–0.24	–0.04	–0.13	–0.15	–0.09	–0.09	–0.07
46	–0.04	–0.15	–0.20	0.26	–0.23	–0.02	–0.16	–0.12	–0.07	–0.08	–0.03
47	–0.10	–0.11	–0.21	0.24	–0.18	–0.12	–0.12	–0.15	–0.09	–0.06	–0.10
48	–0.10	–0.15	–0.21	0.26	–0.24	–0.04	–0.13	–0.15	–0.09	–0.05	–0.07
49	–0.01	–0.11	–0.21	0.26	–0.24	–0.01	–0.13	–0.15	–0.06	–0.07	–0.11
50	–0.12	–0.15	–0.21	0.26	–0.24	–0.04	–0.13	–0.10	–0.08	–0.09	–0.07
51	–0.08	–0.17	–0.20	0.26	–0.24	–0.06	–0.13	–0.15	–0.09	–0.05	–0.05
52	0.03	–0.15	–0.21	0.26	–0.24	–0.03	–0.13	–0.15	–0.08	–0.09	–0.02
53	–0.10	–0.16	–0.21	0.26	–0.24	–0.04	–0.13	–0.15	–0.09	–0.04	–0.07
54	–0.10	–0.15	–0.21	0.26	–0.24	–0.04	–0.13	–0.15	–0.09	–0.04	–0.07
55	–0.10	–0.15	–0.21	0.26	–0.24	–0.04	–0.13	–0.10	–0.08	–0.09	–0.07
56	–0.10	–0.15	–0.21	0.26	–0.24	–0.04	–0.13	–0.15	–0.09	–0.09	–0.07

Table III *continued.*

Compound	CHC11	CHC12	CHC13	CHC14	CHC15	CHC16	CHC17	VOL_3	VOL_6A	VOL_7A	VOL_11 TOT
1	–0.15	–0.14	–0.03	–0.11	–0.15	–0.18	0.20	7.00	1.00	22.08	5.52
2	–0.15	–0.14	–0.09	–0.10	–0.16	–0.15	0.09	7.00	3.00	2.00	5.61
3	–0.16	–0.14	–0.10	–0.10	–0.16	–0.15	0.22	7.00	29.40	1.00	5.66
4	–0.15	–0.14	–0.05	–0.10	–0.16	–0.16	0.04	7.00	27.09	1.00	5.52
5	–0.15	–0.14	–0.08	–0.10	–0.15	–0.15	0.21	7.00	26.95	1.00	5.62
6	–0.15	–0.15	–0.04	–0.10	–0.16	–0.17	0.04	7.00	0.00	0.00	5.45
7	–0.15	–0.14	–0.04	–0.10	–0.15	–0.11	–0.15	7.00	1.00	1.00	5.56
8	–0.15	–0.14	–0.04	–0.10	–0.16	–0.16	–0.15	7.00	3.00	3.00	5.46
9	–0.15	–0.14	–0.03	–0.11	–0.15	–0.18	0.20	7.00	18.98	1.00	5.52
10	–0.10	–0.14	–0.08	–0.10	–0.16	–0.15	0.21	7.00	2.00	2.00	23.33
11	–0.15	–0.14	–0.07	–0.10	–0.16	–0.15	0.21	7.00	2.00	2.00	5.19
12	0.04	–0.18	–0.08	–0.10	–0.15	–0.14	0.21	7.00	1.00	2.00	33.89
13	–0.09	–0.11	–0.09	–0.10	–0.15	–0.15	0.20	7.00	3.00	2.00	17.82
14	–0.10	–0.14	–0.08	–0.11	–0.16	–0.15	0.21	7.00	2.00	1.00	47.20
15	–0.09	–0.14	–0.08	–0.10	–0.15	–0.15	0.20	7.00	2.00	1.00	46.50
16	–0.09	–0.15	–0.08	–0.10	–0.15	–0.15	0.20	7.00	2.00	3.00	57.86
17	–0.07	–0.13	–0.08	–0.10	–0.15	–0.15	0.21	7.00	1.00	2.00	41.71
18	0.02	–0.13	–0.09	–0.10	–0.15	–0.15	0.22	7.00	2.00	4.00	31.46
19	–0.09	–0.14	–0.08	–0.11	–0.15	–0.15	0.21	7.00	3.00	3.00	54.90
20	–0.13	–0.13	–0.05	–0.10	–0.16	–0.15	0.21	7.00	3.00	2.00	2.00
21	–0.12	–0.15	–0.04	–0.11	–0.15	–0.10	–0.15	7.00	2.00	2.00	5.81

Table III *continued.*

Compound	CHC11	CHC12	CHC13	CHC14	CHC15	CHC16	CHC17	VOL_3	VOL_6A	VOL_7A	VOL_11 TOT
22	–0.10	–0.11	–0.08	–0.10	–0.15	–0.15	0.21	7.00	1.00	22.37	21.49
23	0.04	–0.18	–0.08	–0.10	–0.15	–0.14	0.21	7.00	1.00	25.42	32.75
24	–0.13	–0.14	–0.07	–0.10	–0.15	–0.14	0.21	7.00	2.00	2.00	2.00
25	–0.15	–0.14	–0.09	–0.10	–0.16	–0.15	0.09	7.00	2.00	3.00	5.62
26	–0.15	–0.14	–0.08	–0.08	–0.15	–0.15	0.24	7.00	2.00	4.00	5.42
27	–0.10	–0.11	–0.08	–0.08	–0.15	–0.15	0.24	7.00	1.00	3.00	21.63
28	–0.15	–0.14	–0.08	–0.08	–0.15	–0.15	0.23	7.00	2.00	1.00	5.74
29	–0.15	–0.14	–0.09	–0.09	–0.11	–0.15	0.21	7.00	2.00	1.00	5.63
30	–0.12	–0.09	–0.04	–0.10	–0.16	–0.15	0.21	7.00	3.00	4.00	5.22
31	–0.15	–0.09	–0.06	–0.10	–0.16	–0.15	0.21	7.00	3.00	2.00	5.53
32	–0.10	–0.14	–0.04	–0.10	–0.16	–0.17	0.04	7.00	2.00	3.00	19.04
33	0.02	–0.18	–0.04	–0.10	–0.15	–0.17	0.04	7.00	1.00	2.00	13.49
34	0.00	–0.14	–0.09	–0.10	–0.16	–0.15	0.22	7.00	3.00	3.00	35.09
35	–0.07	–0.13	–0.08	–0.10	–0.15	–0.15	0.22	7.00	2.00	22.20	45.66
36	–0.10	–0.13	–0.08	–0.10	–0.15	–0.15	0.21	7.00	3.00	22.14	59.03
37	0.00	–0.14	–0.09	–0.10	–0.15	–0.15	0.21	7.00	2.00	22.76	35.61
38	–0.15	–0.14	–0.08	–0.10	–0.16	–0.15	0.12	7.00	2.00	2.00	5.59
39	–0.14	–0.13	–0.09	–0.10	–0.16	–0.15	0.22	7.00	2.00	3.00	40.59
40	–0.09	–0.14	–0.08	–0.11	–0.16	–0.15	0.21	7.00	3.00	22.13	35.74
41	–0.10	–0.11	–0.08	–0.10	–0.15	–0.15	0.21	7.00	2.00	22.43	79.20
42	0.02	–0.14	–0.10	–0.10	–0.15	–0.15	0.22	7.00	1.00	22.53	34.05
43	–0.11	–0.11	–0.08	–0.10	–0.15	–0.15	0.21	7.00	3.00	3.00	32.27
44	–0.15	–0.14	–0.08	–0.10	–0.15	–0.15	0.21	7.00	1.00	22.49	5.58
45	–0.15	–0.14	–0.09	–0.10	–0.16	–0.15	0.21	7.00	1.00	3.00	5.53
46	–0.13	–0.13	–0.05	–0.11	–0.16	–0.13	–0.15	7.00	0.00	0.00	5.87
47	–0.15	–0.14	–0.03	–0.11	–0.15	–0.18	0.20	7.00	1.00	1.00	5.52
48	–0.10	–0.11	–0.08	–0.10	–0.15	–0.15	0.21	7.00	2.00	3.00	20.16
49	–0.12	–0.14	–0.04	–0.10	–0.16	–0.15	–0.10	7.00	4.00	2.00	5.57
50	–0.15	–0.14	–0.09	–0.10	–0.16	–0.15	0.09	7.00	2.00	20.97	5.82
51	–0.10	–0.12	–0.06	–0.08	–0.16	–0.17	0.23	7.00	3.00	2.00	17.97
52	–0.15	–0.14	–0.10	0.09	–0.19	–0.19	0.03	7.00	3.00	3.00	5.67
53	–0.21	–0.10	–0.08	–0.10	–0.15	–0.15	0.21	7.00	1.00	2.00	29.15
54	–0.21	–0.10	–0.08	–0.10	–0.16	–0.15	0.21	7.00	2.00	3.00	32.42
55	–0.09	–0.14	–0.08	–0.10	–0.15	–0.15	0.20	7.00	1.00	20.06	54.11
56	–0.10	–0.14	–0.07	–0.10	–0.16	–0.15	0.21	7.00	2.00	2.00	43.89

Table III *continued.*

Compound	SUR_11 TOT	VOL_11B	SUR_11B	VOL_11X	SUR_11X	VOL_13B	VOL_17A	SUR_17A	VOL_17B	SUR_17B	MOLWGT
1	11.89	2.00	5.69	0.00	0.00	20.48	26.35	33.44	14.29	19.18	312.30
2	10.64	3.00	5.69	0.00	0.00	22.14	27.19	33.30	14.25	20.24	302.00
3	10.49	3.00	5.69	0.00	0.00	22.38	39.76	42.68	14.32	17.07	340.00
4	10.82	3.00	5.69	0.00	0.00	22.28	48.55	46.29	32.95	38.69	386.00
5	10.62	1.00	5.69	0.00	0.00	22.14	27.61	34.33	13.24	19.02	312.30
6	10.89	2.00	5.69	0.00	0.00	20.19	37.46	44.05	39.02	42.18	404.50
7	11.07	2.00	5.69	0.00	0.00	20.04	3.00	5.69	40.79	44.31	344.30
8	10.93	1.00	5.69	0.00	0.00	20.13	1.00	5.69	36.77	40.19	314.00

Table III *continued.*

Compound	SUR_11 TOT	VOL_11B	SUR_11B	VOL_11X	SUR_11X	VOL_13B	VOL_17A	SUR_17A	VOL_17B	SUR_17B	MOLWGT
9	11.89	3.00	5.69	0.00	0.00	20.48	26.35	33.44	14.29	19.18	312.30
10	30.05	20.95	25.61	0.00	0.00	23.26	25.32	32.61	13.86	19.90	312.30
11	11.15	1.00	5.69	0.00	0.00	29.98	29.25	33.49	13.98	19.37	312.30
12	39.05	29.40	32.45	0.00	0.00	22.74	29.32	34.31	13.87	18.43	328.30
13	23.30	0.00	0.00	17.82	23.30	42.02	27.93	32.64	12.93	18.68	324.30
14	46.69	43.91	42.79	0.00	0.00	23.04	29.60	32.71	13.71	19.98	346.50
15	44.81	33.20	40.27	0.00	0.00	23.05	26.88	32.90	13.27	19.38	326.00
16	52.86	51.59	53.06	0.00	0.00	23.03	26.64	32.72	13.27	19.38	360.50
17	42.27	35.11	36.26	0.00	0.00	20.52	33.83	34.83	12.07	18.46	324.00
18	38.53	30.48	34.10	0.00	0.00	22.01	32.17	34.78	11.37	17.15	322.00
19	48.33	45.82	47.30	0.00	0.00	23.03	28.61	33.58	13.22	19.40	342.00
20	5.69	0.00	0.00	2.00	5.69	34.26	26.91	31.55	12.33	17.79	308.00
21	9.81	3.00	5.69	0.00	0.00	23.60	1.00	5.69	46.92	46.61	342.00
22	26.20	0.00	0.00	21.49	26.20	22.77	27.10	32.80	14.13	18.51	324.00
23	38.30	26.16	31.77	0.00	0.00	22.41	30.83	34.69	13.93	18.45	342.00
24	5.69	0.00	0.00	2.00	5.69	18.83	27.22	32.00	12.39	17.76	294.00
25	10.61	2.00	5.69	0.00	0.00	22.05	69.33	58.04	14.17	20.14	330.00
26	10.98	1.00	5.69	0.00	0.00	42.32	31.38	34.52	13.00	19.27	310.00
27	26.22	0.00	0.00	21.63	26.22	22.15	27.06	31.96	12.43	18.09	308.00
28	10.09	1.00	5.69	0.00	0.00	21.19	26.09	32.06	12.51	17.33	296.00
29	10.58	1.00	5.69	0.00	0.00	22.21	32.22	33.21	12.29	18.28	312.00
30	11.13	1.00	5.69	0.00	0.00	43.22	26.26	33.11	14.06	20.00	324.00
31	10.79	1.00	5.69	0.00	0.00	38.86	30.34	34.28	14.00	20.14	326.00
32	27.22	22.33	27.05	0.00	0.00	22.74	48.27	46.99	37.92	41.87	372.00
33	21.18	13.58	18.50	0.00	0.00	22.17	35.61	43.33	33.65	40.26	406.00
34	38.04	29.93	32.13	0.00	0.00	22.19	30.12	33.16	12.75	17.31	323.00
35	43.14	40.70	37.66	0.00	0.00	22.19	32.92	34.87	12.62	17.67	338.00
36	53.72	53.44	45.86	0.00	0.00	22.15	31.07	34.85	13.46	18.85	356.00
37	39.31	28.84	31.64	0.00	0.00	22.21	27.31	34.04	12.22	17.55	337.00
38	10.69	2.00	5.69	0.00	0.00	19.88	34.63	35.69	14.26	20.17	300.00
39	41.48	31.00	34.49	0.00	0.00	20.30	31.08	34.19	12.11	18.42	348.00
40	33.11	20.37	26.48	0.00	0.00	22.58	29.04	33.65	13.76	19.96	326.00
41	68.89	0.00	0.00	79.20	68.89	20.17	28.62	33.67	13.33	18.83	380.00
42	38.20	30.52	34.16	0.00	0.00	22.32	29.35	34.14	14.64	17.74	336.00
43	37.38	0.00	0.00	32.27	37.38	20.13	26.86	34.22	14.04	20.00	324.00
44	10.69	1.00	5.69	0.00	0.00	22.13	27.61	32.14	13.17	19.01	312.30
45	10.77	2.00	5.69	0.00	0.00	22.29	30.13	33.17	12.59	17.85	298.30
46	10.17	2.00	5.69	0.00	0.00	23.86	1.00	5.69	43.32	42.37	312.30
47	11.89	2.00	5.69	0.00	0.00	20.48	26.36	33.44	14.29	19.18	298.30
48	25.45	0.00	0.00	20.16	25.45	20.17	27.49	33.99	14.18	20.18	310.30
49	11.58	2.00	5.69	0.00	0.00	18.70	29.85	28.65	37.52	40.71	312.00
50	9.99	1.00	5.69	0.00	0.00	19.08	50.29	50.67	13.65	17.41	328.00
51	23.63	0.00	0.00	17.97	23.63	42.31	31.32	33.17	14.00	18.23	322.00
52	10.49	1.00	5.69	0.00	0.00	22.76	57.73	55.89	39.98	44.04	386.00
53	30.43	0.00	0.00	29.15	30.43	21.12	29.62	34.13	14.57	17.71	328.00
54	31.62	0.00	0.00	32.42	31.62	21.00	28.81	33.83	14.62	17.87	328.00
55	46.85	33.66	40.00	0.00	0.00	22.12	27.57	32.43	13.20	19.02	340.00
56	43.95	31.80	36.20	0.00	0.00	18.72	30.21	33.95	13.95	20.16	324.00

Table III *continued.*

Compound	DIPOLE	DIP11	DIP17	VOLTOT	SURTOT	HOMO	LUMO	ELEC	HEAT	VOL_10B
1	0.50	0.00	0.56	241.83	142.60	−9.32	0.86	−30940.18	−29.80	0.00
2	0.60	0.00	0.40	268.14	150.28	−10.06	0.11	−29925.84	−97.92	2.00
3	0.43	0.00	0.96	296.79	164.35	−9.83	0.03	−35806.19	−11.16	20.31
4	0.62	0.00	0.97	309.31	174.34	−10.08	0.11	−43971.71	−142.71	20.96
5	0.46	0.00	0.57	254.45	164.16	−10.07	0.11	−30755.17	−29.98	2.00
6	0.74	0.00	0.90	397.12	187.07	−9.58	−0.53	−43008.58	−120.46	20.98
7	0.55	0.00	0.62	245.81	157.98	−10.09	0.08	−35899.30	−132.20	1.00
8	0.59	0.00	0.54	217.05	146.81	−10.08	0.11	−31652.87	−76.98	21.02
9	0.50	0.00	0.56	275.52	151.70	−9.32	0.86	−30707.64	−27.02	0.00
10	0.48	0.07	0.57	263.08	168.38	−10.06	0.11	−31025.38	−23.57	1.00
11	0.52	0.00	0.61	222.32	150.47	−10.08	0.11	−30870.97	−28.27	1.00
12	0.55	0.35	0.58	300.96	158.88	−9.99	0.16	−33890.28	−58.70	2.00
13	0.54	0.11	0.52	258.48	151.40	−9.97	0.07	−32912.02	90.50	1.00
14	0.47	0.34	0.56	265.28	179.99	−10.19	−0.02	−33491.86	−30.86	3.00
15	0.48	0.09	0.55	290.15	163.89	−10.06	0.10	−33372.64	−25.32	2.00
16	0.63	0.40	0.53	219.11	137.58	−10.14	0.03	−35636.08	−32.49	2.00
17	0.50	0.06	0.53	257.69	160.61	−10.02	0.11	−32631.50	2.07	2.00
18	0.46	0.11	0.47	317.56	161.16	−10.10	0.09	−31979.46	48.38	1.00
19	0.41	0.22	0.56	268.07	160.42	−10.11	0.05	−35940.25	−60.83	1.00
20	0.63	0.00	0.59	211.87	136.42	−8.95	−0.48	−29123.36	29.36	0.00
21	0.32	0.00	0.65	326.58	167.26	−9.28	−0.29	−34954.21	−109.15	0.00
22	0.60	0.07	0.67	286.61	177.11	−9.91	0.08	−32711.62	1.09	1.00
23	0.55	0.36	0.56	307.82	179.86	−9.99	0.19	−36749.31	−54.90	3.00
24	0.62	0.00	0.59	214.69	141.71	−8.96	−0.48	−26775.63	29.63	0.00
25	0.59	0.00	0.42	269.23	155.97	−10.06	0.11	−34239.34	−104.38	1.00
26	0.47	0.00	0.57	304.88	166.65	−10.03	0.09	−30064.26	10.70	1.00
27	0.71	0.07	0.57	307.27	175.67	−9.99	0.05	−29242.75	39.57	2.00
28	0.43	0.00	0.56	195.68	140.66	−10.05	0.08	−27679.15	9.77	2.00
29	0.45	0.00	0.52	294.35	148.23	−10.12	0.09	−30961.66	−27.90	1.00
30	0.56	0.00	0.60	299.45	164.79	−9.90	0.08	−32612.86	6.69	2.00
31	0.53	0.00	0.58	313.58	166.12	−10.09	0.11	−33499.50	−24.08	2.00
32	0.75	0.08	1.02	336.10	176.21	−10.08	0.08	−41383.17	−142.17	1.00
33	0.67	0.29	0.97	328.14	178.11	−10.10	0.08	−48466.46	−215.81	22.97
34	0.60	0.31	0.49	288.87	167.98	−10.26	−0.05	−32087.79	24.30	2.00
35	0.55	0.07	0.24	287.69	169.51	−10.06	0.12	−35724.73	51.42	3.00
36	0.45	0.21	0.52	250.66	144.73	−10.12	0.06	−38946.79	−57.74	1.00
37	0.67	0.32	0.58	257.87	142.15	−10.19	−0.01	−34571.06	10.53	1.00
38	0.54	0.00	0.46	275.82	151.23	−10.07	0.10	−29202.79	−65.57	2.00
39	0.90	0.39	0.20	248.05	162.43	−10.17	0.03	−37190.88	−106.44	1.00
40	0.41	0.08	0.54	301.74	171.38	−9.99	0.13	−33667.64	−24.73	2.00
41	0.38	0.17	0.57	377.84	213.28	−9.54	0.09	−41898.25	−19.07	2.00
42	0.55	0.15	0.32	240.17	150.84	−9.80	−0.01	−34834.55	82.85	2.00
43	0.52	0.16	0.57	213.30	135.23	−9.55	0.08	−32461.18	−4.44	2.00
44	0.38	0.00	0.56	280.98	157.17	−10.06	0.11	−31027.30	−29.41	2.00
45	0.42	0.00	0.50	281.85	151.68	−10.13	0.08	−28565.72	−25.94	2.00
46	0.51	0.00	0.61	289.75	162.15	−9.47	−0.33	−30915.53	196.04	0.00
47	0.51	0.00	0.56	244.27	142.27	−9.33	0.85	−28370.70	−28.60	0.00
48	0.53	0.12	0.62	272.66	154.60	−9.96	0.08	−30053.66	0.42	2.00

Table III *continued.*

Compound	DIPOLE	DIP11	DIP17	VOLTOT	SURTOT	HOMO	LUMO	ELEC	HEAT	VOL_10B
49	0.55	0.00	0.54	274.70	162.67	–9.22	–0.28	–30610.42	–53.78	0.00
50	0.56	0.00	0.41	278.07	155.87	–10.02	0.12	–33970.65	–70.61	3.00
51	0.51	0.12	0.60	279.05	159.37	–9.94	0.18	–32007.55	102.36	1.00
52	0.45	0.00	0.88	337.54	169.69	–10.05	0.14	–45173.76	–137.53	22.58
53	0.40	0.20	0.58	289.01	156.83	–9.50	–0.01	–32882.00	–34.82	2.00
54	0.65	0.20	0.59	313.55	166.11	–9.50	0.13	–32999.00	–32.27	2.00
55	0.57	0.10	0.42	279.36	156.02	–10.03	0.11	–36161.17	–26.33	1.00
56	0.49	0.18	0.49	340.16	170.87	–10.05	0.11	–32916.33	–20.45	2.00

'Whole molecule' properties: MME (molecular mechanics energy), MOLWGT (molecular weight), DIPOLE (dipole moment), VOLTOT (total molecular volume), SURTOT (total surface area of molecule), HOMO (energy of HOMO (AM1)), LUMO (energy of LUMO (AM1)), ELEC (total electronic energy (AM1)), HEAT (heat of formation (AM1)).
Atom and substituent properties: The numbers refer to the position of the substituents and stereochemistry is indicated by A and B for alpha and beta respectively. Substituents connected to a sp^2 carbon in the steroid skeleton are indicated by X. 'TOT' refers to all atoms in all substituents connected to the corresponding carbon atom in the steroid skeleton.
CHC*nr* (atomic charge of steroid ring atoms C1 to C17), CH_SKEL (sum of atomic charges of all steroid ring atoms), CH_*subs* (sum of atomic charges of atoms in substituent *subs*), VOL_*subs* (total volume of atoms in substituent *subs*), SUR_*subs* (total surface area of atoms in substituent *subs*), DIP*subs* (dipole moment of atoms in substituent *subs*).

Genetic function approximation

The genetic function approximation (GFA) as implemented in the Drug Discovery Workbench of the Cerius^2 molecular modelling package (Cerius^2, 1994) was used to derive equations that relate the calculated properties to the RBA. A detailed description of the GFA has been published previously (Rogers and Hopfinger, 1994). Basically, GFA starts with a population of equations that relate one or more properties to the observed activity. A fitness function (e.g. lack-of-fit (LOF) or correlation coefficient) is used to estimate the quality of each individual equation. The equations with the best scores are most likely to be chosen to mate and to propagate their genetic material (properties) to offspring through the cross-over operation, in which pieces of genetic material are taken from each parent and recombined to create the child. After many mating steps, the average fitness of the individual equations increases as good combinations of properties are discovered and spread through the population. It is claimed that by replacing regression analysis with the GFA, models comparable with or superior to standard techniques can be constructed and that information is made available which is not provided by other techniques.

For the data set under study the GFA was configured to use default values where possible. However, preliminary experiments indicated that some settings should be changed in order to obtain a better sampling of all properties. Therefore, the GFA was started with 250 randomly generated equations with a randomized equation length of 7, which were evolved for 20 000 generations. The types of basis functions not only included linear functions, but also square functions, spline functions and squared splines. A spline term is denoted by angle brackets and partitions the data samples into two classes, depending on the value of its feature. For example, <f(x) – a> is equal to zero if the value of (f(x) – a) is negative; otherwise it is equal to (f(x) – a). The constant 'a' is called the knot of the spline.

Neural networks

The Stuttgart Neural Network Simulator (SNNS, 1994) with the default back-propagation algorithm without momentum was applied for simulating neural networks. The learning rate was set to 0.05 and all weights were randomly initialized between –0.3 and 0.3. Biases are added automatically by the program and are represented as a unit parameter, which is in contrast to other network simulators where the bias is simulated by an extra link weight. All networks consisted of an input layer with 10 input units, a single hidden layer with 1 to 10 units and an output layer with one unit. All units were given a *tanh* activation function which takes values from –1 to +1. We allowed for small extrapolations in our models by scaling all properties to a range from –0.8 to +0.8.

A well-known problem in the application of neural networks is overtraining. Generally, the performance of a neural network on the training set increases monotonously during training, but the predictive performance on new data (the test set) shows an optimum after a certain number of epochs after which the performance steadily decreases indicating that the network is overtrained (Zupan and Gasteiger, 1993). This effect of overtraining may be avoided by model validation which can take place by using cross-validation procedures or by applying a test set with compounds that are not used during model building.

In QSAR studies, especially in the field of CoMFA, cross-validation is extensively used as a method to assess the predictive power of a model. Although leave-one-out (LOO) cross-validation tends to overestimate the predictive power this method is still widely used because it is easy to apply and gives a reproducible indication of the predictive power. The use of a test set is only possible if sufficient compounds are available and requires a careful selection of compounds in order to obtain meaningful results. The use of such a test set leads to a model that has an optimized performance on this particular test set. The true predictive power on really unknown compounds requires an additional (external) test set. In many QSAR studies, and also in our case, there are not enough compounds (i.e. observables) to create two test sets. Therefore, we will apply cross-validation to validate the network and we will use our test set of 13 new compounds as a true test of the predictive power on unknown compounds.

In order to avoid overtraining in our experiments, a LOO cross-validation procedure was implemented to determine the optimum number of epochs. In the first step of this procedure the first compound is omitted from the training set and the remaining 42 are used to train a network for 2000 epochs. After every 20 epochs the training is suspended and the RBA of the missing compound is predicted. In the next step the second compound is omitted from the data set and a new network is built and initialized with random weights to start training from scratch. These steps are then repeated for each of the compounds in the training set resulting in a total of 100 predictions of the RBA of each compound.

All predictions that have been made after an identical number of epochs are now collected and they are compared with the observed values. In this way 100 data sets of predicted versus observed activities are obtained and the correlation coefficients of these data sets are plotted as a function of the number of epochs, so that network performance can be monitored while training.

RESULTS

The analysis of a data set as described here comprises three issues: data reduction, model building, and model validation. Data reduction is required to reduce the ratio of variables over compounds and to eliminate variables with a distorted distribution. Model building and validation initially take place by using cross-validation procedures. The resulting predictive power is tested by using the external data set of 13 new compounds.

The complete data set contains several properties that are expected to correlate with RBA but also properties that are less likely to affect RBA. Data reduction should lead to a set of properties that can distinguish between different compounds. A close inspection of the compounds in the training set shows that most structural variation is found at positions 11α, 11β, 17α, and 17β and to a lesser extent at positions 7α and 13β. Therefore, it is anticipated that properties related to these positions should be chosen for model building. On the other hand, the choice of properties should not be biased by the author's opinion because other unexpectedly selected properties may lead to valuable new insights in the relationship between structure and RBA.

Data reduction

Although neural networks have been used for data reduction or selection (Wikel and Dow, 1993) our efforts with this data set remained unsuccessful because the Hinton diagrams we obtained for our data set depended strongly on the number of training cycles and choice of starting weights. Moreover, the interpretation of the diagrams remains rather subjective. Recently, a more

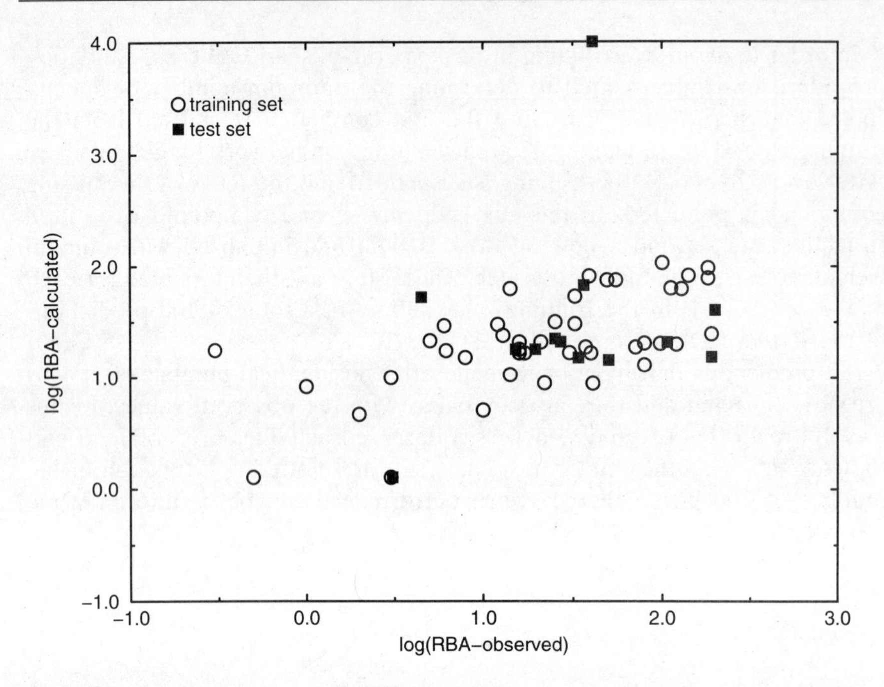

Figure 3 Observed versus predicted activity as calculated by a stepwise linear regression model (Eq. (2)). Open circles: training set. Solid squares: test set.

sophisticated application of neural networks for data selection has been described using pruning of weights (Maddalena and Johnston, 1995). We did not have access to software including a pruning option and therefore we tried to apply an alternative technique to select properties to be used as input for a neural network.

The selection of properties can be done using a technique such as stepwise regression which selects a combination of properties that correlates best with the activity. This method leads to the model in Eq. (2) ($r^2_{\text{training set}} = 0.42$, $r^2_{\text{test set}} = 0.02$). The corresponding plot of calculated versus observed values (Figure 3) indicates that predictions made by this model are very inaccurate. The poor performance of this model is probably caused by the fact that properties are selected that do not provide an accurate description of the structural variation in the data set.

$$\text{ACT} = -15.99 \times \text{CHC13} + 35.86 \times \text{CHC14} + 0.06 \times \text{SUR\_17B} + 2.50 \quad (2)$$

Techniques like PLS reduce the number of variables in a data set by extracting components that are a linear combination of the original properties. This technique results in a one-component model with similar correlation coefficients ($r^2_{\text{training set}} = 0.37$, $r^2_{\text{test set}} = 0.06$), as shown in Figure 4. Two compounds in the test set are known to possess highly deviating structures

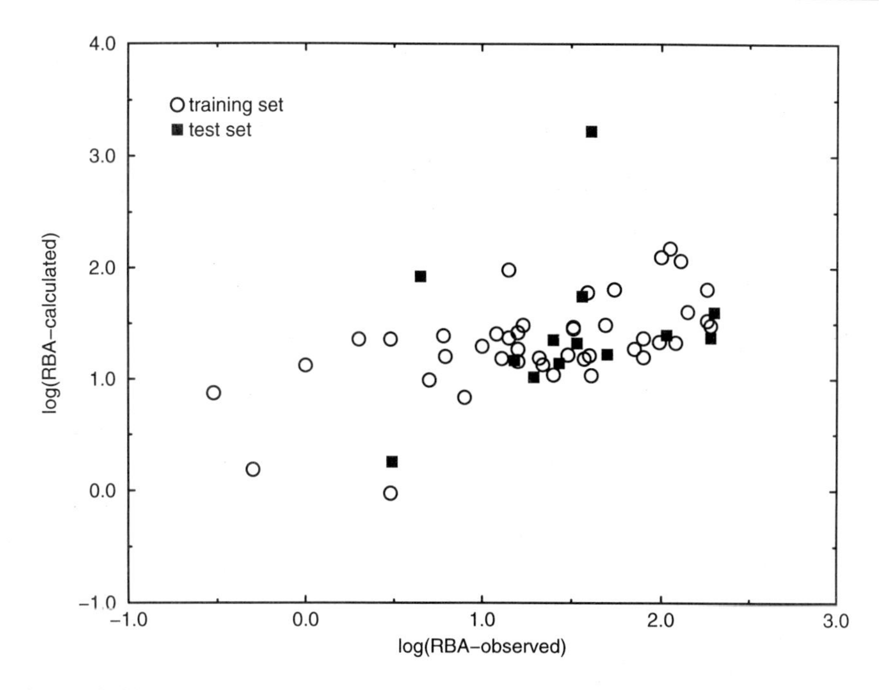

Figure 4 Observed versus predicted activity as calculated by PLS. Open circles: training set. Solid squares: test set.

(as discussed below) and if these are omitted from the plot the correlation coefficient for the test set increases to 0.59. At first sight, this figure seems to indicate that we have obtained a fairly good model. A closer inspection, however, leads to another conclusion. Firstly, Figure 4 shows that most of the predictions for the compounds from the test set fall in a relatively narrow range around 1.3, whereas the observed values range from –0.5 to 2.2. Thus, the predictions more or less fall on a straight horizontal line, which explains the relatively high r^2 observed. Secondly, the predictions for the test set are actually better than those for the training set, suggesting that the former result might be fortuitous. An analysis of the loadings of the PLS component (Table IV) shows that once more the most important properties in the model do not correspond to properties that provide an accurate description of the structural variation in the data set. We therefore concluded that PLS does not lead to a satisfactory model.

In order to understand the poor performance of these statistical methods all individual properties were plotted against the observed activity and correlation coefficients were calculated. Table V lists the 10 properties that correlated best with the activity. CHC3 is the property with the highest correlation coefficient (r^2 = 0.17) but this property is useless as can be seen in Figure 5: CHC3 has a distorted distribution. The next two properties are

Table IV *Ten properties with the highest loadings in the PLS model.*

Property	Loading
CHC7	–0.269
CHC4	–0.254
VOL_7A	–0.245
LUMO	–0.217
VOL_17B	0.247
SUR_17B	0.259
CHC5	0.265
CH_17B	0.267
CHC14	0.283
CHC3	0.309

Table V *Correlation coefficients of individual properties with log RBA.*

Property	r^2
CHC3	0.17
CHC14	0.14
CHC7	0.13
CH_17B	0.12
CHC5	0.12
SUR_17B	0.12
CHC4	0.11
VOL_7A	0.10
VOL_17B	0.10
LUMO	0.08

also of limited value because they do not relate to substituents with much variation in the training set. In fact, only the properties related to position 17B and VOL_7A in this list are likely to be important for describing activity. From earlier in-house studies it is known that the nature of the substituent at position 11 considerably influences the progestagenic activity but according to the correlation coefficients none of the properties of the 11-substituent correlates well to the activity. VOL11B, for instance, only has a correlation coefficient of 0.04. The plot of VOL11B versus the activity (Figure 6) shows that the relationship between VOL11B and activity is highly non linear. For small substituents (hydrogen atoms) there is no correlation but larger substituents show a fair negative correlation of volume with activity. These results explain why standard (linear) statistical techniques lead to poor models. As these results suggest that nonlinear methods are required for the analysis of this data set it was decided to apply a GFA to this data set.

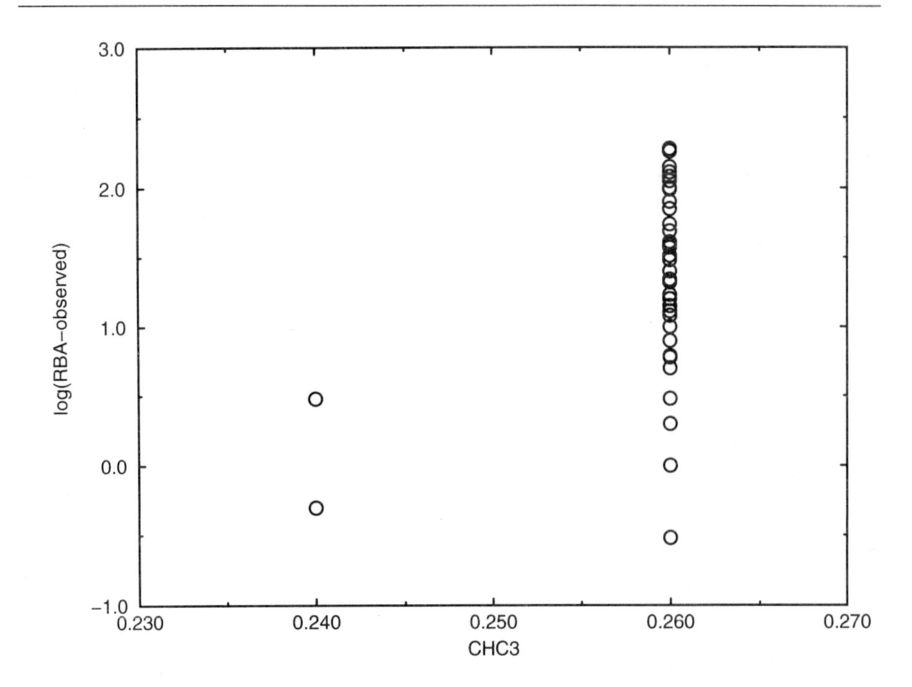

Figure 5 Relationship between activity and CHC3.

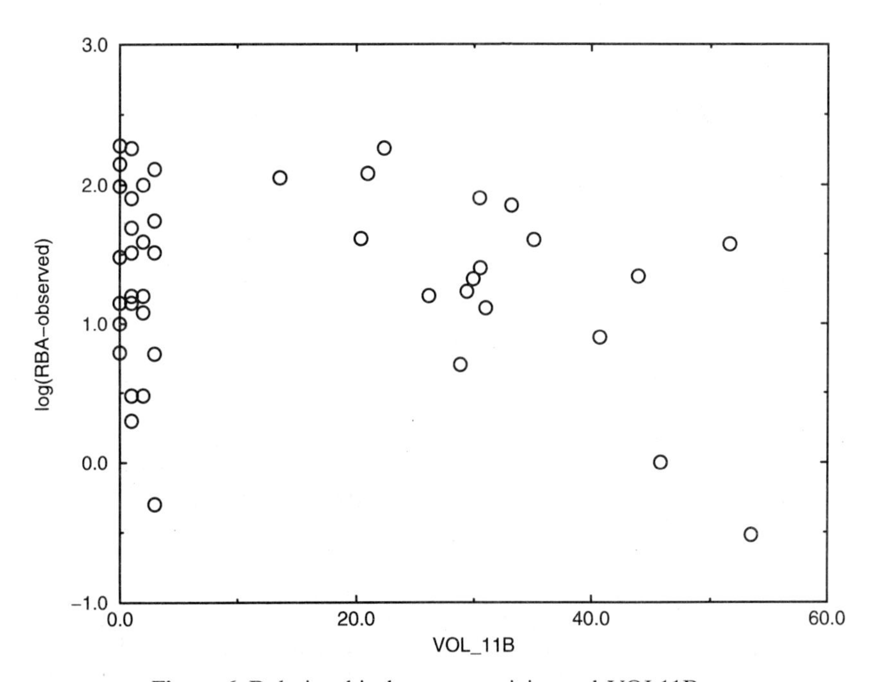

Figure 6 Relationship between activity and VOL11B.

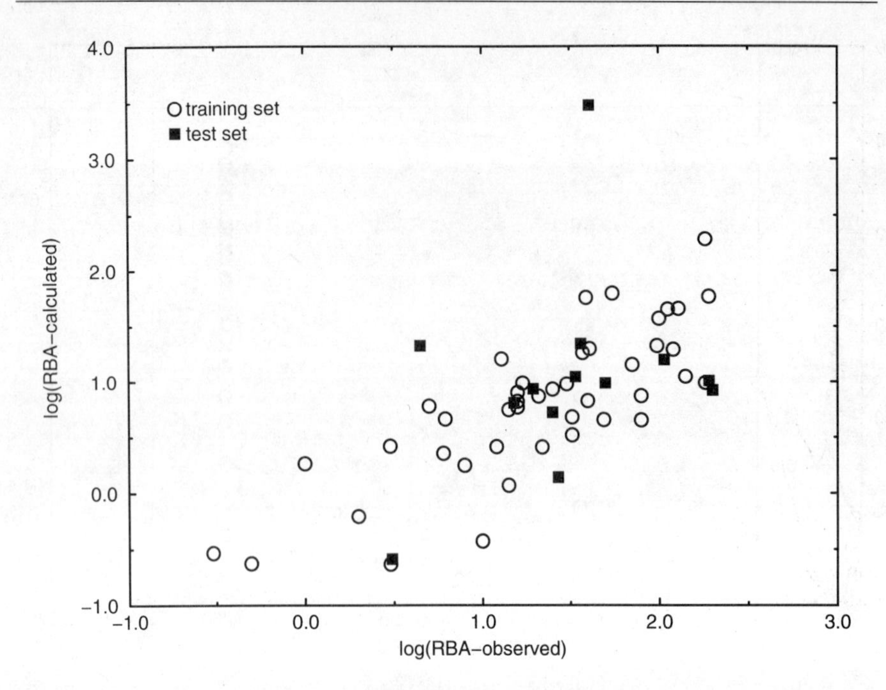

Figure 7 Observed versus predicted activity as calculated by the best GFA model (Eq. (3)). Open circles: training set. Solid squares: test set.

The genetic procedure described in the experimental section resulted in a set of 250 (nonlinear) equations with a maximum r^2 of 0.64 for the training set. In principle, one of these equations could be chosen as a model for predicting the activity. As an example the best equation (in terms of the LOF score) is given below, and the corresponding plot of observed versus calculated values is shown in Figure 7.

$$\begin{aligned}\text{ACT} = {} & -2.079 + 0.119 \times \text{SUR\_17B} + 0.014 \times \text{MME} + 0.247 \\ & \times <\text{SUR\_11B} - 47.3> - 0.065 \times <\text{VOL\_11B} - 30.48> + 1.8 \times 10^{-9} \\ & \times (\text{ELEC} + 43971.7)^2 - 18.885 \times (\text{CH\_17A} + 0.17)^2 - 31.142 \\ & \times <\text{CHC13} + 0.08> - 5.05 \times 10^{-4} \times (\text{SUR\_11TOT} - 37.38)^2 \end{aligned} \quad (3)$$

The quality of this model in terms of the correlation coefficient of the training set is higher than the quality of the stepwise and PLS models shown in Figures 3 and 4, but the correlation coefficient of the test set is still very low. The GFA model, however, is intuitively preferred because it uses a set of properties that properly reflects the structural variation in the data set.

It is interesting to note that the GFA has applied a spline function to VOL_11B. Figure 6 suggests that such a function is indeed required to describe the relationship between VOL11B and the RBA. However, a knot of approximately 10 would be more appropriate than the value of 30.48

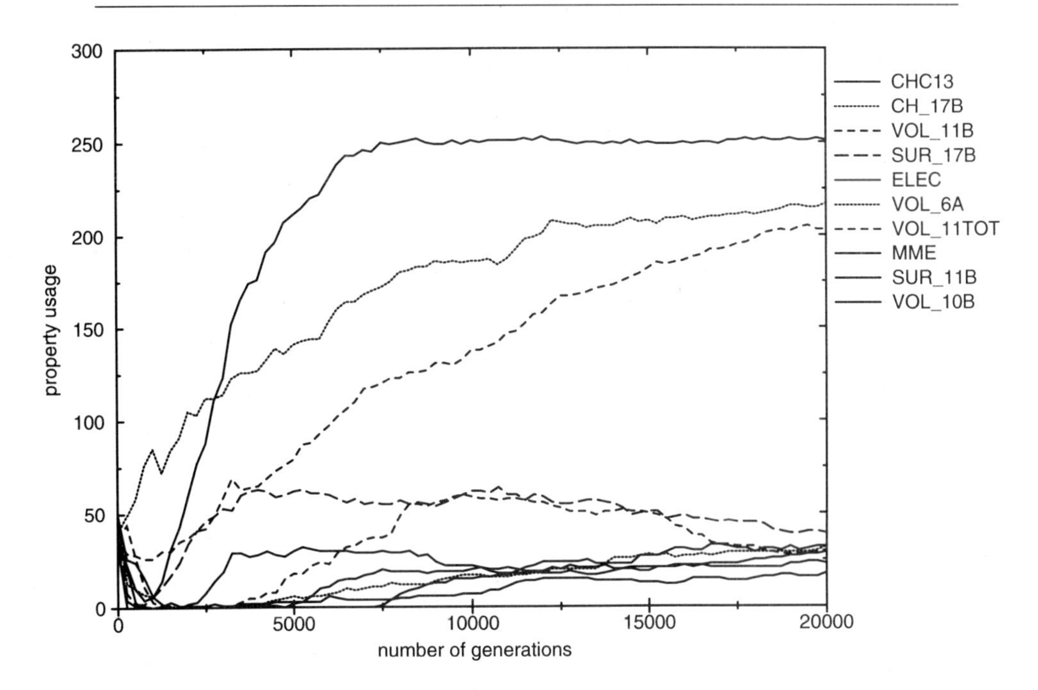

Figure 8 Property usage during GFA evolution. The order of the properties in the legend is equal to the order of usage after 20 000 generations.

which is found by the GFA. Nevertheless, the fact that a property related to position 11 is selected in this model is an improvement over standard statistical methods. Furthermore, most other properties also correspond to substituents known to be important.

Eq. (3) is just a single example from a population of equations found by the GFA. Other equations use other properties and these should also be taken into account. This can be accomplished by monitoring the property usage during the genetic evolution. Figure 8 shows which 10 properties are used most frequently during the formation of new generations. After 20 000 mutations three properties have emerged as the most important for the prediction of activity: CHC13, CH_17B, and VOL_11B. This is an encouraging result since it has already been noted that properties related to substituents 11 and 17 represent the bulk of structural variation in the compounds of this study. Furthermore, CHC13 also relates to a substituent that is known to influence RBA. Seven other properties are selected somewhat less frequently by the GFA but again some of these properties relate to important regions in the compounds. It is interesting to note that the GFA selected three properties related to position 11 of which VOL_11B and SUR_11B may look very similar at first sight since they both describe the size of this substituent. However, surface and volume descriptors may discriminate between spherical and elongated substituents. As the properties selected by GFA provide a good description of the structural variation in the data set they were applied as input variables for the neural network as will be discussed in the next paragraph.

Determining the optimal network architecture

Some initial analyses of the data set using neural networks showed that the default learning parameters of SNNS were appropriate for training our networks. The only two parameters that remained to be evaluated are the number of training cycles (epochs) and the number of hidden neurons. In order to determine the optimal number of epochs we applied the cross-validation procedure as described in the experimental section and monitored the correlation coefficient between predicted and observed values while training proceeded. The results for a 10–5–1 network are shown in Figure 9. The solid line shows the correlation coefficient of the training set. Obviously, this value should increase as the network is trained. After 2000 cycles an r^2 of 0.96 is obtained. The dashed line shows the results from cross-validation. Here the correlation coefficient reaches maximum after about 400 to 500 epochs. Further training results in loss of predictive power; in other words: the net is now overtrained. Thus, cross-validation indicates that training should be stopped after about 400 to 500 epochs to avoid overtraining.

As explained earlier, the test set was not used to determine when to stop training because it represents a set of new (yet unknown) compounds. Nevertheless, we have also included the results for the test set in this graph to allow for comparison with the cross-validation results. Figure 9 shows that the results from the test set are fairly similar to those from cross-validation with a maximum around 400 epochs. Ideally, the cross-validation and test set should have a maximum at the same number of epochs but that depends strongly on the contents of the training and test sets. It can be concluded from this graph that similar conclusions would have been drawn if a test set was used instead of cross-validation. Thus, cross-validation provides a good measure for detecting overtraining with this steroid data set.

Similar analyses were made for networks with 1, 2, 3, 4, 6, 8, and 10 hidden neurons and these resulted in plots very similar to those in Figure 9. The highest correlation coefficient obtained for each of these networks was plotted as a function of the number of hidden neurons (Figure 10) to determine the optimum network architecture.

Figure 10 indicates that an increase in the number of hidden neurons increases the correlation coefficient of the training set. Apparently, networks with one or two hidden neurons have an insufficient number of weights to store all information. At about five neurons the correlation coefficient reaches a plateau. More than five hidden neurons make the network more complicated (with a higher risk of overtraining) without a significant improvement of the results.

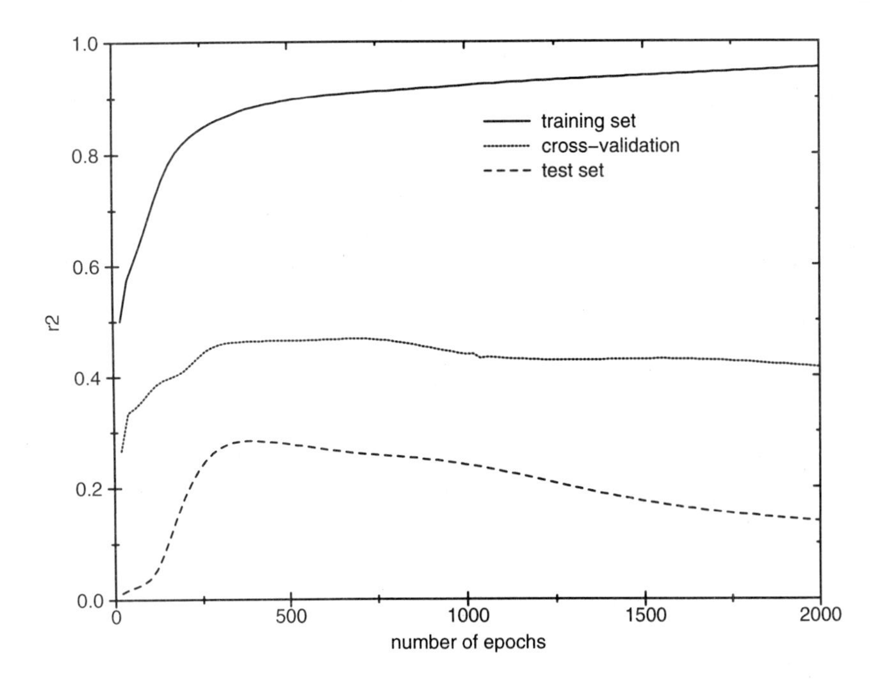

Figure 9 Training of a 10–5–1 neural network. The solid line represents the r^2 for the training set, the dotted line represents the r^2 found by cross-validation and the results for the test set are presented with a dashed line.

The results of the cross-validation are somewhat difficult to interpret because some cross-validation curves do not show a clear maximum during training. Therefore, the maximum r^2 values in Figure 10 may not be accurate. Despite these problems, the plot suggests that the predictive power does not improve significantly with more than 6 hidden neurons.

By combining the results of training set and cross-validation we concluded that a network with fewer than five hidden neurons may not lead to an optimal model but that a further increase of hidden units does not lead to a better model in terms of fitting and generalizing. Therefore, we have chosen to continue with five hidden neurons.

Again for comparison only, we have also included the results of the test set in Figure 10. With fewer than five neurons the network behaves unpredictably. With five hidden neurons the predictions are almost optimal and with a higher number of hidden neurons the quality of the predictions decreases again. This suggests that a network architecture with five or six hidden neurons results in the highest predictive power for this particular problem which is in agreement with the conclusion drawn from the cross-validation procedure.

It is especially interesting to note that in all networks the optimum of the cross-validated predictions closely corresponds to the position of the kink in

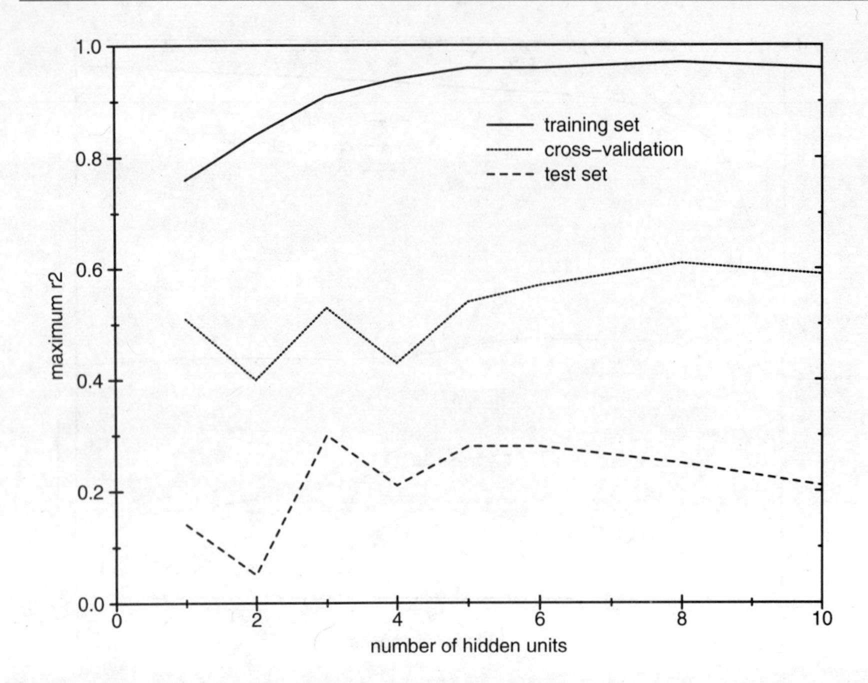

Figure 10 Maximum correlation coefficient (r^2) as a function of the number of hidden neurons. The solid line represents the r^2 for the training set, the dotted line represents the r^2 found by cross-validation and the results for the test set are presented with a dashed line.

the curve for the training set. This suggests that the kink in the training curve may represent the point where memorizing starts. Further experiments are required to validate this hypothesis but if it is true it would be unnecessary to perform the time-consuming cross-validation procedure!

In a paper on neural networks (Andrea and Kalayeh, 1991) a parameter ρ has been introduced that gives the ratio of the number of compounds to the number of network connections. It was found that ρ values larger than 2.2 result in networks with poor predictive power and that ρ values smaller than 1.8 may result in overfitting. This ρ value has been used by several other authors but it has also been concluded that it is not possible to give general guidelines since the optimum ρ value depends on the data set under study (Manallack *et al.*, 1994). In our study both cross-validation and the application of a test set indicate that a ρ value of 0.70 (five hidden neurons, $\rho = 43/61$) leads to a network with excellent predictive power. It should be noted that overtraining is likely to occur (as Figure 9 shows); however, the present study indicates that this can be avoided by applying the cross-validation procedure.

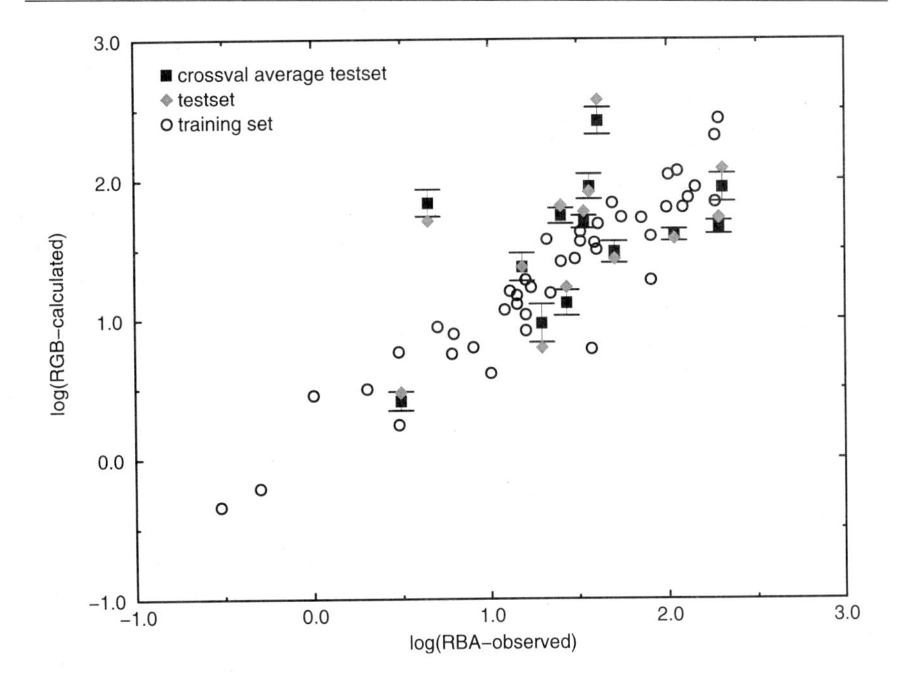

Figure 11 Observed versus predicted activity as calculated by a 10–5–1 network. Open circles: training set. Gray diamonds: test set. Solid squares represent average predictions from 43 networks trained during the cross-validation procedure. The standard deviations of these averages are given by the error bars.

Data analysis with a 10–5–1 network

As the previous results indicate that a 10–5–1 network that is trained approximately 500 epochs provides the best balance between increasing predictive power and avoiding overtraining, a network was trained with these parameters using all 43 compounds in the training set. Predictions were made for the compounds in the test set as shown in Figure 11. Most of these predictions have errors comparable to the predictions of the compounds in the training set. The two outliers in Figure 11 are compounds **46** and **52** which are structurally different from the training set as will be discussed below. It can be concluded that this neural network performs very well for compounds that fall within the scope of the training set but that extrapolations are not possible. Obviously, this is not a disadvantage particular to neural networks, but of QSAR methods in general. In fact, the poor predictions of these two unknown compounds might also be interpreted as further evidence that the neural network really extracted relevant information from the training set.

A closer look at the individual predictions of compounds in the test set reveals some interesting features of the neural network (Table VI). As

Table VI *Neural network predictions of log RBA for compounds in the test set.*

Compound	log RBA (observed)	log RBA (calculated)
44	1.18	1.37
45	1.41	1.81
46	0.65	1.71
47	0.49	0.47
48	2.04	1.58
49	1.57	1.91
50	1.44	1.23
51	2.31	2.08
52	1.62	2.57
53	1.54	1.77
54	2.29	1.73
55	1.30	0.80
56	1.71	1.43

discussed in the introduction three compounds (**46**, **52**, and **56**) were added to the test set to assess the ability to extrapolate beyond the scope of the training set. **46** and **52** are very different from all compounds in the training set and they are predicted very badly: these two compounds are the outliers in Figure 11. Compound **56** has a 11β–13β bridge and is predicted reasonably well despite the fact that two somewhat similar compounds in the training set (**11** and **15**) are known to have a conformation not possible in **56**. This conformational difference, however, is not represented by any of the properties used and apparently this is not really required for predicting the RBA of this compound. Compounds **44** and **45** differ only by a methyl group at position 7α and the latter is correctly predicted to have a higher RBA. Compounds **45** and **47** only differ by the position of a double bond and again the relative RBAs are predicted correctly. Compound **51** contains a series of structural features (11-methylene, 13-ethyl, and Δ^{15}) each of which is known to enhance RBA. However, the observed RBA is not a summation of all these enhancements but instead the actual RBA is somewhat lower than might be expected. The neural network correctly predicts this compound not to be extremely active. Finally, the neural network is not able to distinguish compounds **53** and **54** which differ in the position of a fluoro atom on the 11-methylene group. This is not surprising since no properties are included as input that might describe this difference and as a result the neural network predicts these two compounds to be almost equally active.

A matter of concern in neural network applications is the reproducibility of the results. A network can easily fall into a local minimum during training so that the result would depend on the starting configuration. Thus, initialization with other weights may lead to different results. Despite the fact that no special measures were taken to avoid local minima during the training of our networks we obtained useful results which seemed to be independent of changes in starting configuration. In order to test the reproducibility of the results more thoroughly, each of the 43 cross-validation networks was used to predict the test set. As these networks were trained independently with different compounds and new random weights this provides a way to assess the reproducibility of our networks. Figure 11 shows the average predictions of these networks together with the standard deviation in the predictions. The relatively small standard deviations indicate that the results presented in this paper are reproducible.

CONCLUDING REMARKS

In this paper we have derived a model for the prediction of progesterone receptor binding of steroids. It has been shown that conventional statistical techniques are inadequate because the data set contains nonlinear relationships. The GFA finds models with nonlinear terms but predictions of training set and test set are not significantly better than the results of statistical techniques. A major advantage of GFA, however, is the ability to select relevant properties even if these are not linearly correlated with the RBA. The selected properties provide a good starting point for neural network simulations and the final predictions of a 10–5–1 network proved to be superior to all other methods tested in this study. The r^2 values as summarized in Table VII already strongly point in this direction; however, it should be stressed that visual inspection of plots such as depicted in Figure 11 should also be taken into consideration when judging the validity of a particular model.

It may be argued that the 10–5–1 network contains too many connections compared with the number of compounds and that the model is overtrained. Our results indicate that overtraining may occur but that it can be prevented by applying cross-validation to determine the optimum number of epochs for training. It should be noted that we presented results on a test set to assess the predictive power of the NN, but that this test set has not been

Table VII *Summary of r^2 values obtained with different models in this study. As the test set contains two structurally unique molecules (**46** and **52**) that are predicted inaccurately by all models, the table also includes a column for the r^2 that is obtained if these two compounds are omitted from the test set.*

Model	r^2 Training set	r^2 Test set (all 13 compounds)	r^2 Test set (11 compounds)
Stepwise regression	0.42	0.02	0.42
PLS	0.37	0.06	0.59
GFA (best model)	0.64	0.09	0.49
NN	0.88	0.28	0.57

used to determine which are the optimal parameters. Thus, the test set provides independent evidence for the predictive power of the NN.

Analysis of the predictions of certain couples of compounds in the test set proved that the model correctly predicts the relative RBA of similar compounds. This indicates that the neural network managed to learn relevant relationships in the data set. Steroids that are very different from the compounds in the training set are predicted inaccurately, but this is a problem of QSAR in general. By averaging over 43 independent networks we have proven that the results are reproducible and that our simulations do not suffer from local minima. A disadvantage of the application of neural networks may be the black-box-character of the model. Although the RBA of new, but structurally related, compounds can be predicted with some accuracy, it is difficult to design new and more active compounds because it is not obvious how the individual properties affect the RBA. However, there may be ways to extract such information from a NN and these will be the subject of a future study.

Another topic for further study is the combination of GFA and NNs which was applied in this study. We used an existing GFA in which the fitness of functions is evaluated by (linear) regression models. Although we have shown that the resulting properties can be used successfully to train a neural network, it would be preferable to evaluate the fitness of a certain combination of properties by training a network with these properties. Such a direct coupling may lead to truly nonlinear models.

REFERENCES

Andrea, T.A. and Kalayeh, H. (1991). Applications of neural networks in QSAR of dihydrofolate reductase inhibitors. *J. Med. Chem.* **34**, 2824–2836.

Belaisch, J. (1985). Chemical classification of synthetic progestogens. *Rev. Fr. Gynecol. Obstet.* **80**, 473–477.

Bergink, E.W., van Meel, F., Turpijn, E.W., and van der Vies, J. (1983). Binding of progestagens to receptor proteins in MCF-7 cells. *J. Steroid Biochem.* **19**, 1563–1570.

van den Broek, A.J., Broess, A.I.A., van den Heuvel, M.J., de Jongh, H.P., Leemhuis, J., Schönemann, K.H., Smits, J., de Visser, J., van Vliet, N.P., and Zeelen, F.J. (1977). Strategy in drug research. Synthesis and study of the progestational and ovulation inhibitory activity of a series of 11β-substituted-17α -ethynyl-4-estren-17β-ols. *Steroids* **30**, 481–510.

Broess, A.I.A., van Vliet, N.P., Groen, M.B., and Hamersma, H. (1992). Synthetic approaches toward total synthesis of 12β-methyl- and 12-methylene-19-norpregnanes. *Steroids* **57**, 514–521.

Cerius^2 release 1.5 (1994). Molecular Simulations Inc., 200 Fifth Avenue, Waltham, MA 02154, US.

Chem-X Jan 94 release, Chemical Design Ltd, Roundway House, Cromwell Park, Chipping Norton, UK.

Cramer III, R.D., Patterson, D.E., and Bunce, J.D. (1988). Comparative molecular field analysis (CoMFA). *J. Am. Chem. Soc.* **110**, 5959–5967.

Coburn, R.A. and Solo, A.J. (1976). Quantitative structure–activity relationships among steroids. Investigations of the use of steric parameters. *J. Med. Chem.* **19**, 748–754.

Doré, J.C., Gilbert, J., Ojasoo, T., and Raynaud, J.P. (1986). Correspondence analysis applied to steroid receptor binding. *J. Med. Chem.* **29**, 54–60.

Duax, W.L., Griffin, J.F., and Rohrer, D.C. (1984). Steroid conformation, receptor binding, and hormone action. In, *X-Ray Crystallography and Drug Action, Course Int Sch. Crystallogr. 9th,* (A. S. Horn and C. J. de Ranter, Eds). Oxford Univ. Press, Oxford, pp. 405–426.

Free, S.M. and Wilson, J.W. (1964). A mathematical contribution to structure activity studies. *J. Med. Chem.* **7**, 395–399.

Gasteiger, J., Rudolph, C., and Sadowski, J. (1990). Automatic generation of 3D atomic coordinates for organic molecules. *Tetrahedron Comp. Meth.* **3(6C)**, 537–547.

van Geerestein, V.J. (1988). Structure of the ethanol solvate of 13-ethyl-11β-methyl-18-norlynestrenol. *Acta. cryst. Section C* **44**, 376–378.

Hansch, C. and Fujita. T. (1964). ρ-σ-π Analysis. A method for the correlation of biological activity and chemical structure. *J. Am. Chem. Soc.* **86**, 1616–1626.

van Helden, S.P. and Hamersma, H. (1995). 3D QSAR of the receptor binding of steroids: A comparison of multiple regression, neural networks and comparative molecular field analysis. In, *Proceedings of the 10th Symposium on Structure–Activity Relationships: QSAR and Molecular Modelling*. Barcelona, September 1994.

Hopper, H.O. and Hammann, P. (1987). The influence of structural modification on progesterone and androgen receptor binding of norethisterone. Correlation with nuclear magnetic resonance signals. *Acta Endocrinol. (Copenhagen)* **115**, 406–412.

Lee, D.L., Kollman, P.A., Marsh, F.J., and Wolff, M.E. (1977). Quantitative relationships between steroid structure and binding to putative progesterone receptors. *J. Med. Chem.* **20**, 1139–1146.

Loughney, D.A. and Schwender, C.F. (1992). A comparison of progestin and androgen receptor binding using the CoMFA technique. *J. Comput.-Aided Mol. Des.* **6**, 569–581.

Maddalena, D.J. and Johnston, G.A.R. (1995). Prediction of receptor properties and binding affinity of ligands to benzodiazepine/$GABA_A$ receptors using artificial neural networks. *J. Med. Chem.* **38**, 715–724.

Manallack, D.T., Ellis, D.D., and Livingstone, D.J. (1994). Analysis of linear and non-linear QSAR using neural networks. *J. Med. Chem.* **37**, 3758–3767.

Moriguchi, I., Komatsu, K., and Matsushita, Y. (1981). Pattern recognition for the study of structure–activity relationships. *Anal. Chim. Acta* **133**, 625–636.

Ojasoo, T., Doré, J.C., Gilbert, J., and Raynaud, J.P. (1988). Binding of steroids to the progestin and glucocorticoid receptors analyzed by correspondence analysis. *J. Med. Chem.* **31**, 1160–1169.

Pitt, C.G., Rector, D.H., Cook, C.E., and Wani, M.C. (1979). Synthesis of 11β-,13β- and 13β-,16β-propano steroids. Probes of hormonal activity. *J. Med. Chem.* **22**, 966–970.

Rohrer, D.C., Hazel, J.P., Duax, W.L., and Zeelen, F.J. (1976). 11β-Methyl-19-nor-17α-pregn-4-en-20-yn-17β-ol ($C_{21}H_{30}O$). *Crystal Struct. Commun.* **5**, 543–546.

Rogers, D. and Hopfinger, A.J. (1994). Application of genetic function approximation to quantitative structure activity relationships and quantitative structure property relationships. *J. Chem. Inf. Comput. Sci.* **34**, 854–866.

Rozenbaum, H. (1982). Relationships between chemical structure and biological properties of progestogens. *Am. J. Obstet. Gynecol.* **142**, 719–724.

Stewart, J.J.P. (1990). MOPAC: A semiempirical molecular orbital program. *J. Comput-Aided Mol. Des.* **4**, 1–105.

SNNS version 3.2 (1994). IPVR, University of Stuttgart, Breitwiesenstrasse 20-22, 70565 Stuttgart, Germany.

Teutsch, G., Weber, L., Page, G., Shapiro, E.L., Herzog, H.L., Neri, R., and Collins, J.A. (1973). Influence of 6-azido and 6-thiocyanato substitution on progestational and corticoid activities and a structure–activity correlation in the Δ^6-6-substituted progesterone series. *J. Med. Chem.* **16**, 1370–1376.

Wikel, J.H. and Dow, E.R. (1993). The use of neural networks for variable selection in QSAR. *Bioorg. Med. Chem. Lett.* **3(4)**, 645–65.

Zupan, J. and Gasteiger, J. (1993). *Neural Networks for Chemists.* VCH Verlagsgesellschaft, Weinheim, p. 142.

8 Genetically Evolved Receptor Models (GERM): A Procedure for Construction of Atomic-Level Receptor Site Models in the Absence of a Receptor Crystal Structure

D.E. WALTERS* and T.D. MUHAMMAD

Department of Biological Chemistry, Finch University of Health Sciences/The Chicago Medical School, 3333 Green Bay Road, North Chicago, IL 60064, USA

The goal of the present research is to produce atomic-level models of receptor sites, based on a small set of known structure–activity relationships. First, a set of five or more active compounds are selected and superimposed in low-energy conformations. Then a 'receptor site' is constructed by placing 40–60 atoms around the surface of the superimposed active compounds. A genetic algorithm is used to alter and optimize the atom types of the receptor site in such a way as to maximize the correlation between calculated drug–receptor binding and measured drug activity. The genetic algorithm rapidly finds large numbers of receptor models with correlation coefficients typically > 0.95, even for non-homologous series.

These receptor models can be used in several ways. First, they can serve the purpose of a QSAR, correlating calculated binding with measured bioactivity. Second, they can be used to predict the bioactivities of new compounds, by docking the compounds, calculating their binding energies, and calculating their bioactivities. Third, the receptor models can substitute for protein crystal structures in software which searches 3-D databases or does de novo drug design. In this chapter we describe the examination of alternate binding modes for ligands. In recent years this has become an important consideration when designing ligands to interact specifically with a receptor site. We also

* Author to whom all correspondence should be addressed.

In, *Genetic Algorithms in Molecular Modeling* (J. Devillers, Ed.)
Academic Press, London, 1996, pp. 193–210.
ISBN 0-12-213810-4

demonstrate the utility of carrying out sequence analysis on the genes describing the evolved models. Sequence analysis permits identification of the most important structural features in determining the affinity and, particularly, the selectivity of the ligand–receptor interaction.

KEYWORDS: *genetic algorithm; receptor model; drug design; predictive models; alternate binding modes.*

INTRODUCTION

When the structure of a receptor is known from X-ray crystallography or from NMR spectroscopy, it provides a great deal of information which can aid in the design of selective new ligands. The receptor site structure can be used in screening 3-dimensional structure databases (Kuntz *et al.*, 1982; Bartlett *et al.*, 1989), or it can be used in conjunction with *de novo* design programs (Moon and Howe, 1991; Nishibata and Itai, 1991; Böhm, 1992; Lawrence and Davis, 1992; Verlinde *et al.*, 1992; Rotstein and Murcko, 1993), or newly designed ligands can be docked into the site to assess their fit prior to their synthesis. Unfortunately, such structural information is unavailable in a large percentage of real-world ligand design problems.

In the absence of such structural information, one could conceptually construct a model of the receptor site in a very simplistic way, as illustrated in Figure 1: superimpose a series of known ligands, in low-energy conformations, in such a way as to present a common pattern of key functional groups; then place complementary atoms or functional groups in proximity to the ligands, building up a 'shell' of atoms to represent the nearest atoms in the receptor binding site. For example, we could place receptor model atoms with partial negative charge near ligand atoms with partial positive charge; we could place hydrogen bond donors adjacent to hydrogen bond acceptors and vice versa; we could place hydrophobic receptor atoms close to hydrophobic substituents on the ligands. In fact, using a set of atom types

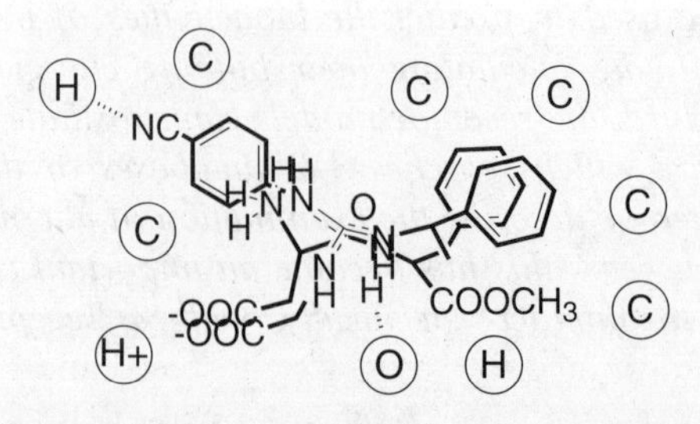

Figure 1 Simplistic approach to receptor model construction. Active ligands are superimposed, and complementary atoms are placed in proximity.

commonly found in proteins (such as those listed in Table I), we can very easily make such models. The problem is this: if we construct a shell of 50 atoms, and each atom can be chosen from a set of 14 atom types, then we could construct 2×10^{57} different models.

The genetic algorithm approach is well suited to highly multidimensional search problems of this sort if we are able to satisfy two criteria:

- We must find an appropriate way to represent a potential solution to the problem as a string or linear code.
- We must find an appropriate way to quantitatively evaluate how 'good' any given solution is.

How do we decide what constitutes a 'good' model? And once we have established our criteria, how do we find 'good' models in an essentially infinite multidimensional search space? If we take one position at a time and try all possible atom types, we could clearly come up with the model which has the highest affinity for the ligands. But real receptors do not necessarily have the maximum possible affinity for their ligands–receptors must have a relatively high affinity for their ligands, and they must be able to discriminate between more active and less active ligands.

In a recent paper (Walters and Hinds, 1994), we have described a method to generate good receptor models using a genetic algorithm, using a program which we call Genetically Evolved Receptor Models (GERM). The starting point is a series of active compounds (typically 5–10 are used) with experimentally determined bioactivities. It is important to use compounds which are known or suspected to act at a common receptor site. The structures are superimposed in low-energy conformations in such a way as to present a common pattern of steric and electronic properties to the receptor site. A shell of atoms representing the receptor surface is constructed around the superimposed ligands, spacing the atoms as evenly as possible, and placing the atoms in van der Waals contact with the ligands or moving them away from the ligands by 0.1–0.5 angstroms (to compensate for the fact that we are constructing a rigid receptor site model and treating the ligands as rigid molecules). Once the atoms are placed in 3-dimensional space, their positions are fixed for the remainder of the calculation.

METHODS

To satisfy the first requirement for using a genetic algorithm, a model is encoded in a linear string as illustrated in Figure 2. *Position* in the string corresponds to position in three-dimensional space, and the *numerical value* at each position corresponds to a specific atom type. The 'genetic code' which we use is shown in Table I. It consists of 14 typical protein atom types taken from the CHARMm force field (Brooks *et al.*, 1983). We also include a null

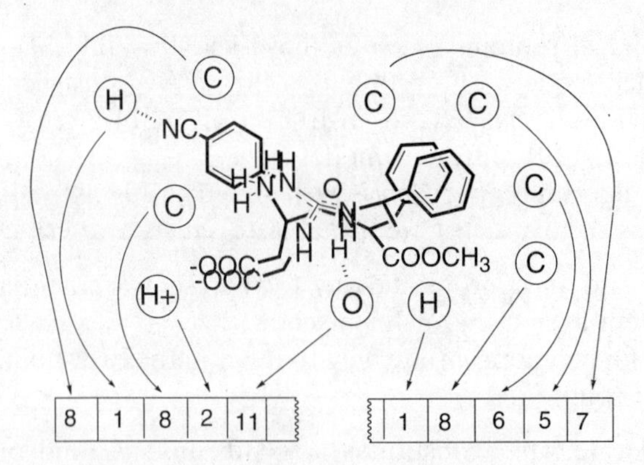

Figure 2 Encoding a receptor model into a linear string or code. The *position* in the string corresponds to the position in three-dimensional space, and the *value* at each position corresponds to the atom type at that position.

atom type (type zero), corresponding to no atom at all in a given position. This allows receptor models to have an open face if necessary.

The second requirement (quantitative evaluation of fitness) is satisfied as follows. The ligands are docked into a receptor model, one at a time. For each ligand, we calculate an intermolecular interaction energy (all ligand atoms interacting with all receptor atoms; no intramolecular terms are used). We then calculate the statistical correlation coefficient between log (1/calculated interaction energy) and log (experimentally measured bioactivity). The *correlation coefficient* serves as the quantitative fitness measure, with the underlying working assumption that bioactivity is proportional to binding affinity. Thus, a model is considered to be a 'good' one if it provides a good correlation between *calculated binding energies* and *experimentally measured bioactivities*. The fitness function, log (experimentally measured bioactivity) $\propto$ (1/calculated interaction energy), is mathematically similar to the equation which relates binding constants to Gibbs free energy, $K = \exp(-\Delta G/RT)$. Clearly, though, we are calculating only the enthalpy component of Gibbs free energy.

The genetic algorithm is then implemented by generating an initial population of 1000–5000 models. The size of the initial population will, of course, depend on the number of positions in the models and on the number of compounds in the training set. We have typically used a population of 2000 models for sets of six compounds, and a population of 5000 for sets of 11 compounds. We also use a rather ruthless selection process: the weakest member of the population is always replaced by any offspring with higher fitness. This results in fairly rapid loss in genetic diversity, for which we have compensated with larger population sizes. It is likely that a less ruthless

Table I *The 'genetic code' and parameters used in this work consisted of a list of atom types and their corresponding force field parameters. Atom types were chosen from the CHARMm force field (Brooks et al., 1983), and charges are values which approximate those found in the standard 20 amino acids in the commercially distributed version of the CHARMm force field (Molecular Simulations, Inc., Burlington, MA 01803, USA). Type 0 corresponds to having no atom at all in a given position.*

Atom type code	CHARMm type	E_{min} (kcal/M)	R_{min} (Ångstroms)	Partial atomic charge
0	–	–	–	–
1	H (H on polar atom)	–0.0498	0.800	0.25
2	HC (H on charged N)	–0.0498	0.600	0.35
3	HA (aliphatic H)	–0.0450	1.468	0.00
4	C (carbonyl C)	–0.1410	1.870	0.35
5	CH1E (CH group)	–0.0486	2.365	0.00
6	CH2E (CH_2 group)	–0.1142	2.235	0.00
7	CH3E (CH_3 group)	–0.1811	2.165	0.00
8	CT (aliphatic C)	–0.0903	1.800	0.00
9	NP (amide N)	–0.0900	1.830	–0.40
10	NT (amine N)	–0.0900	1.830	–0.30
11	O (carbonyl O)	–0.2000	1.560	–0.50
12	OT (hydroxyl O)	–0.2000	1.540	–0.60
13	OC (carboxylate O)	–0.1591	1.560	–0.55
14	S	–0.0430	1.890	–0.20

method (for example, replacement of the weaker parent, as was suggested to us by Dr Brian Luke) would enable us to use smaller populations.

Initially, each position of each model is randomly assigned an atom type from 0 to 14, corresponding to the types listed in Table I. The fitness of each model is calculated. Pairs of models are then chosen to serve as 'parents'. Parent selection is currently done on a fitness-weighted basis, using a roulette wheel scheme (a model with fitness of 0.90 is ten times more likely to be selected than one with a fitness of 0.09). Pairs of 'offspring' models are generated by cross-over at a point which is randomly selected each time (Figure 3). Mutation is carried out by selecting one or more random positions on the offspring and assigning an atom type between 0 and 14 (corresponding to the atom types listed in Table I). We have typically used a frequency of 1 mutation per offspring, but we have applied it with a Poisson distribution so that any given offspring may have 0, 1, or more mutations, and the *average* occurrence of mutation is 1 per offspring. The fitness of each offspring is evaluated. If it is better than the weakest member of the population, then it replaces that member; if it is not better than any member, it is discarded. The selection/crossover/evaluation process is repeated for a fixed number of generations, or until there is no significant change in the population's mean fitness over some number of generations. As the population evolves, the

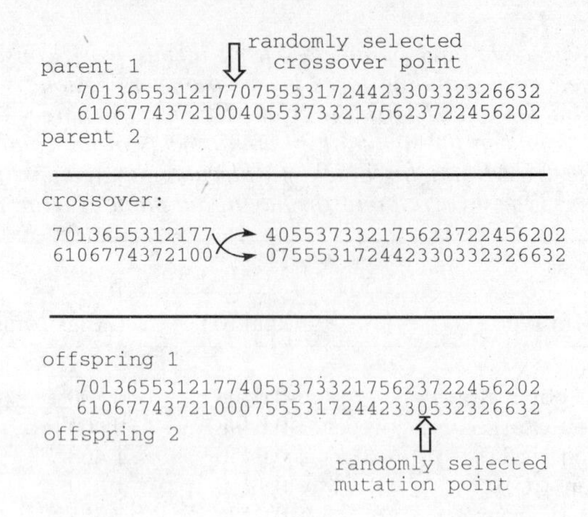

Figure 3 Illustration of the crossover and mutation operators.

mean fitness score may change from a very low (or even negative) number to a very high level. Ultimately, the method produces a set of several thousand models, *each* of which has a high correlation (typically $r^2 > 0.9$) between *calculated* ligand–receptor binding energy and *measured* bioactivity. Thus, these are *discriminative* models – they are able to quantitatively distinguish between highly active and less active ligands.

Our initial studies with this method have involved a series of 22 sweet-tasting molecules with potencies ranging from 50 to 200 000 times that of sucrose (Figure 4 and Table II). We selected this set for several reasons. First, we have previously studied these structures extensively (Culberson and Walters, 1991) and have carried out the prerequisite conformational analysis and alignment steps. Second, this set provides considerable structural diversity, including aspartic peptides, arylureas, and guanidine-acetic acid derivatives. Third, the measured bioactivity in this series is *in vivo* data which is not complicated by questions about bioavailability, absorption, metabolism, or excretion. Taste results came from trained human taste panels (DuBois *et al.*, 1991), where the receptors are freely accessible on the surface of the tongue.

In our previous paper (Walters and Hinds, 1994), we carried out several series of cross-validation calculations, including a number of leave-n-out experiments. Ultimately we found it possible to leave out half of the data set. Models constructed around 11 of the 22 compounds could make very reasonable estimates of the potencies of the other 11 compounds. We also tested for over-fitting by re-running the calculations on the 8 aspartic peptides in the set with bioactivity data scrambled. Whereas the compounds with real bioactivity gave a mean r^2 of 0.975, with scrambled data the mean r^2 was 0.344. Thus, the method is not over-fitting to the training set.

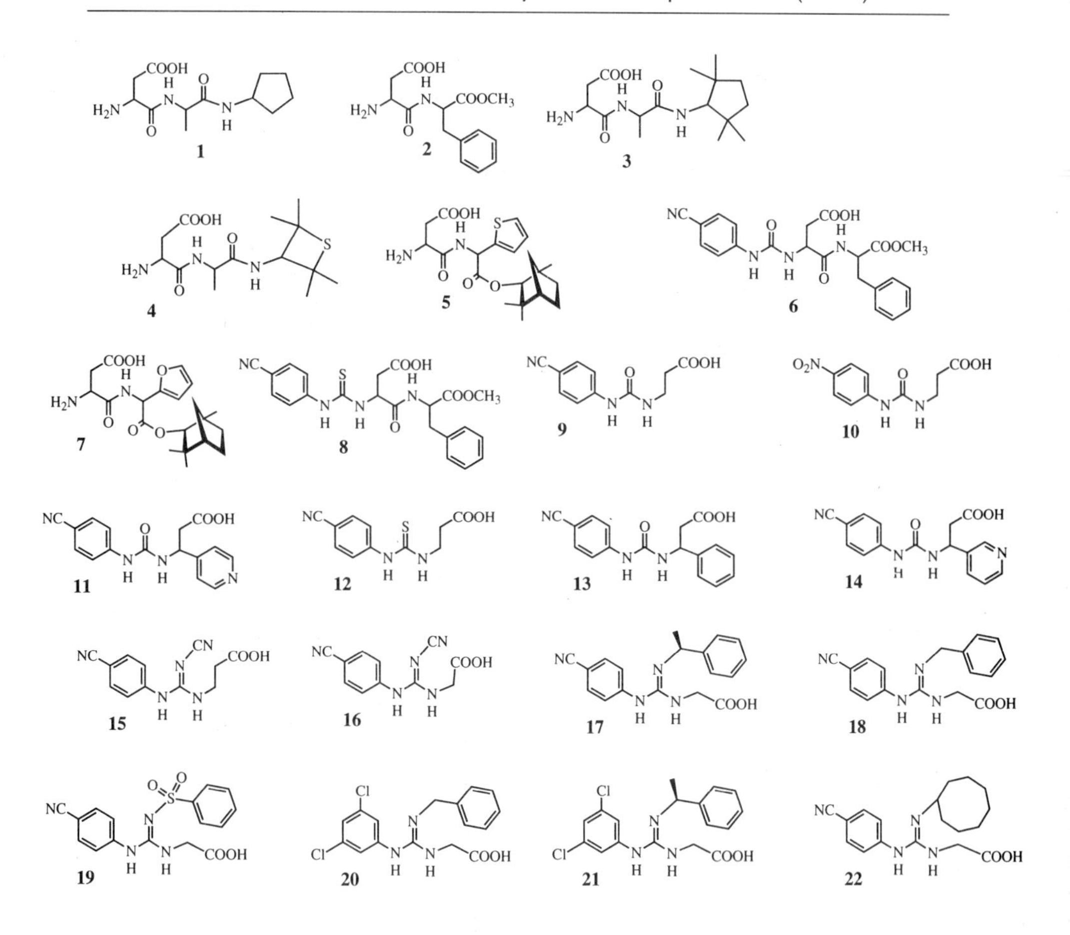

Figure 4 Structures of the initial set of 22 sweetener structures. Compounds 1–8 are aspartic acid derivatives; compounds 6 and 8–14 are arylurea derivatives; compounds 15–22 are guanidine derivatives. For sources of the compounds and bioactivity data, see Walters and Hinds (1994).

In this paper we describe further studies using the GERM method. First, we illustrate the utility of the evolved models in evaluating alternative binding modes for several compounds. It is not uncommon, when using protein crystallography to study a series of ligands bound to a receptor protein, to find a ligand which binds in an orientation which is different and unexpected, relative to the other ligands in the series (for example, see Kester and Matthews, 1977). Because of such occurrences, medicinal chemists have begun to consider the possibility of alternate binding modes. In earlier work on sweeteners (Culberson and Walters, 1991), we identified a number of key structural features which contribute to bioactivity of the aspartic peptides,

Table II *Experimentally determined log (potency) values for the 22 sweet-tasting compounds of Figure 4.*

Compound	log (potency)
1	**1.70**
2 (aspartame)	2.26
3	2.90
4 (alitame)	**3.30**
5	**3.38**
6 (superaspartame)	**4.00**
7	4.22
8	**4.70**
9	2.65
10 (suosan)	**2.85**
11	3.30
12	3.32
13	**4.00**
14	4.30
15	2.95
16	**3.85**
17	**4.45**
18	4.48
19	4.48
20	**4.90**
21	5.08
22	**5.23**

arylureas, and guanidine-acetic acid series of sweeteners. Despite their structural diversity, each of the compounds in Figure 4 contains a carboxylate group, two or more polar N-H groups, and a large hydrophobic substituent; many also contain an aryl ring with electron-withdrawing substituent(s). These structures can be superimposed in low-energy conformations in such a way as to place these functional groups in a consistent pattern. Figure 5 shows the qualitative receptor model surface which we reported previously (Culberson and Walters, 1991), with representative aspartic, arylurea, and guanidine structures in place. For highly potent structures such as the guanidine-acetic acid derivatives (structures 15–22) and superaspartame (structure 6), it is easy to identify each of the key structural features and to dock it properly into the receptor model. For many simpler structures, however, not every feature is present, and it is not necessarily clear how to dock the structure properly. Below, we describe the evaluation of alternate binding modes of three different sweeteners using evolved receptor models.

Second, we show the value of carrying out sequence analysis of the genes describing the evolved models. Sequence analysis points out which atoms in the receptor model are highly conserved and which ones are variable. The locations and types of conserved atoms give information about the ways in

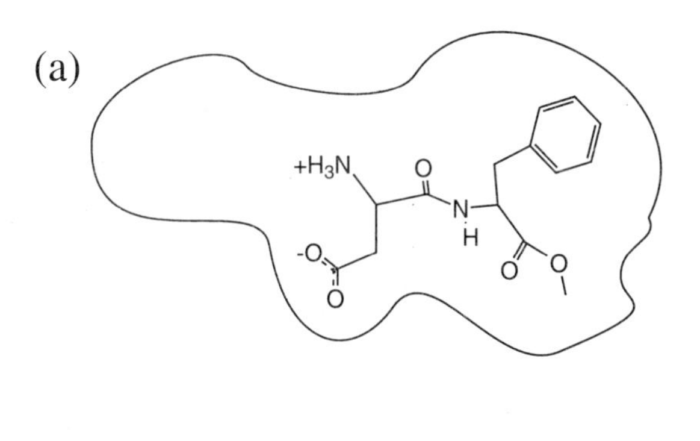

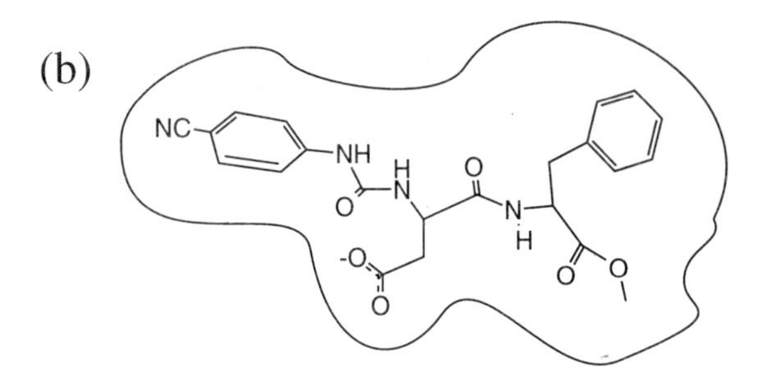

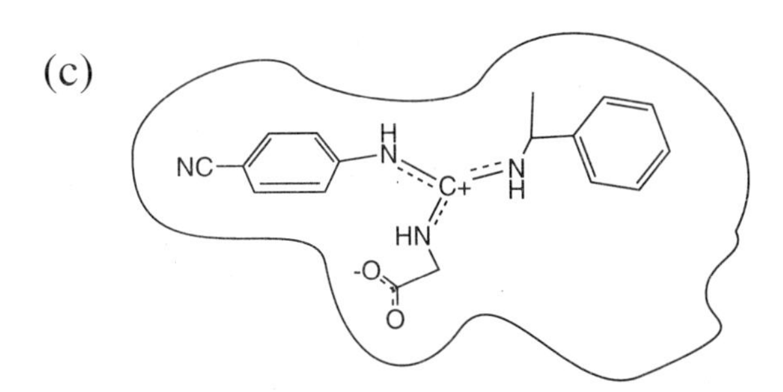

Figure 5 Schematic picture of receptor model (Culberson and Walters, 1991) showing the orientation of representative members of all three structural classes of compounds: (a) orientation of aspartame, compound 2 of Figure 4; (b) orientation of superaspartame, compound 6 of Figure 4; (c) orientation of compound 17 of Figure 4. Note that in each case there is a carboxylate group in the lower left lobe; there is a polar NH group near the top-center; there is a polar NH group just below and to the right of the center; there is a large hydrophobic group in the upper right lobe; in (b) and (c) there is an aromatic ring with electron-withdrawing substituent in the upper left lobe.

which the models discriminate between high and low activity compounds, and it points out the key features of the ligands with respect to receptor recognition. Since we have the ability to generate populations of several thousand models, each of which has a high correlation between calculated binding energy and experimentally measured bioactivity, we considered the possibility that there may be some sites in the receptor model which consistently have the same atom type (or very similar atom types), while other sites may be highly variable. If this is the case, it may be an indication as to which sites are most important in determining the *selectivity* of the models. Further, if type 0 (null atom type) is strongly favored at a given position, it may be an indication as to the likelihood of a particular face of the receptor site being open. In the Results and Discussion section, we describe the analysis of gene sequences from three large populations of evolved sweetness receptor models.

RESULTS AND DISCUSSION

Evaluation of alternate binding modes

Often, when a receptor site has several binding groups and a potential ligand has several functional groups, there is ambiguity about the best way to dock the ligand into the receptor site. In the case of suosan (structure 10), for example, we initially aligned the structure in such a way that the nitrophenyl ring occupies the upper-left region of the model usually occupied by aromatic rings with electron-withdrawing substituents (Figure 6a). This alignment places one of the urea NH groups in good alignment with the other NH groups at the top of the model, but the other NH group is not optimally placed with respect to the second NH site near the center of the model. However, we can consider an alternate orientation (Figure 6b) which does a better job of matching both NH sites and which places the nitrophenyl ring in the large hydrophobic region.

Using the population of 5000 models described previously (Walters and Hinds, 1994), we calculated log (potency) for suosan in both orientations. Orientation (a) produces a calculated value of 3.2, while orientation (b) produces a value of 1.6; the experimentally determined value is 2.9. Thus, the alternate mode (b) appears unlikely in this case.

The previous example could have been biased, since suosan in orientation (a) was one of the structures in the training set. As a further test, we evaluated different binding modes for two structures which were not in the training set: D-tryptophan and 6-chloro-D-tryptophan (Figure 7). These structures are of moderate potency: D-tryptophan, log (potency) = 1.5 (Berg, 1953), and 6-chloro-D-tryptophan, log (potency) = 2.1 (Suarez *et al.*, 1975). Both structures have a carboxylate group, two different NH groups, and an

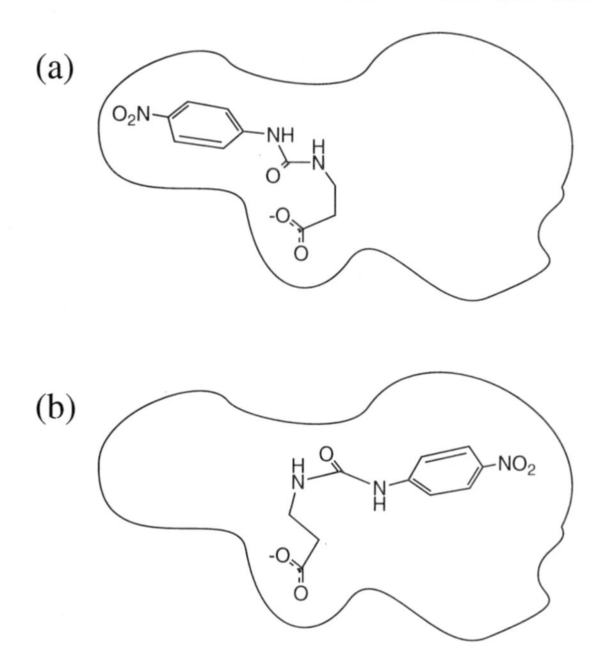

Figure 6 Alternate binding modes for suosan. (a) Nitrophenyl ring in aryl region. (b) Nitrophenyl ring in hydrophobic region.

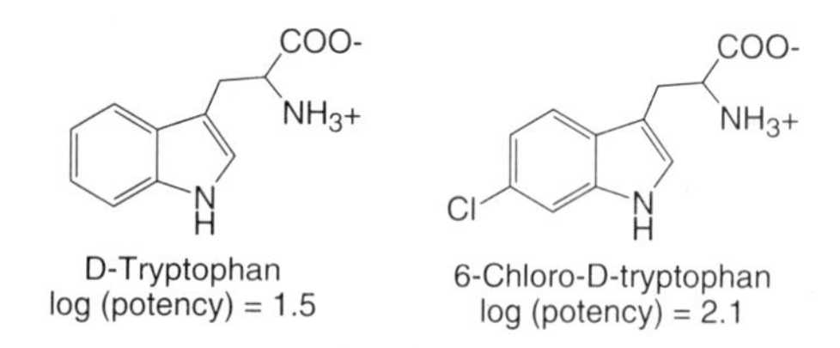

Figure 7 D-Tryptophan and 6-chloro-D-tryptophan.

aromatic indole ring system; in one case, the aryl ring has a chloro-substituent. Should the indole ring be docked into the aryl region (Figure 8a) or the hydrophobic region (Figure 8b)? Should the amino group occupy the upper NH site or the central NH site? Can the indole NH group satisfactorily occupy the remaining NH site? Does the chloro-substituent make a difference in the preferred binding mode? Since there were no D-amino acids included in the model generation, this test should not be biased by the training set. As shown in Table III, the models do not show any difference in calculated potency for D-tryptophan in the two different orientations. Both orientations produce a calculated log potency of 1.6, very close to the experimentally observed value. 6-Chloro-tryptophan, on the other hand, shows a

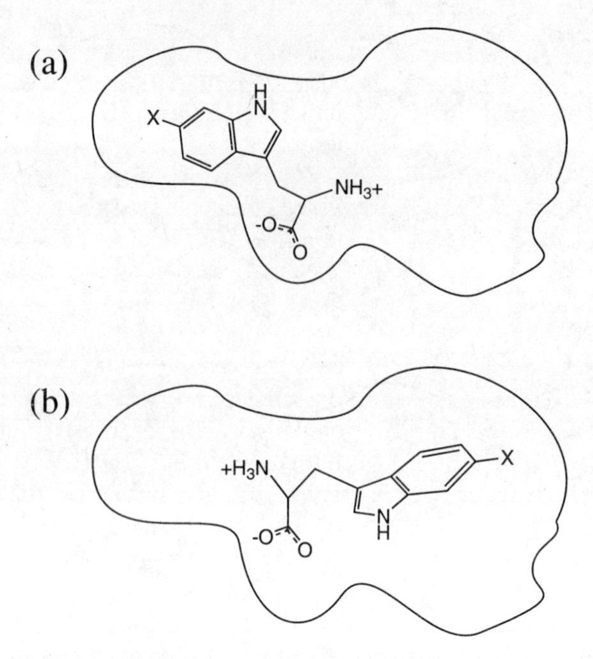

Figure 8 Alternate binding modes for D-tryptophan (X = H) and 6-chloro-D-tryptophan (X = Cl). (a) Indole ring in aryl region, indole NH in upper NH site, amino group near central NH site. (b) Indole ring in hydrophobic region, amino group near upper NH site, indole NH in central NH site.

modest preference for binding in orientation (a), with calculated log potency of 2.3 vs. 1.9. The preference is probably not sufficiently strong to allow us to rule out orientation (b) for chloro-tryptophan, but the results make us aware that we must consider both possibilities when looking at analogs of D-tryptophan. Note also that the predicted potencies are quite close to the experimentally measured ones, even though there were no D-amino acids in the training set.

Table III *Experimental and calculated log (potency) values for D-tryptophan and 6-chloro-D-tryptophan, using the two different orientations shown in Figure 8a and 8b.*

Compound	Experimental log (potency)	Calculated log (potency) in orientation (a)	Calculated log (potency) in orientation (b)
D-Tryptophan	1.5	1.6	1.6
6-Chloro-D-tryptophan	2.1	2.3	1.9

Sequence analysis

We previously described a set of 5000 models based upon structurally diverse sweet peptides, ureas, and guanidines (Walters and Hinds, 1994). In generating these models, we used compounds 1, 4, 6, 7, 8, 10, 13, 16, 17, 18, and 22 of Figure 4 for the training set. This provided both a diversity of structures and a broad range of bioactivities. We generated models containing 46 atom sites, evolved for 50 000 generations. The final mean fitness score for this populations was r = 0.944. Conservation at a single position within a single population could indicate either that the atom type at that position is important, or that, because of inbreeding, a single atom type came to dominate just by chance. Thus, we considered it important to compare several populations which used the same training compounds but which had independently randomized starting populations. Therefore, we generated two additional sets of 5000 models each (summarized in Table IV). For each set, we analysed the frequency of occurrence of each atom type at each position. Figures 9–11 shows scatter plots (as bubble diagrams) for the three sets. In each case, the x-axis corresponds to the position in the model; the y-axis corresponds to the atom type; the area of the bubble is proportional to the fraction of atoms at that position having the indicated atom type. For a given population, we can see that some positions are highly variable. In set 1, for example, position 20 has significant amounts of types 4, 6, 7, 10, 11, 12, and 13, and lesser amounts of types 1, 3, 5, and 9. There is no clear preference in atom type at this position. We can also see some positions which are highly conserved; in set 1, positions 2 and 3 are almost always a hydrophobic atom type (especially types 5 and 6).

Position 39 has a high percentage of negatively charged atom types in set 1, more hydrophobic types in set 2, and more positively charged types in set 3. We can consider this position to be variable, even though in a single population it might appear conserved. On the other hand, position 3 is 100% hydrophobic in all three sets, and position 4 is 87%, 88%, and 84% hydrophobic, respectively, in the three sets. Clearly, these positions have a strong preference for hydrophobic atom types.

Table V summarizes some of the most highly conserved positions in the three populations. Hydrophobic types (3, 5–8) predominated at positions 3, 4, and 10. As shown in Figure 12, position 3 is located at the top of the large

Table IV *Summary of three populations of models used in sequence analysis. Set 1 was described previously (Walters and Hinds, 1994), and sets 2 and 3 were generated for this work. All three sets had population size = 5000 and were run for 50 000 generations.*

	Set 1	Set 2	Set 3
Mean r (fitness)	0.944	0.954	0.946

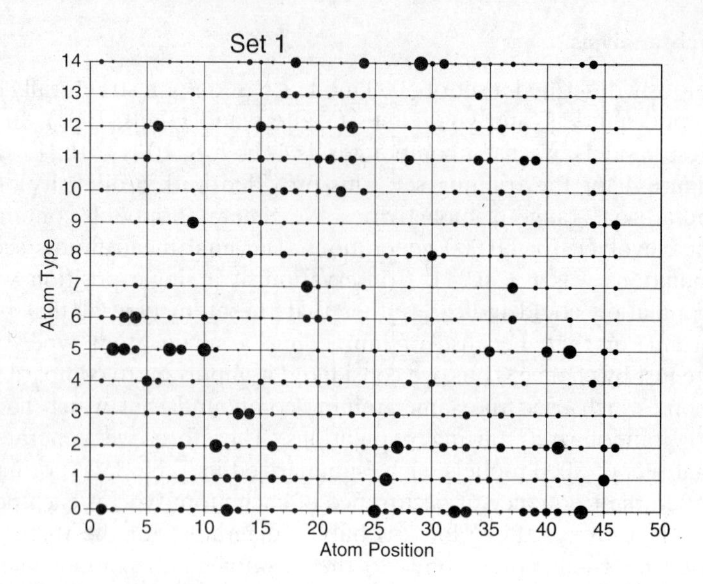

Figure 9 Bubble graph showing the frequency of occurrence of specific atom types as a function of position in the model for set 1. The area of the bubble is proportional to the frequency of occurrence.

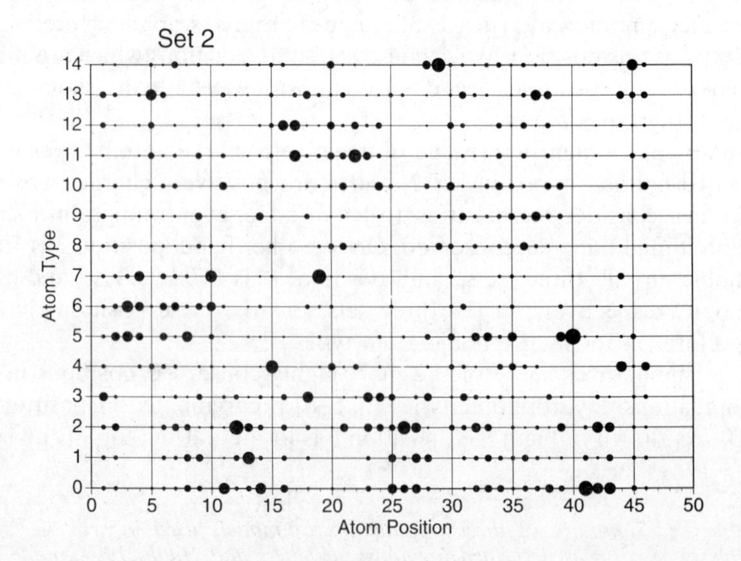

Figure 10 Bubble graph showing the frequency of occurrence of specific atom types as a function of position in the model for set 2. The area of the bubble is proportional to the frequency of occurrence.

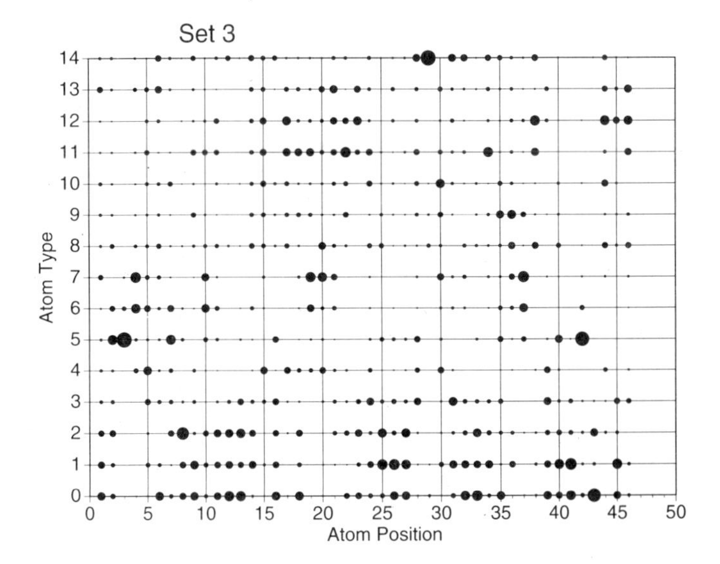

Figure 11 Bubble graph showing the frequency of occurrence of specific atom types as a function of position in the model for set 3. The area of the bubble is proportional to the frequency of occurrence.

hydrophobic region; position 4 is on the front surface near the aryl-ring region; position 10 is on the back surface, close to the methyl ester group of the aspartame-type structures.

Two sites showed a strong preference for a negatively charged atom type. Position 22 is located in close proximity to the NH binding site near the top

Table V *Some of the most highly conserved positions in the three populations of receptor models studied.*

Atom types	Position #	Set 1	Set 2	Set 3	Mean
Hydrophobic	3	100%	100%	100%	100%
	4	88	87	84	86
	10	93	95	62	83
Negative	22	88	88	76	84
	29	79	83	93	85
Positive	26	73	72	50	65
	27	65	66	64	65
Zero	43	70	43	78	64
Small (zero or H)	12	100	99	90	96
	13	94	93	98	95
	25	90	85	81	85
	26	89	95	89	91
	27	97	98	100	99
	41	99	98	100	99

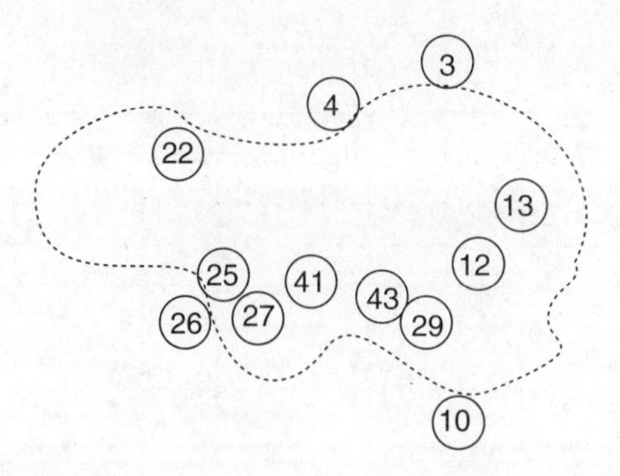

Figure 12 Schematic view of the composite surface over the sweetener structures (dotted line). Compare with Figures 5, 6, and 8 for orientation of the ligands. Locations of receptor model atom positions having conserved atom types are illustrated relative to this surface. Positions 22, 25, 27, 41, 43, 29, 12, and 13 are on the back face of the composite surface.

of the receptor model. This is consistent with our expectation of complementarity for a major ionic/polar interactions. Position 29 is located on the back surface near the methyl ester group of aspartame. In the model, this negatively charged atom may aid in discriminating between aspartame (which has an ester oxygen atom nearby) and alitame (which has a methyl group in this region). The negative charge on the receptor model should make aspartame binding less favorable, consistent with its observed lower potency.

There are also two sites which have a strong tendency to carry a partial positive charge. These are positions 26 and 27, located on either side of the region where the carboxylate group interacts. Again, this is consistent with our expectation that the major ionic functional group should have one or two atoms with which to interact favorably.

The highest occurrence of the zero (null) atom type appears at position 43, located at the back of the large hydrophobic region. With an overall frequency of 64%, we might question whether this is pointing toward an open face or a region where steric bulk is accommodated well. However, many nearby sites show some tendency toward a zero atom type. And when we consider together atom types which are either zero or small (hydrogen, regardless of charge), we find several atoms in this region with a strong tendency toward small/zero atom types. Interestingly, these positions show no tendency to discriminate between neutral and positive atom types. These so-called 'small' atoms form a broad band across the back of the model surface (positions 13, 41, 43, 25, 27).

Overall, the sequence analysis identifies conserved regions of the receptor model which correspond to known key features in the ligands: the carboxylate group, the NH group in the upper part of the model, the large hydrophobic group, and the electronegatively substituted aryl group. In addition, this analysis identifies previously unsuspected features such as the one which enables the models to distinguish aspartame and alitame. Thus, the models apparently are able to capture steric and electronic field information from a structure–activity series and incorporate it into a quantitative model.

CONCLUSION

We have previously shown that a genetic algorithm can be used to generate receptor site models which quantitatively discriminate among compounds with varying biological activities. In this paper, we extend the utility of these models in two ways. First we demonstrate the application of evolved models to the evaluation of alternate binding modes for ligands which can adopt different orientations or conformations. Second, we show that sequence analysis can provide further insight into the probable nature of the receptor and can aid in identifying key features of the ligand which are required for ligand–receptor recognition.

Further studies in progress relate to the use of these models in conjunction with three-dimensional database searching and with *de novo* design software. We anticipate that evolved receptor models will continue to be useful in cases where there is no available structural information about the receptor being targeted.

ACKNOWLEDGMENTS

D.E. W. thanks Procter and Gamble Co. for financial support, G.D. Searle Company and The NutraSweet Company for gifts of computer equipment, and Molecular Simulations, Inc., for providing the Quanta/CHARMm and Cerius^2 software used in these studies.

REFERENCES

Bartlett, P.A., Shea, G.T., Telfer, S.J., and Waterman, S. (1989). CAVEAT: A program to facilitate the structure-derived design of biologically active molecules. In, *Molecular Recognition in Chemical and Biological Problems* (S.M. Roberts, Ed.). Royal Society of Chemistry, London, Vol. 78, pp. 182–196.

Berg, C.P. (1953). Physiology of the D-amino acids. *Physiol. Rev.* **33,** 145–189.

Böhm, H.-J. (1992). The computer program LUDI: A new method for the *de novo* design of enzyme inhibitors. *J. Comput.-Aided Mol. Design* **6**, 61–78.

Brooks, B.R., Bruccoleri, R.E., Olafson, B.D., States, D.J., Swaminathan, S., and Karplus, M. (1983). CHARMm: A program for macromolecular energy, minimization, and dynamics calculations. *J. Comput. Chem.* **4**, 187–217.

Culberson, J.C. and Walters, D.E. (1991). Three-dimensional model for the sweet taste receptor: Development and use. In, *Sweeteners: Discovery, Molecular Design, and Chemoreception* (D.E. Walters, F.T. Orthoefer, and G.E. DuBois, Eds.). American Chemical Society, Washington, DC, Symposium Series Vol. 450, pp. 214–223.

DuBois, G.E., Walters, D.E., Schiffman, S.S., Warwick, Z.S., Booth, B.J., Pecore, S.D., Gibes, K., Carr, B.T., and Brands, L.M. (1991). Concentration-response relationships of sweeteners. A systematic study. In, *Sweeteners: Discovery, Molecular Design, and Chemoreception* (D.E. Walters, F.T. Orthoefer, and G.E. DuBois, Eds.). American Chemical Society, Washington, DC, Symposium Series Vol. 450, pp. 261–276.

Kester, W.R. and Matthews, B.W. (1977). Crystallographic study of the binding of dipeptide inhibitors to thermolysin: Implications for the mechanism of catalysis. *Biochem.* **16**, 2506–2516.

Kuntz, I.D., Blaney, J.M., Oatley, S.J., Langridge, R., and Ferrin, T.E. (1982). A geometric approach to macromolecule-ligand interactions. *J. Mol. Biol.* **161,** 269–288.

Lawrence, M.C. and Davis, P.C. (1992). CLIX: A search algorithm for finding novel ligands capable of binding proteins of known three-dimensional structure. *Proteins: Struct. Funct. Genet.* **12**, 31–41.

Moon, J.B. and Howe, W.J. (1991). Computer design of bioactive molecules: A method for receptor-based *de novo* ligand design. *Proteins: Struct. Funct. Genet.* **11**, 314–328.

Nishibata, Y. and Itai, A. (1991). Automatic creation of drug candidate structures based on receptor structure. Starting point for artificial lead generation. *Tetrahedron* **47**, 8985–8990.

Rotstein, S.H. and Murcko, M.A. (1993). GroupBuild: A fragment-based method for *de novo* drug design. *J. Med. Chem.* **36**, 1700–1710.

Suarez, T., Kornfeld, E.C., and Sheneman, J.M. (1975). Sweetening Agent. U.S. Patent 3,899,592.

Verlinde, C.L.M.J., Rudenko, G., and Hol, W.G.J. (1992). In search of new lead compounds for trypanosomiasis drug design: A protein structure-based linked-fragment approach. *J. Comput.-Aided Mol. Design* **6**, 131–147.

Walters, D.E. and Hinds, R.M. (1994). Genetically evolved receptor models: A computational approach to construction of receptor models. *J. Med. Chem.* **37**, 2527–2536.

9 Genetic Algorithms for Chemical Structure Handling and Molecular Recognition

G. JONES*[†], P. WILLETT[†], and R.C. GLEN[‡§]
[†]Krebs Institute for Biomolecular Research and Department of Information Studies, University of Sheffield, Western Bank, Sheffield S10 2TN, UK
[‡]Department of Physical Sciences, Wellcome Research Laboratories, Beckenham, Kent BR3 3BS, UK

Genetic algorithms (GAs) are novel optimization algorithms which emulate the process of Darwinian evolution to solve complex search problems. Guided by the mechanics of evolution, successive generations of populations of artificial creatures called chromosomes search the fitness landscape of a problem to determine optimum solutions. The application of the GA to problems in computational chemistry is the subject of much investigation.

In the realm of chemical structure handling a particular problem of interest is substructure searching of databases of 3-dimensional (3-D) compounds. A GA was found to be highly efficient and effective in searching the conformational space of small 3-D molecules for pharmacophoric patterns. Problems in molecular recognition are yet more demanding. Not only are powerful search engines, capable of solving multiple minima problems, required, but an appreciation of the process of molecular recognition is required to generate suitable target functions. A GA for docking flexible ligands into partially flexible protein sites has been developed. In order to identify binding modes successfully, an attempt has been made to quantify the ability of common substructures to displace water from the receptor surface and form hydrogen bonds. Excellent results have been obtained on a number of receptor–ligand complexes. Using similar techniques a second GA has been used to superimpose flexible molecules automatically, such that common functional groups are overlaid. The algorithm has proved successful in reproducing experimental

* Author to whom all correspondence should be addressed.
§ Current address: Tripos Inc., St Louis, MO 63144, USA.

In, *Genetic Algorithms in Molecular Modeling* (J. Devillers, Ed.)
Academic Press, London, 1996, pp. 211–242.
ISBN 0-12-213810-4

and structure–activity relationship results and should provide an invaluable tool for pharmacophore elucidation. The design of these algorithms provides insight into the mechanism of molecular recognition.

KEYWORDS: *genetic algorithm; pharmacophore elucidation; protein docking; three-dimensional search.*

INTRODUCTION

Genetic algorithms (GAs) are a class of non-deterministic algorithms that provide good, though not necessarily optimal, solutions to combinatorial optimization problems at a low computational cost (Goldberg, 1989; Davis, 1991). The GA mimics the process of evolution by manipulating data structures called *chromosomes*. A *steady-state-with-no-duplicates* GA (Davis, 1991) was used in the experiments reported here. Starting from an initial, randomly-generated population of chromosomes the GA repeatedly applies two genetic operators, *crossover* and *mutation,* these resulting in chromosomes that replace the least-fit members of the population. Crossover combines chromosomes while mutation introduces random perturbations. Both operators require *parent* chromosomes that are randomly selected from the existing population with a bias towards the fittest, thus introducing an evolutionary pressure into the algorithm. This selection is known as *roulette-wheel-selection*, as the procedure is analogous to spinning a roulette wheel with each member of the population having a slice of the wheel that is proportional to its fitness. This emphasis on the survival of the fittest ensures that, over time, the population should move towards the optimum solution.

Here we describe the application of the GA to a number of problems in chemical structure handling: three-dimensional pharmacophore search (Clark *et al.*, 1994); docking flexible ligands (Jones *et al.*, 1995a); and the superimposition of flexible molecules (Jones *et al.*, 1995b).

3-D CONFORMATIONAL SEARCH

Introduction to conformational search

The searching of databases of flexible 3-D molecules for the presence of *pharmacophores* is of considerable importance in the area of drug design. The pharmacophore is a geometric arrangement of structural features (normally atoms) in a drug molecule that is necessary for biological activity at a receptor site. Thus in order to identify candidate drug structures from a database of 3-D chemical structures, the database should be searched to determine structures that match a query pharmacophore.

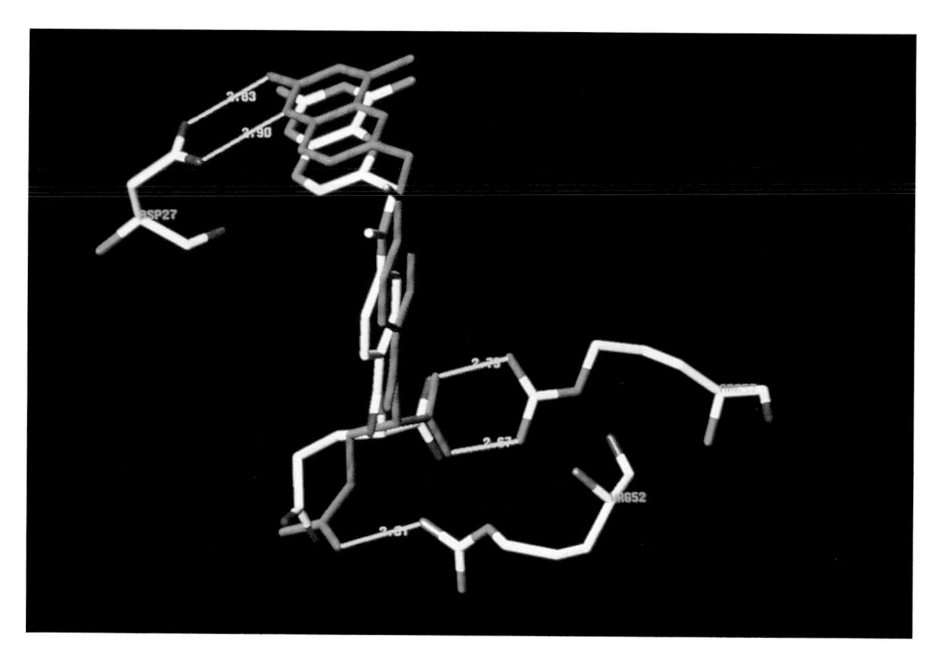

Plate 1. Docking of methotrexate into dihydrofolate reductase.

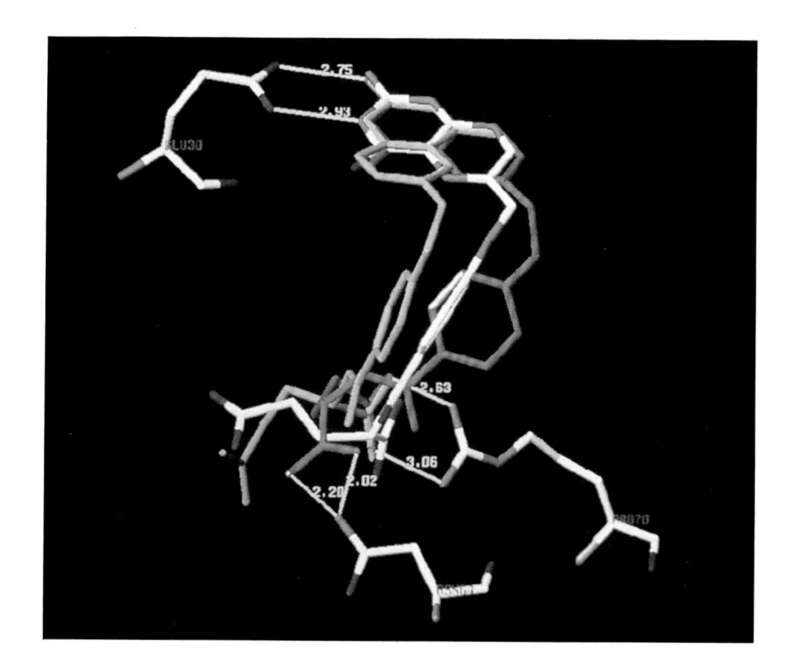

Plate 2. Docking of folate into dihydrofolate reductase.

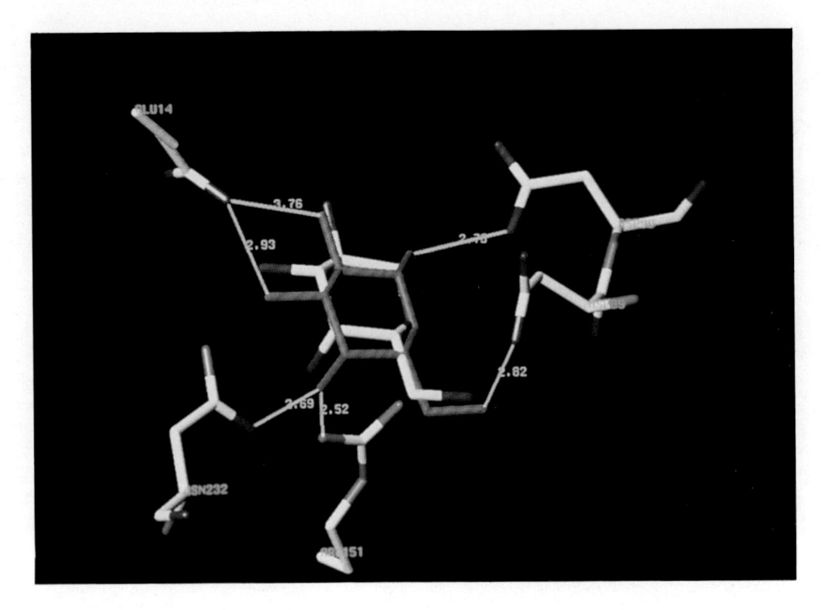

Plate 3. Docking of D-galactose into an L-arabinose binding protein.

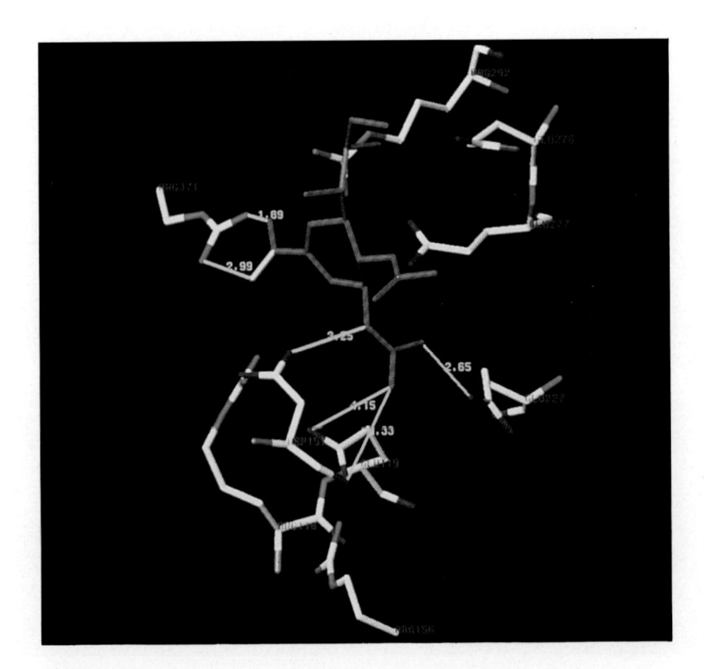

Plate 4. Docking of 4-guanidino-2-deoxy-2,3-di-dehydro-D-N-acetylneuramic acid into influenza sialidase.

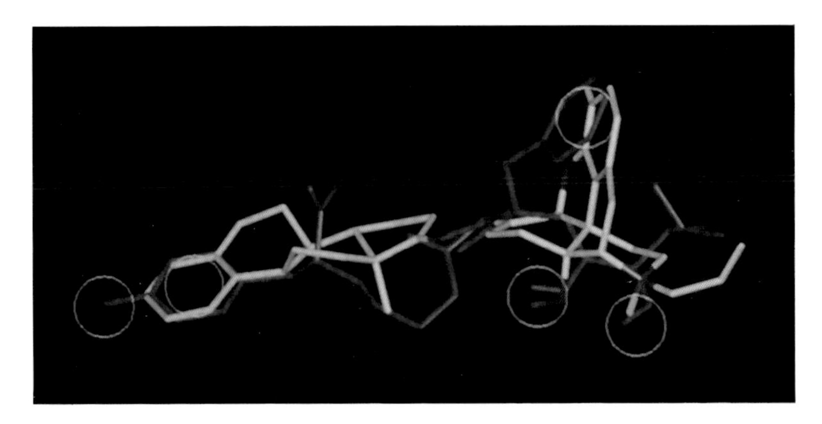

Plate 5. Overlay of leu-enkephalin and hybrid morphine. The hybrid morphine, EH-NAL is shown coloured by atom type and leu-enkephalin is shown coloured purple. The elucidated pharmacophore is indicated by the yellow circles.

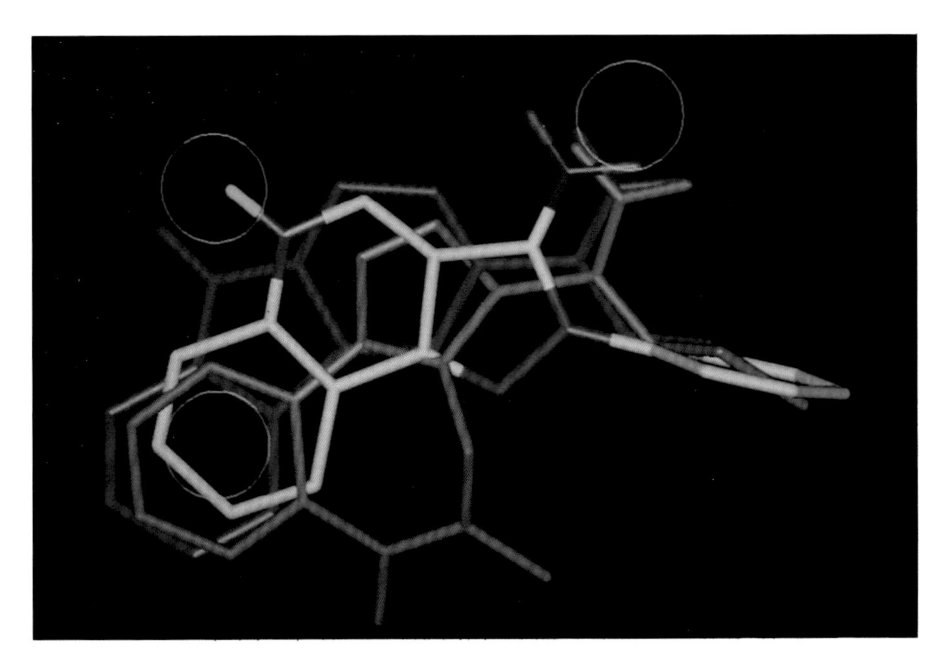

Plate 6. Overlay of benzodiazepine receptors ligands. CGS-8216 is coloured by atom type, Rol15-1788 is in orange and methyl-beta-carboline-3-carboxylate is in purple. Points of interest are indicated by the yellow circles (see the text for discussion).

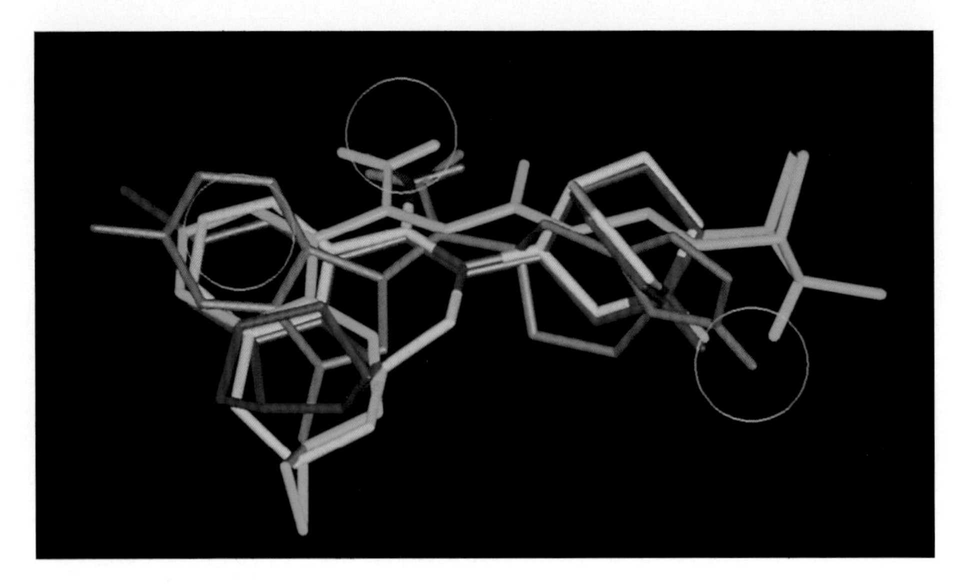

Plate 7. Overlay of 5-HT_3 antagonists. Structure 37 is coloured by atom type, structure 44 is in purple, structure 45 is cyan and structure 47 is orange. The elucidated pharmacophore is indicated by the yellow circles.

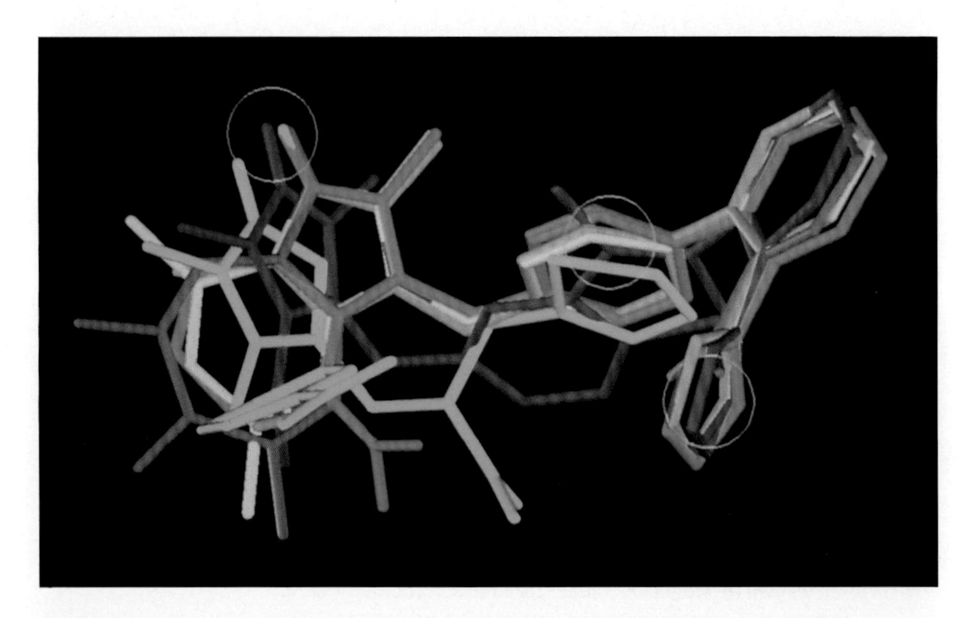

Plate 8. Overlay of six angiotensin II receptor antagonists. The base molecule L-158809, is shown coloured by atom type, GLAXO is coloured magenta. SEARLE is orange, SKB 108566 is cyan, TAK is green and DuP is yellow. Points of interest are indicated by the yellow circles (see the text for discussion).

Flexible searching can be envisaged as a three-stage process (Clark *et al.*, 1992, 1994):

1. A *screen* or *key search* which determines if the screen set of the query is contained in the screen set of the database structure.
2. A *geometric search* that uses the Ullmann subgraph isomorphism algorithm to determine potential hits based on the bounded distance matrices of query and database structure.
3. A *conformational search* that examines potential hits from stage 2 to determine if a structure conformation of low energy exists that matches the query.

The screen search rapidly eliminates all structures that are definitely not possible matches to the query. The geometric search uses the Ullmann relaxation and backtrack algorithm to assign query atom nodes to structure atom nodes, such that the bounded distances of the query are compatible with those of the structure. Not all hits from the geometric search correspond to conformations of the database molecule that match the query and are of low energy, thus necessitating the final, conformational search. There are several techniques available for conformational search (Howard and Kollman, 1988; Leach, 1991), all of which are computationally very expensive: we have used a GA for the conformational search stage of 3-D flexible search, as detailed below.

The genetic algorithm for conformational search

Molecular conformational analysis has been a promising application area for GAs, with several published results (Lucasius *et al.*, 1991, Blommers *et al.*, 1992; Payne and Glen, 1993). In particular, Payne and Glen (1993) have suggested the use of a GA for conformational search.

Conformational-analysis GAs work by representing angles of rotation around bond vectors as binary numbers. A molecular conformation can then be encoded as a binary string chromosome composed of a linear sequence of rotation angles. A steady-state-with-no-duplicates GA was used (Davis, 1991) with a fixed population size of 35. Parent selection was on linear normalized penalty values (Davis, 1991). The individual with the highest penalty score (the least fit individual) was given a fitness score of 10. Other individuals then added 10 to the fitness of the previous individual to obtain their assigned fitness values. The GA terminated after the application of 1000 genetic operators, or if the penalty score associated with the best individual had not decreased by 0.1 in the past 100 operations. The GA is illustrated in Figure 1.

The chromosome representation

Chromosomes were binary strings composed of linear sequences of binary representations of rotatable bonds, where a rotatable bond was defined as

1. A set of reproduction operators (crossover, mutation) are chosen. Each operator is assigned a weight.
2. An initial population is randomly created and the fitnesses of its members determined.
3. An operator is chosen using roulette wheel selection based on operator weights.
4. The parents required by the operator are chosen using roulette wheel selection based on linear normalized penalty values.
5. The operator is applied and child chromosomes produced. Their fitness is evaluated.
6. The children replace the least fit members of the population.
7. If the conditions for termination have been fulfilled stop otherwise goto 3.

Figure 1 Operator based GA.

any single freely-rotatable chain bond that was not the terminating bond in a chain. Each angle was encoded in eight bits as Gray-coded integers (Goldberg, 1989), since Payne and Glen (1993) have shown that Gray coding is of particular advantage for conformational analysis.

For any structure it was possible that a particular rotatable bond lay outside the pharmacophoric pattern. Normally, manipulation of such a bond was not required to generate a conformation that matched the pharmacophore. The GA was also able to make use of a *reduced representation* whereby any rotatable bonds that did not influence the distances in the pharmacophore were discarded.

The penalty function

Given a chromosome, the conformation represented by its bit-string could be obtained by decoding the chromosome. The rotation around each bond involved multiplying the position vectors for all atoms on one side of the rotatable bond by a transformation matrix for rotation about an axis defined by the bond vector (Newman and Sproull, 1981).

Each chromosome was assigned a penalty value. The most fit chromosomes were those with the smallest penalty value. The penalty value comprised two terms: a distance penalty for how close the conformation was to matching the pharmacophoric pattern and an energy penalty. The final penalty value was the weighted sum of both penalty terms. The distance penalty was obtained by comparing the distances between pharmacophore atoms in the conformation with the desired distance bounds specified in the pharmacophore. If the calculated distance lay within the allowed bounds no penalty was given, otherwise a penalty value equal to the modulus of the difference between the distance and the closest bound, in Å, was assigned. An energy penalty was incorporated into the final penalty score in order that the most fit conformations generated by the GA be energetically viable. The energy was expressed as the difference in steric energy between the initial low energy reference conformation and the GA conformation. The steric energy term was a 6–12 term identical to that used in the General Purpose Tripos 5.2 Forcefield (Clark *et al.*, 1989; Tripos Inc., 1994). The contribution of the energy to the penalty of the conformation was scaled by the factor *VDW_ENERGY_WT*, where *VDW_ENERGY_WT* was a user-defined parameter that controlled the weighting of the energy penalty term relative to the pharmacophore match penalty term.

Genetic operators

Two genetic operators were available: mutation and crossover. Roulette wheel selection based on operator weights was used to select a particular operator (each operator had an equal chance of selection) (Davis, 1991). The mutation operator used simple bit mutation with a mutation probability of 0.05 per bit, while the crossover operator performed one-point crossover (Goldberg, 1989).

Comparison of conformational search methods

The GA was used as part of a three-stage flexible search system. An extensive comparison of conformational search methods was undertaken in order to evaluate the effectiveness of the GA as a tool for flexible search. A summary of the results and conclusions of this work are presented here and full details can be found in Clark *et al.* (1994).

Testing the GA on a subset of the POMONA89 database

A subset of the POMONA89 database (as used by Clark *et al.* (1992)) was chosen to evaluate the efficiency and effectiveness of the GA. 1538 2-D structures were selected from this database and converted to 3-D structures using the CONCORD program (Rusinko *et al.*, 1989). Bounded distance matrices were then created for these structures. Using a test set of eight pharmacophore queries the test structures were passed through the screening and geometric stages of the flexible search process. Hits from the bounded distance matrix stage were then passed to the GA. Each query had approximately 100–200 potential hits.

Table I *GA conformational search results with a van der Waals energy term for the full and reduced definition. The table shows percentages of the database that were found to be hits over ten GA runs and also the mean percentage of hits per run. CPU times are for an ESV 3 or 30 workstation.*

	Reduced definition			Full definition		
Query	Total hits	Hits/run	Time/structure	Total hits	Hits/run	Time/structure
1	0.0	0.0	1.84	0.0	0.0	3.17
2	43.5	39.7	1.47	43.5	38.8	2.45
3	0.0	0.0	2.26	0.0	0.0	3.42
4	32.9	28.9	2.14	32.9	27.1	3.12
5	37.4	20.2	2.22	34.2	15.2	3.25
6	37.5	35.3	1.03	37.0	34.6	1.63
7	32.6	13.0	3.00	25.6	7.77	4.27
8	16.6	13.8	0.75	15.3	11.8	1.39
Mean	25.1	20.1	1.84	23.6	16.9	2.84

In order to take account of the possible variation in GA results, the GA was run 10 times on each structure-query pair. An experiment was performed to see if more hits were retrieved over 20 runs. Little difference was found in the total number of hits and it was concluded that 10 runs were sufficient to sample the whole solution space. Experiments were performed with and without a van der Waals energy calculation on both reduced and full definitions of the rotatable bonds (Clarke *et al.*, 1994). All run times given are in CPU seconds on an Evans and Sutherland ESV3 or ESV30 workstation.

Table I shows the results obtained using reduced and full definitions with a van der Waals energy penalty (*VDW_ENERGY_WT* = 0.02). The number of hits is recorded as a percentage of possible hits, i.e., the number of structures passing through the screening and geometric search stages. The total percentage of unique hits obtained over the 10 runs and the average percentage of hits retrieved in each run is also displayed. It is clear from Table I that the reduced definition is much quicker in execution than the full definition; the fact that the reduced definition recovers more hits than the full definition is somewhat surprising. It would appear that the possible overlap of side chains which remain fixed under the reduced definition was not a problem (at least in the case of the database studied here).

A second set of experiments was performed without a van der Waals energy term in the penalty function (*VDW_ENERGY_WT* = 0.0), using both the reduced and full definitions. Table II shows the results of these experiments. Again the reduced definition is superior in terms of the mean number of hits retrieved and speed of execution. In fact, in the absence of van der Waals calculations, there is no advantage in using the full definition over the reduced definition.

Table II *GA conformational search results without a van der Waals energy term for the full and reduced definition. The table shows percentages of the database that were found to be hits over ten GA runs and also the mean percentage of hits per run. CPU times are for an ESV 3 or 30 workstation.*

	Reduced definition			Full definition		
Query	Total hits	Hits/run	Time/structure	Total hits	Hits/run	Time/structure
1	0.0	0.0	0.90	0.0	0.0	1.41
2	44.1	43.1	0.65	44.1	42.0	0.96
3	0.0	0.0	1.00	0.0	0.0	1.47
4	32.9	28.6	0.92	32.9	28.9	1.24
5	41.1	24.2	0.98	40.5	20.4	1.43
6	37.0	36.2	0.50	37.0	36.0	0.72
7	32.6	16.0	1.24	33.7	12.3	1.75
8	15.3	14.1	0.41	15.3	13.1	0.70
Mean	25.4	20.3	0.83	25.4	19.1	1.21

Comparing these results with the previous set it can be seen that a greater number of hits are retrieved when the energy term is omitted from the penalty function. Presumably, the extra hits contained bad contacts. A decrease in execution time is also observed. This is primarily due to the computational load in determining 6–12 energies.

This GA was compared with the following techniques: distance geometry (Blaney *et al.*, 1990), systematic search (Tripos Inc., 1994), and directed tweak (DT) (Hurst, 1994; Moock *et al.*, 1994). The DT algorithm uses a pseudo-energy function similar to the pharmacophore match penalty used by the GA. The derivatives of this energy with respect to rotatable bonds are determined and a standard minimization technique can then be used to fit the molecule to the pharmacophore. Of these techniques the GA and DT methods were found to be superior both in terms of efficiency and effectiveness and these algorithms were selected for further comparison: full details of these experiments on the POMONA89 database are presented by Clark *et al.* (1994).

Experiments on a larger database

The GA and DT algorithms for the execution of the conformational search stage of a pharmacophore search were then tested on a much larger database of 10000 structures extracted from the Chemical Abstracts Service database. Following the geometric and screening stages there were approximately 750 structures for each query pharmacophore that were required to undergo the conformational search.

Table III summarizes the results of these experiments. The GA was run using the reduced definition with no van der Waals energy calculation. It

Table III *GA and DT search results on the larger database. The table shows percentages of the database that were found to be hits over ten GA runs, the mean percentage of hits per GA run and the percentage of hits per DT run. CPU times are for an ESV 3 or 30 workstation.*

	GA			DT	
Query	Total hits	Hits/run	Time/structure	Hits	Time/structure
1	0.0	0.0	0.77	0.0	0.39
2	28.7	27.1	0.67	29.2	0.56
3	0.0	0.0	0.49	0.0	0.39
4	45.4	39.4	0.97	42.8	0.97
5	18.0	10.3	0.77	17.8	0.71
6	43.6	43.0	0.40	46.8	0.39
7	19.3	9.9	0.83	19.3	0.82
8	0.0	0.0	0.32	0.0	0.35
Mean	19.4	16.2	0.65	19.5	0.57

Table IV *Analysis of hitlists from GA and DT searches on the larger database.*

	Number of hits found by		
Query	GA and DT	GA only	DT only
2	174	5	8
4	78	10	5
6	119	11	10
8	82	3	3

was observed that, while the GA retrieved approximately the same number of hits over 10 runs as did the DT program, the DT algorithm was consistently more effective than the GA algorithm over a single run. Furthermore, the DT algorithm is faster than the GA. However, the difference is only 13%. Although Table III also shows that the total number of structures retrieved by the two methods may be nearly identical, Table IV shows that the two hitlists are not identical. This suggests that the two algorithms may be used as complementary tools.

Discussion of search results

This section documents the use of a GA for the conformational search stage of flexible searching. Although the DT algorithm has been shown to be slightly more efficient and effective than the GA, this has been shown to be a particularly successful application area. With proper investment and improvements to the GA it should be possible to increase GA efficiency massively. Secondly, a proper study of GA parameterization should improve the effectiveness of the algorithm. A major advantage of the GA is that it would be simple to expand the fitness function to include additional pharmacophore features, such as torsion angles, vectorial constraints and included and excluded volumes. It is not yet clear whether or not it will be possible to formulate the pseudo-energy derivatives that DT requires for these features. The GA has also been shown to be a successful search strategy for the conformational analysis of large molecules (Lucasius *et al.*, 1991; Blommers *et al.*, 1992; Payne and Glen, 1993), where the DT algorithm is as yet unproved.

In summary, the GA is still ripe for development in this area and it is hoped that future flexible search systems can take advantage of this technology.

FLEXIBLE LIGAND DOCKING

Introduction to docking

Biological receptors exhibit highly selective recognition of small organic molecules. Evolution has equipped them with a complex three-dimensional 'lock' into which only specific 'keys' will fit. This has been exploited by medicinal chemists in the design of molecules to augment or retard key biochemical pathways and so exhibit a clinical effect. X-ray crystallography has revealed the structure of a significant number of these receptors or 'active sites', in the case of enzymes. It would be advantageous in attempting the computer-aided design of therapeutic molecules to be able to predict and explain the binding mode of novel chemical entities (the 'docking' problem) when the active site geometry is known.

The process of automated ligand docking has been the subject of a major effort in the field of computer-aided molecular design (Blaney and Dixon, 1993). Early efforts, including the original DOCK program (Kuntz *et al.*, 1982), only consider orientational degrees of freedom, treating both ligand and receptor as rigid (DesJarlais *et al.*, 1986; Lawrence and Davis, 1992; Shoichet *et al.*, 1993).

More recently a number of algorithms have been reported that take into account the conformational degrees of freedom of the ligand. These include docking fragments of the ligand and recombining the molecule (DesJarlais *et al.*, 1986), systematic search (Leach and Kuntz, 1992) and clique detection (Smellie *et al.*, 1991). These algorithms are all inherently combinatorial. Another approach is simulated annealing (Goodsell and Olsen, 1990). Recently the use of GAs to perform conformational search of the ligand has been reported. Oshiro *et al.* (1995) have used a GA to incorporate ligand flexibility into the DOCK program and Judson *et al.* (1994) used a GA to minimize the energy of a flexible ligand. Leach (1994) has developed an algorithm that takes into account conformational flexibility of protein side-chains as well as ligand flexibility. His results highlight the need for appropriate methods to estimate the strength of ligand binding.

Inspection of proteins with associated *high affinity* ligands reveals that they appear to conform closely to the shape of the receptor cavity and to interact at a number of key hydrogen-bonding sites. The docking process probably involves the exploration of an ensemble of binding modes with the most energetically stable being observed. These exhibit low temperature factors and little crystallographic disorder. We believe that the key requirements for high affinity may simply be the complementarity of the three-dimensional shape of the ligand and protein and the ability of the ligand to displace water and form hydrogen-bonds at key hydrogen-bonding sites when associated with the active site. Thus sufficiently accurate simulation of these interactions may be enough to explain and to predict the binding mode of the majority of high affinity ligands.

The docking process involves an exploration of the allowed conformational space of the ligand and protein as they merge together, resulting in a mutually complementary shape. Thus, both the ligand and the protein exhibit flexibility, the so-called rubber lock and key. The GA described here utilizes a novel representation of the docking process. Each chromosome encodes not only an internal conformation of the ligand and protein active site but also includes a mapping from hydrogen-bonding sites in the ligand to hydrogen-bonding sites in the protein. On decoding a chromosome, a least-squares fitting process is employed to position the ligand within the active site of the protein in such a way that as many of the hydrogen-bonds suggested by the mapping are formed. To enable the GA to reliably estimate binding affinities we have attempted to quantify the ability of common substructures of drug molecules to displace water from the receptor (protein) surface. This, in conjunction with a steric term, allows the GA to find the most energetically favourable combination of interactions.

Preparation of input structures

Known crystal structures of bound ligand complexes were obtained from the Brookhaven Protein Data Bank (Bernstein *et al.*, 1977). Water molecules and ions were removed (including ordered water) and hydrogen atoms were added at appropriate geometry. Groups within the protein were ionized if this was appropriate at physiological pH. The structures were then partially optimized (to relieve any bad steric contacts introduced by hydrogen addition) for ten cycles of molecular mechanics (using the SYBYL MAXIMIN2 energy minimizer (Tripos Inc., 1994)). The ligand was extracted from the structure and then fully optimized using molecular mechanics. The remaining structure then comprised the protein active site.

A flood-fill algorithm determined which protein atoms in the active site were solvent accessible (Ho and Marshall, 1990). Protein donor and acceptor atoms that were able to interact with the ligand were determined by searching these solvent accessible atoms. Lone pairs were added to acceptors at a distance of 1.0 Å. Any single bond that connected a donor or acceptor to the protein was selected as rotatable provided that (except for hydrogens and lone pairs) the donor or acceptor terminated an acyclic chain.

All atoms within the ligand were searched to determine which ligand atoms were hydrogen donors and which were acceptors, and lone pairs were added. All acyclic single non-terminal bonds were marked as rotatable. Prior to docking a random translation was applied to the ligand and random rotations were applied to all rotatable bonds.

The genetic algorithm for flexible docking

As in conformational search a steady-state-with-no-duplicates GA (Davis, 1991) was used. Each docking run consisted of the application of 100000 operators. However, if an improvement in the best fitness score had not been observed in the past 6500 operations the GA terminated. As before, parents were selected using the technique of roulette-wheel selection with rank-based fitness values (Davis, 1991). The rank-based fitness scores were scaled in such a way that the most-fit member of the population was 1.1 times more likely to be chosen as a parent than an individual with average fitness (that is, a *selection pressure* of 1.1 was chosen).

The chromosome representation

Each chromosome in the GA comprised four strings. Conformational information was encoded by two binary-strings: one for the protein and one for the ligand (using the same encoding as described for conformational search). An integer-string encoded a mapping from lone pairs in the ligand to hydrogens in the protein, such that if V was the integer value at position P on the string, then the chromosome mapped the Pth lone pair in the active site to the Vth donor hydrogen in the ligand. In a similar manner, a second integer-string encoded a mapping from hydrogens in the ligand to lone pairs in the protein. On decoding a chromosome, the GA used the binary strings to produce conformations of the ligand and protein active site and the integer strings as input to a least-squares routine that tried to form as many of the hydrogen bonds as possible.

The fitness function

The GA used the following model to form bonds. For every hydrogen, a virtual point was created at 1.45 Å from the donor in the same direction as the hydrogen. Similarly, for every lone pair a virtual point was created at 1.45 Å from the acceptor. When the GA decoded a chromosome it obtained a set of possible hydrogen bonds between pairs of hydrogens and lone pairs. The GA attempted to form bonds by performing a least-squares fit between the virtual points associated with each hydrogen and lone pair.

A Procrustes rotation (Digby and Kempton, 1987), with a correction to remove inversion, was then applied. This generated a geometric transformation that minimized the distance between each ligand virtual point and the protein virtual point to which it was mapped within the chromosome. After the application of this rotation, a second Procrustes rotation was applied that minimized the distance between those pairs of virtual points that were less than 5 Å apart. Following the application of the least-squares fitting procedure the ligand was docked within the protein and a fitness score was assigned to the bound ligand.

Each hydrogen-bonding pair contributes to the overall energy of binding. Initially the donor (d) and acceptor (a) are in solution but on coming together (da), water (w) is stripped off. Therefore, to simulate this, the interaction energy is composed of four terms:

$$\text{Epair} = (\text{Eda} + \text{Eww}) - (\text{Edw} + \text{Eaw}) \quad (1)$$

Six hydrogen-bond donor and twelve hydrogen-bond acceptor atom types were defined (based on SYBYL molecular mechanics atom types (Clark *et al.*, 1989; Tripos Inc., 1994)) which resulted in seventy-two possible pairwise interactions. These were embedded in model fragments, e.g. $CH_3C{=}O$ ... $HN^+(CH_3)_3$. The calculation of the interaction energy was performed by a number of different approaches. These included gas-phase semi-empirical (MOPAC, AM1, PM3) (Stewart, 1990), solvent based (AMSOL, SM3) (Cramer *et al.*, 1994), VAMP (T. Clark, 1994, personal communication) with a self consistent reaction field (Rinaldi *et al.*, 1993), molecular mechanics with MOPAC (AM1 and PM3) Coulson and Mulliken atom charges (Stewart, 1990). Full details of these experiments may be found in Jones *et al.* (1995a), and Jones (1995). Currently, gas-phase molecular mechanics (Clark *et al.*, 1989; Tripos Inc., 1994) with PM3 Mulliken charges and a dielectric constant of 1.0 in the receptor cavity gives results consistent with the observed binding modes. This may be surprising, however recent nuclear magnetic resonance spectroscopic studies imply that the protein–ligand hydrogen-bond interactions in the receptor cavity may have energetics more consistent with gas-phase values (Beeson *et al.*, 1993) than has previously been appreciated.

Each hydrogen bond in the docked conformation was assigned a weight (between 0 and 1) that was used to scale the full bonding energy. The weighting scheme was chosen so as to favour those hydrogen bonds that were linear and whose virtual points overlapped. Poor bond angles or long interaction distances gave rise to weights of zero. Let the total bonding energy for the ligand be *H_Bond_Energy*.

Following the placement of the ligand into the active site of the protein, a Lennard-Jones 6–12 potential (Hirschfelder *et al.*, 1964) was used to determine the steric energy of interaction between the ligand and protein, and the internal energy of the ligand. The energy of interaction between atoms forming hydrogen bonds was reduced. Let the interaction energy between ligand and protein be *Complex_VDW_Energy* and let *Internal_VDW_Energy* be the internal energy of the ligand.

The final fitness score was a weighted sum of energies

$$H\_Bond\_Energy - (5.0e^{-4} \times Complex\_VDW\_Energy) - (0.5 \times Internal\_VDW\_Energy) \quad (2)$$

The weights were determined by empirical adjustment to best reproduce known bound ligand structures. These weights allowed the GA to form strong hydrogen bond motifs without generating high energy solutions.

Genetic operators

The GA made use of two genetic operators: mutation and crossover (Davis, 1991). Operators were chosen randomly using a weighting scheme that meant that, on average, four mutations were applied for every crossover, so that the GA emphasized more the exploration of the search space than the optimization of the solutions that were obtained. Each operator was randomly applied to one of the four chromosome strings.

The crossover operator performed two-point crossover on the integer-strings and one-point crossover on the binary-strings. The mutation operator performed binary-string mutation on binary-strings and integer-string mutation on integer-strings. Each bit in the binary-string had a probability of mutation of 0.05. The integer-string mutation performed a single mutation on an integer value. A position was randomly chosen on the integer-string and the value at that position was mutated to a (different) new value.

Applying the docking GA to known complexes

The performance of the GA docking procedure was determined by comparing the results obtained from its use with crystallographic data for several diverse ligand–protein complexes, four of which are described here. None of the proteins contained metal ions in the active site, although it is intended to extend the algorithm to these cases at a later date. To ensure that most of the high affinity binding modes were explored the GA was run fifty times for each of the examples examined (probably less runs are required as the top 20 results in most cases were very similar). The result with the best fitness score was used as the solution (although other results, with poorer fitness scores, might have had better fits to the crystal structures). On average, each run of the GA took approximately 5–8 minutes of CPU time on a Silicon Graphics R4000 Iris computer.

The GA docking procedure was applied to the problems illustrated in Plates 1–4. In these plates, ligand crystal structures (where available) and important protein close contacts have been coloured by atom type. For reasons of clarity, not all close contacts with the protein are displayed. The GA solutions are shown in red.

Methotrexate and dihydrofolate reductase

Dihydrofolate reductase (DHFR) is an enzyme of vital importance to DNA synthesis and preferential inhibition of the enzyme has been exploited to produce a range of potent anti-bacterial drugs and anti-cancer drugs. The geometry of binding of the anti-cancer agent methotrexate with *Escherichia coli* DHFR has been elucidated by X-ray crystallography (Bolin *et al.*, 1982).

Bound methotrexate is protonated at N1 and the binding mode of the pteridine ring results in a number of close contacts: pteridine N1 and 2-amino. . .Asp27 carboxylate, 2-amino group. . .water, 4-amino group. . .Ile5 and Ile94 carbonyls, benzoylcarbonyl. . .Arg52 guanidinium (and water), L-glutamate α-carboxylate. . .Arg57, L-glutamate γ-carboxylate. . .Arg52 and various waters. Not all of these are shown for reasons of clarity.

The GA solution (red) is shown overlaid upon the experimental result (coloured by atom type) in Plate 1. Key hydrogen-bond interactions are shown in yellow. The similarity between the predicted and experimental results is excellent, with a root mean square (RMS) deviation for the atoms of 0.99 Å. The pteridine ring is correctly oriented with the hydrogen-bond (pteridine N1 and 2-amino ... Asp27) the main feature. The benzoylcarbonyl is also correctly oriented, interacting with Arg52. The L-glutamate γ-carboxylate is rotated slightly from the X-ray solution (where it is solvated), but the GA twists the side chain to interact with Arg52. The L-glutamate α-carboxylate is correctly positioned.

Folate and dihydrofolate reductase

Vertebrate DHFR catalyzes the NADPH linked reduction of folate to 7,8-dihydrofolate and on to 5,6,7,8-tetrahydrofolate. The structure of the human recombinant enzyme has been determined by X-ray diffraction (Davis *et al.*, 1990). At physiological pH, folate appears to bind as the neutral species and displays a number of close contacts in the active site: pteridine ring 2-amino. . .Glu30, pteridine ring 2-amino. . .water. . . Thr136, pteridine ring N3. . .Glu30, the pteridine ring carbonyl O4. . .water. . .Glu30, the carbonyl of the ρ-aminobenzoic acid. . .Asn64, the folate α-carboxylate of the glutamate. . .Arg70, the folate γ-carboxylate of the glutamate. . .waters (not all shown for reasons of clarity).

In Plate 2 the GA solution is shown in red with the X-ray data coloured by atom type. The GA correctly predicts the binding mode of the pteridine ring with the pteridine ring distances 2-amino. . .Glu30 and N3. . .Glu30 similar to the X-ray result, with an RMS deviation for the atoms in the pteridine ring of 0.63 Å. The benzene ring is rotated relative to the crystal structure due the interaction of the folate glutamate γ-carboxylate with Arg70. Solutions having the alternative interaction (the folate α-carboxylate of the glutamate. . .Arg70) were found by the GA; however, the illustrated solution had the best GA score with an RMS deviation for the atoms in the complete structure of 2.48 Å. The X-ray result has one carboxylic acid in solvent while the GA solution has both carboxylic acids interacting with the protein.

Folate binding to DHFR has been studied by nuclear magnetic resonance spectroscopy (NMR). These studies have suggested that three forms exist in equilibrium (Cheung *et al.*, 1993). The predominant form at low pH (form-1) (enol, N1 protonated) appears to bind in a similar manner to methotrexate while the neutral form (form-2) (keto form with N1 unprotonated) binds with the pteridine ring rotated 180°. Also, a third protonated form (form-3) (enol N1 protonated) coexists with form-2 (the protonation state of folate in the protein in forms-2 and -3 is relatively insensitive to buffer pH) and this also binds in a similar mode to methotrexate. The GA solution for form-2 has previously been described. Form-3 is predicted by the GA to bind with the pteridine ring rotated by 180°; this prediction is shown in yellow in Plate 2 and is consistent with the NMR results.

Galactose and L-arabinose binding protein

The structure of the carbohydrate-binding protein, L-arabinose binding protein mutant (Met[108]Leu) has been determined by X-ray crystallography (Vermersch *et al.*, 1991). This is a mutant of a periplasmic transport receptor of *Escherichia coli* with high affinity for the sugar, L-arabinose. The ligand (D-galactose) position was determined from difference Fourier analysis. There are many close contacts with the protein: O1. . .Asp90, O2. . .water, O3. . . Glu14, O4. . .Asn232, Lys10. . .O2, Asn205. . .O3, Asn232. . .O3, Arg151. . . O4, Arg151. . .O5, water. . .O5 (not all of these are shown for reasons of clarity).

The GA solution (Plate 3) is almost identical to the experimental result, with the RMS deviation of the atoms being as low as 0.67 Å with all of the experimentally-observed close contacts being correctly reproduced. Again, the GA predicts the correct binding mode despite the absence of bound water in the model. The alternative epimer (L-galactose, results not shown in diagram for clarity) was also docked by the GA. The solution that was obtained was again very close to the experimental result, with an RMS deviation of 0.66 Å.

4-Guanidino-2-deoxy-2,3-di-dehydro-D-N-acetylneuramic acid and influenza sialidase

Potent inhibitors of influenza virus replication have been reported based upon inhibition of influenza sialidase. The binding mode of the inhibitor 4-guanidino-2-deoxy-2,3-di-dehydro-D-*N*-acetylneuramic acid (4g-neu5Ac2en) to sialidase was reported based upon X-ray crystallography analysis of the protein with bound ligand (von Itzstein *et al.*, 1993). The ligand binds with a number of reported close contacts: terminal guanidinyl nitrogens. . . . Glu227 (2.7 Å), terminal guanidinyl nitrogens. . . .Glu119 (3.6 and 4.2 Å), carboxylate. . .Arg371.

The GA solution (red) is overlaid on the crystal structure of sialidase (coloured by atom type) in Plate 4. This solution has been compared with the reported intermolecular distances, difference map and diagrams (von Itzstein *et al.*, 1993). It appears from these that the GA correctly predicts the binding mode, closely mirroring the experimental result. The GA solution is shown in red with close intermolecular contacts in yellow.

Discussion of docking results

The results demonstrate the effectiveness of the algorithm in reproducing crystallographic studies of the bound complexes, even in the case of highly flexible ligands with variable protonation states. The accuracy with which the binding modes of ligands are predicted suggests that the representation and fitness function used here provides a useful description of the most important features involved in binding, at least for these high-affinity ligands. If this is indeed so, our results would suggest that the mechanism of molecular recognition requires the ligand not only to exhibit shape complementarity with the receptor cavity (as has been assumed by previous docking programs) but also to displace water at key points so forming hydrogen-bonds in preference to water. The absence of bound water from the GA model (all water molecules are removed from the protein before docking) seems not to affect the ability of the GA to correctly predict the binding mode. Perhaps the bound water (in these examples) is a consequence of binding rather than a major contributory factor to the stabilization of the bound ligand.

Several potential improvements to the algorithm are under investigation (such as ring conformational search, the inclusion of metal ions and accounting for possible variations in dielectric and solvent competition across the active site) and the results of these developments will be reported elsewhere.

FLEXIBLE MOLECULAR OVERLAY AND PHARMACOPHORE ELUCIDATION

Introduction to pharmacophore elucidation

It is often the case, in a drug discovery project, that there is little or no information about the 3-D structure of the receptor. In such cases, methods such as the active analogue approach (Marshall *et al.*, 1979) must be invoked that seek to rationalize the ligand–receptor interaction on the basis of structural characteristics of those active molecules that have been identified thus far. These approaches involve aligning the active molecules to identify common structural features with the aim of elucidating the pharmacophore that is responsible for the observed activity. There are many ways in which the alignment can be carried out, as reviewed recently by Klebe (1993). However, these procedures typically have one or more of the following limitations: they require intervention to specify at least some points of commonalty, thus biasing the resulting overlays; they encompass conformational flexibility by considering some number of low-energy conformations, rather than considering the full conformational space of the molecules that are to be overlaid; or they are extremely time-consuming in operation. Here we describe the use of a GA for the overlay of sets of molecules that seeks to overcome these limitations.

The first report of the use of a GA for the superimposition of flexible molecules, and the starting point for the work reported here, was a paper by Payne and Glen (1993). However, that GA was controlled by fitting to constraints or by minimizing the distance between known pharmacophore points in the two molecules that were being compared. The GA presented here encodes not only conformational information but also intermolecular mappings between important structural features (such as lone pairs, hydrogen-bond donor protons and aromatic rings) that may be required for activity; in addition, the algorithm does not require any prior knowledge regarding either the constraints or the nature of the pharmacophoric pattern. Indeed, one of the main applications of the procedure we shall describe is the identification of such patterns, since these can then be used to search a database of 3-D structures, as described earlier in this paper.

The GA utilized a novel representation in attempting to tackle this superimposition problem. Each chromosome contained binary strings that encoded angles of rotation about the rotatable bonds in all of the molecules, and integer strings that mapped hydrogen-bond donor protons, acceptor lone pairs and ring centres in one molecule to corresponding sites in each of the other molecules. A least-squares fitting process was used to overlay molecules in such a way that as many as possible of the structural equivalencies suggested by the mapping were formed. The fitness of a decoded chromosome was then a combination of the number and similarity of overlaid features, the volume integral of the overlay and the van der Waals energy of the molecular conformations. It will be realised from this brief description that the GA exploits methods that were developed in the flexible docking algorithm described previously.

Preparation of input structures

In the absence of refined crystallographic co-ordinates, an input structure was normally created using the SYBYL BUILD module (Tripos Inc., 1994) and hydrogen atoms were added to all atoms with free valences. Groups within the input structure were ionized if this was appropriate at physiological pH (e.g., alkyl amine, carboxylic acid) and specific atoms were protonated if this was indicated by pK_a or NMR data. A low energy conformation was generated using molecular mechanics (the SYBYL MAXIMIN2 energy minimizer with Gasteiger-Marsilli charges (Tripos Inc., 1994)). Following this procedure, each input structure was written out from SYBYL as a MOL2 file. All of the rings in each structure were identified using a smallest-set-of-smallest-rings (SSSR) algorithm (Zamora, 1976), and each structure was then analyzed to determine the *features* that were present, where a feature was a hydrogen-bond donor proton, a lone pair or a ring. Given a set of active molecules, the GA selects one of them as a *base molecule*, to which the other molecules are fitted. The base molecule was defined to be that with the smallest number of features.

As in the docking GA, hydrogen-bond donor and acceptor atoms were identified in each of the input structures using the SYBYL atom-type characterization. Donor hydrogens were identified, and lone pairs added to acceptors at a distance of 1.0 Å from the acceptor. All freely rotatable single bonds that were not connected to terminating atoms were selected as rotatable. Prior to superimposition, a random translation was applied to each input structure (including the base molecule) and random rotations were applied to all rotatable bonds.

The genetic algorithm for flexible molecular overlay

As described previously a steady-state-with-no-duplicates GA (Davis, 1991) was used. Each overlay run consisted of the application of 50000 operators. However, if an improvement in the best fitness score had not been observed in the past 6500 operations the GA terminated. As before, parents were selected using the technique of roulette-wheel selection with rank-based fitness values (Davis, 1991). As in the docking experiments, the rank-based fitness scores were scaled in such a way that the most-fit member of the population was 1.1 times more likely to be chosen as a parent than an individual with average fitness (i.e. a *selection pressure* of 1.1 was chosen).

To exploit the fact that GAs are very well suited to implementation in a distributed environment an *island* model was implemented. This involves separate sub-populations and the migration of individual chromosomes between the sub-populations (Starkweather *et al.*, 1990; Tanese, 1989). The island model has attracted growing interest, not only because it represents a practical and efficient method of parallelizing the GA, and thus reducing the observed run-time, but also because it has been observed that the resulting distributed GA with several small sub-populations often outperforms a GA with a single large population equal in size to the sum of the distributed sub-populations.

A simple island model was implemented using a serial algorithm. Five sub-populations, each comprising 100 individuals, were created by the GA and arranged in a ring, such that each island had two neighbours. Genetic operators were then applied to each sub-population in turn, with parents being selected from that sub-population and children being inserted into that same sub-population. As 50000 operations were performed by the GA in the course of a run, 10000 operations would be applied to each sub-population.

Initial experiments showed that there was no perceptible difference in performance between using five sub-populations of size 100 and a single population of size 500, though the island model showed slightly faster run-times. However, if a parallel version were implemented on five processors (either on a multiprocessor machine or a workstation LAN) up to five times speed-up could be achieved.

The chromosome representation

A chromosome of $2N$–1 strings was used to encode a molecular overlay involving N molecules. This contained N binary strings, each encoding conformational information for one structure (as described above for conformational search) and N–1 integer strings, each encoding a mapping between features in a molecule (other than the base molecule) to features, of the same type, in the base molecule. For example, a lone pair in one molecule could be mapped to a particular lone pair in the base molecule, under the implicit assumption that the lone pairs in both molecules interacted with the same donor-hydrogen in the receptor.

On decoding a chromosome, the fitness function of the GA would attempt to satisfy the specified mapping by using a least-squares fitting technique. In order to make the mapping chemically sensible, the mapping was one-to-one between similar features. For example, it would not make sense if a lone pair was mapped to two different lone pairs in the base molecule, or if a lone pair was mapped onto a hydrogen-bond donor proton. Each integer string was of length L, where L was the number of features in the base molecule. Because the mappings were one-to-one the integer string was constrained to have no duplicate values. Each feature in every molecule was assigned a unique label. The labels of the base molecule features were then arranged in a list of length L. If V was the integer value at position P on the integer string and if B was the Pth element in the list of base molecule labels then the feature with label V was mapped onto the base molecule feature with label B. By associating features in each molecule to base-molecule features, these mappings suggested possible pharmacophoric points. On decoding the chromosome, the GA used a least-squares routine to attempt to form as many points as possible. The fitness function assigned scores based on the quality of the resulting pharmacophore.

The fitness function

A set of molecular conformations was generated by decoding each of the N binary strings in the chromosome to generate a molecular conformation, as described for a conformational search. These conformations were then passed to the least-squares fitting procedure.

A *virtual point*, representing a donor or acceptor atom in the receptor with which the molecules interact, was created for each hydrogen-bond, donor proton and acceptor lone-pair in a molecule at a distance of 2.9 Å from the donor or acceptor, in the direction of the hydrogen or lone pair. Virtual points were also created at the centre of each ring. Consider the superimposition of one molecule, A, onto the base molecule. Let N be the number of base molecule features, so that decoding a chromosome gave rise to N pairs containing a virtual point in the base molecule and a virtual point in A. As in the docking GA, a Procrustes rotation (Digby and Kempton, 1987), with a correction to remove inversion, was used to yield a geometric transformation that, when applied to all the virtual points from molecule A in the N pairs, minimized the least-squared distances between all of the virtual points from molecule A and the corresponding base-molecule virtual points. As not all possible features in the base molecule will necessarily be included in a pharmacophore, a second least-squares fit was applied to minimize the distance between those pairs of points that were less than 3 Å apart.

The internal steric energy for each molecule was calculated using a Lennard-Jones 6–12 potential (Hirschfelder *et al.*, 1964). Let *vdw_energy* be the mean 6–12 energy per molecule. Pairwise common volumes were determined between the base molecule and each of the other molecules. Because an exact determination of common molecular volume would be extremely time-consuming, the calculation was approximated by treating atoms as spheres and summing the overlay between spheres in the two different molecules. The mean volume integral per molecule with the base molecule was determined. Let this term be *volume_integral*.

A similarity score was determined for the overlaid molecules. This score, *similarity_score*, was the sum of three terms: the first term was a score for the degree of similarity in position, orientation and type between hydrogen-bond donors in the base molecule and hydrogen-bond donors in the other molecules; the second term was a score derived from comparing hydrogen-bond acceptors; and the third term was a score that resulted from comparing the position and orientation of aromatic rings. Thus

$$\textit{similarity\_score} = \textit{donor\_score} + \textit{acceptor\_score} + \textit{ring\_score} \quad (3)$$

In order to assign similarity scores the GA required the use of a function that determined how similar two hydrogen-bond donor or acceptor types were. Let *type_sim*[a, b] be a weight between 0 and 1 that was a measure of the similarity between hydrogen-bond types a and b, where a and b were either both donor types or a and b were both acceptor types. This index was determined using the pairwise bonding energies that were calculated for the docking program (Jones, 1995; Jones *et al.*, 1995b).

Each virtual point corresponding to a hydrogen-bond donor proton in the base molecule was used to define a *hydrogen-bonding centre*, with the potential to interact with acceptors within a receptor molecule. A virtual point from every other molecule, corresponding to the hydrogen-bond donor proton that was geometrically closest to the base-molecule virtual point, was added to each of these hydrogen bonding centres. A score, *vec_wt* × *sim_wt*, was then assigned to the hydrogen bonding centre, where *vec_wt* was a measure of the closeness of virtual point positions and positions of hydrogen-bond donors in the hydrogen-bond centre (Jones, 1995; Jones *et al.*, 1995b).

Sim_wt was a measure of the similarity of the donors involved in the hydrogen-bond centre. The similarity index *type_sim* (defined above) was used to determine a score for the hydrogen bond centre. In an overlay of N molecules each hydrogen bond centre will contain $2N-1$ pairs of donors. *Sim_wt* was then set to the smallest pairwise similarity value found in the hydrogen bond centre.

Once *vec_wt* and *sim_wt* had been determined for a given hydrogen-bonding centre the contribution *vec_wt* × *sim_wt* was determined, and *donor_score* was then the sum of all such contributions from all hydrogen-bonding centres containing donor hydrogens.

Each acceptor lone pair in the base molecule was used to define a hydrogen-bonding centre with the potential to interact with donors within a receptor macromolecule. The process used to generate the score *acceptor_score* was entirely analogous to that used when determining *donor_score*.

In a similar fashion, each aromatic ring centre in the base molecule was used to define a hydrogen bond centre. Each such hydrogen bond centre made a contribution to *ring_score* that depended on the closeness of ring centres and ring normals in the hydrogen bond centre.

The final fitness score was determined by a weighted sum of the common volume, similarity score and steric energy. The fitness score was given by

$$volume\_integral + 750 \times similarity\_score - 0.05 \times vdw\_energy \qquad (4)$$

The weights of 750 and 0.05 were determined by empirical adjustment to give reasonable overlays (where the algorithm is driven to generate good pharmacophores without producing high energy structures) over a wide range of examples. The selection of ideal weights is a complex process and is an area of current investigation.

Genetic operators

The GA made use of three genetic operators: mutation, crossover and migration. The crossover operator required two parents and produced two children. The mutation operator required one parent and produced one child (Goldberg, 1989; Davis, 1991). The operator weight for mutation was set equal to the operator weight for crossover.

The crossover operator performed two-point crossover on integer strings and one-point crossover on the binary strings. The one-point crossover was the traditional GA recombination operator (Goldberg, 1989) while the two-point crossover was the PMX crossover operator (including the duplicate removal stage) that is described by Brown *et al.* (1994) and by Goldberg (1989). The mutation operator performed binary-string mutation on binary strings and integer-string mutation on integer strings. The binary-string mutation was identical to that described by Davis (1991). Each bit in the binary string had a probability of mutation equal to $1/L$, where L was the length of the binary string. The integer-string mutation was identical to that described by Brown *et al.* (1994). A position was randomly chosen on the integer string and the value at that position was mutated to a (different) new value that was randomly chosen from the set of allowed integer values. If this new value occurred elsewhere on the integer string, then it was replaced by the original value at the mutated position.

The migration operator required one parent and produced one child. The child was an exact copy of the parent. Let p be the sub-population to which the migration operation was applied and let n be a sub-population randomly selected from the two neighbours of p. Roulette-wheel parent selection was performed on n to produce the parent and the child was then inserted into p. An operator weight was used to determine how many migrations were performed, relative to mutation and crossover, and it was found that a 5% migration rate gave good results.

Results of overlay experiments

This section describes the application of the GA to a number of diverse overlay and pharmacophore-generation problems; further examples of the application of this algorithm are presented by Jones (1995) and Jones *et al.* (1995b). The GA was run ten times for each of the problems, and the resulting fittest solutions from each run ranked in order of decreasing fitness: in the following references to the 'best solution', the 'worst solution', etc. correspond to the position of the solution in the ranked fitness list of final solutions. All CPU times are for a Silicon Graphics R4000 Indigo II Workstation. The experimental results are illustrated in Plates 5–8, where the base molecules are coloured by atom type and where hydrogens and lone pairs are not generally displayed unless they were of particular significance in the overlay.

Overlay of leu-enkephalin on a hybrid morphine

The first superimposition problem to be considered here involved the two very different structures shown in Figure 2, specifically the hybrid morphine molecule EH-NAL, a mixed azide between estrone and naloxone (Kolb, 1991), and leu-enkephalin. Although only two molecules are involved, this is an extremely demanding problem as leu-enkephalin is highly flexible, containing 20 rotatable bonds. EH-NAL, conversely, has just six rotatable bonds in sidechains. The GA was run 10 times to generate 10 possible overlays. The mean run time was 9 minutes 13 seconds.

Plate 5 shows the best solution (ranked by GA fitness score) that was obtained. The base molecule, EH-NAL, is shown coloured by atom type,

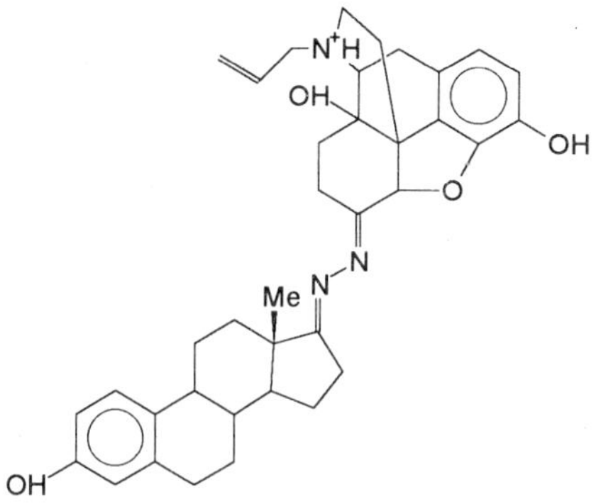

EH-NAL

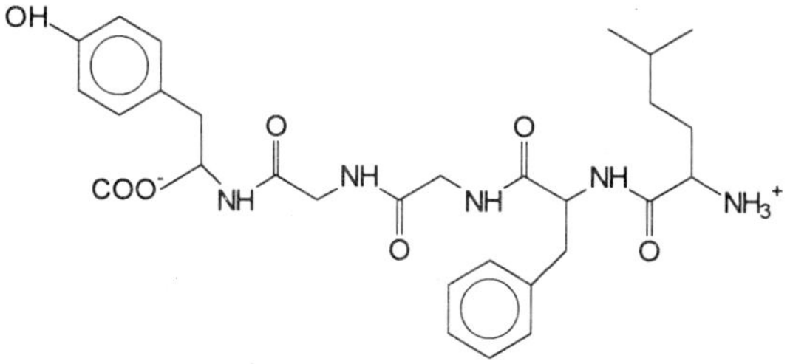

Leu-enkephalin

Figure 2 Hybrid morphine and leu-enkephalin.

while leu-enkephalin is shown in purple. The pharmacophore identified by the GA contains the five features indicated by the yellow circles: two aromatic rings; one phenol group; the protonated nitrogen (for which the connected hydrogens are shown in the plate to illustrate their common directionality); and the GA has also overlaid an sp^3 oxygen in EH-NAL with an sp^2 oxygen in leu-enkephalin, such that a lone pair from each (displayed in the colour plate) is aligned in the same direction. Six of the ten runs (including the five best runs) identified the first four of these five features, with just the fittest identifying the oxygen overlap.

The bound conformations of these molecules are not known so it is not possible to judge the accuracy of the GA. It is, however, interesting to note that Kolb (1991) obtained a very similar fit (superimposing both rings, the phenol group and the protonated nitrogen) using molecular dynamics with simulated annealing. However, this approach is not fully automated, unlike the GA, and would appear to be much more time-consuming in operation (though exact times are not available).

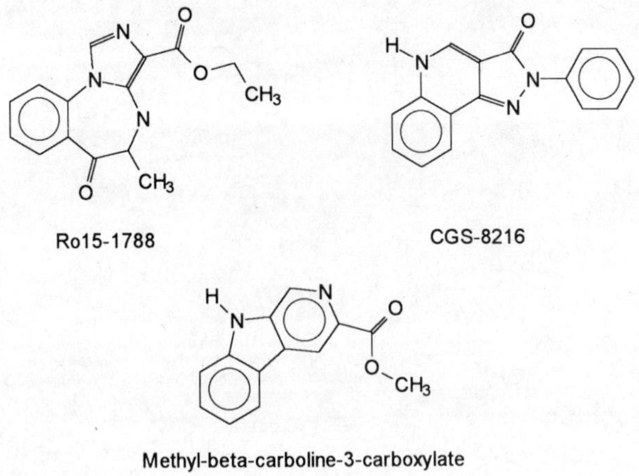

Ro15-1788

CGS-8216

Methyl-beta-carboline-3-carboxylate

Figure 3 Benzodiazepine receptor ligands.

Overlay of three benzodiazepine ligands

Codding and Muir (1985) have produced an overlay of ligands that bind to the benzodiazepine receptor, using structure–activity studies and functional-group similarities. Three of the ligands used in their analysis are shown in Figure 3: two of these molecules are inverse agonists (which promote convulsions on binding) while Ro15-1788 is an antagonist that has no convulsant effects. The mean execution time of the GA was 2 minutes and 10 seconds.

The best solution obtained from the ten runs is shown in Plate 6, where the base molecule, CGS-8216, is shown coloured by atom type, Ro15-1788 is in orange and methyl-beta-carboline-3-carboxylate is shown in purple. The pharmacophore elucidated by the GA comprised a benzene ring and an sp^2 oxygen acceptor: these features are indicated in the figure by yellow circles and the lone pairs connected to the sp^2 oxygens are also displayed. Also of interest is the fact that the donor nitrogens in the two inverse agonists are closely positioned (again this is indicated by a circle and the donor hydrogens are also displayed). It has been suggested that an absence of this donor is required for antagonism (Codding and Muir, 1985). All of the other GA runs produced this overlay, though the worst run produced a very untidy fit.

The GA solution is in fair agreement with Codding and Muir's structure–activity studies, which predicted a binding site that recognizes four features: an aromatic ring, an sp^2 oxygen, a hydrophobic side chain and an N–H donor group, though this last feature is not present in Ro15-1788. Although the GA does not identify hydrophobic regions, the volume overlay term ensured that the side chains that comprised this feature were correctly overlaid.

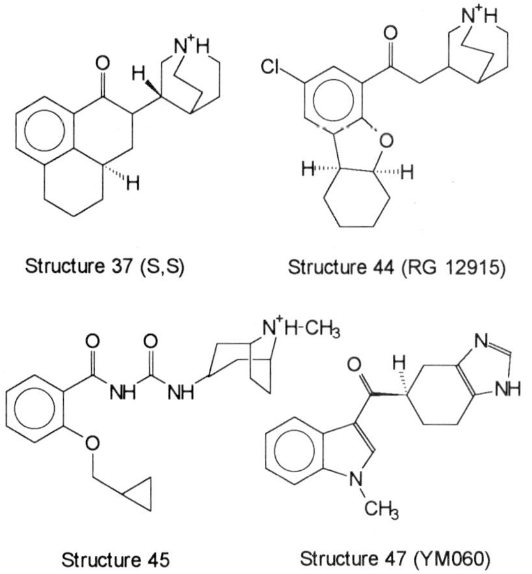

Structure 37 (S,S) Structure 44 (RG 12915)

Structure 45 Structure 47 (YM060)

Figure 4 5-HT_3 receptor agonists (structure numbers from Clark *et al.* (1993)).

Overlay of four 5-HT_3 antagonists

Clark *et al.* (1993) have synthesized several series of *N*-(quinuclidin-3-yl)aryl and heteroaryl-fused pyridones and tested them for 5-HT_3 receptor affinity. An overlay was attempted of the four antagonists shown in Figure 4. The default parameters were used, with the mean run-time being 6 minutes and 9 seconds.

The GA elucidated a pharmacophore consisting of a nitrogen donor, an sp^2 oxygen acceptor and an aromatic ring in all but the least-fit run. Plate 7 shows the superimposition obtained by the GA run that generated the highest fitness score. The base molecule, structure 37, is shown coloured by atom type, structure 44 is coloured purple, structure 45 is cyan and structure 47 is orange. The yellow circles in this plate indicate the three pharmacophore points: the normals of the aromatic ring, the lone pairs of the sp^2 oxygen; and the donor hydrogens bonded to the nitrogens. Although the nitrogens are not overlaid, their donor hydrogens are clearly in a position to interact with the same point in the receptor. This pharmacophore is the same as that identified by Clark *et al.* (1993), although the centres of the aromatic rings in their overlay do not appear to be as close as in the GA solutions.

Overlay of six angiotensin II receptor antagonists

Perkins and Dean (1993) have described a novel strategy for the superimposition of a set of flexible molecules, using a combination of simulated

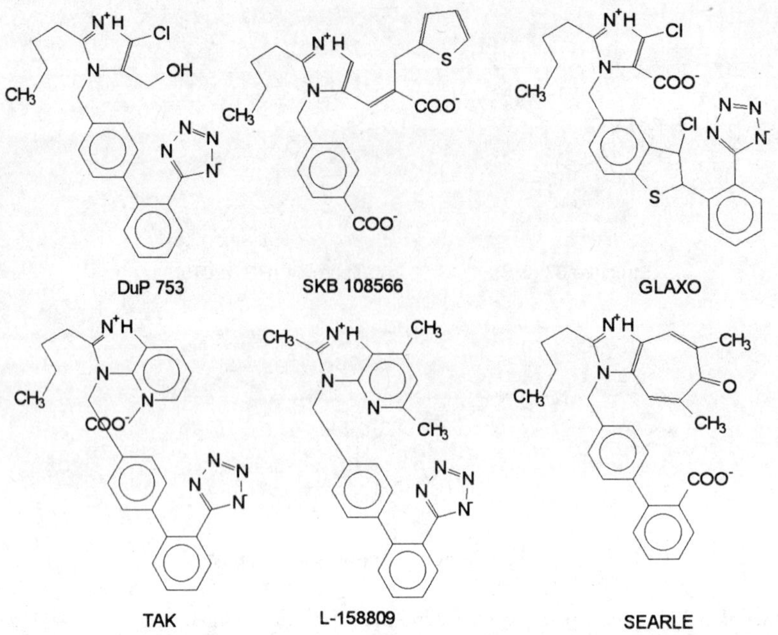

Figure 5 Six angiotensin II antagonists (structures from Perkins and Dean (1993)).

annealing and cluster analysis. The conformational space of each molecule is searched using simulated annealing. Significantly different low-energy conformations are extracted from the conformational analysis history using cluster analysis. For each pair of molecules, every possible combination of conformations found by cluster analysis is matched by simulated annealing, using the difference distance matrix as the objective function. The molecules are then superimposed using the match statistics, either by reference to a base molecule or by a consensus method. The algorithm was tested on the six angiotensin II antagonists shown in Figure 5, and we have also used these structures to evaluate the GA. The butyl side chains in five of the six structures were replaced by methyl groups, as this simplification was also performed by Perkins and Dean (1993). As the GA proved slow to converge, the algorithms was run for 60 000 operations. The average run time of the GA was 7 minutes and 56 seconds.

Six of the ten overlays (including the three fittest solutions) generated a pharmacophore comprising an aromatic ring and a protonated nitrogen. The best overlay is shown in Plate 8. The base molecule, L-158809, is shown coloured by atom type, GLAXO is coloured magenta, SEARLE is orange, SKB 108566 is cyan, TAK is green and DuP 753 is yellow. The yellow circles indicate the two pharmacophore points and the hydrogen-bond donor proton bonded to the protonated nitrogen is also displayed. It was hoped that the GA might have been able to overlay an acidic group from each structure. However, this was not the case, though it was able to overlay acid groups from all structures except SKB 108566 (which is the most dissimilar structure from L-158809). This overlay is indicated by a yellow circle and includes all the tetrazol groups. The fact that this overlay of five acidic groups is present in the final solution suggests that there may have been chromosomes within the population that encoded an overlay of acidic groups from all structures.

The superposition obtained by Perkins and Dean (1993) has some similarity with the GA result in that the imidazole group and benzene rings are also successfully superimposed (due to their structural similarity). However, their procedure is far more time-consuming. Conformational analysis took about 8 minutes (for SKB 108566), while the pairwise matching process took about 6 hours for each pair of molecules (based on the time for matching all conformers of DuP 753 and SKB 108566). These times are for a Sun Microsystems SPARCstation IPX workstation (a CPU that is comparable to that used for the GA): a superposition of the six structures, by reference to a base molecule, should thus take about 31 CPU hours, with a consensus superposition taking considerably longer.

Discussion of overlay results

The design and implementation of a GA for the superimposition of flexible molecules and the use of the resulting overlays to suggest possible pharmacophoric patterns has been described. The experiments reported here demonstrate the effectiveness and versatility of the algorithm, in that it has been possible to superimpose molecular structures in structurally-diverse test systems with results that are both intuitively acceptable and often in agreement with overlays suggested by alternative means. That said, there are several additions and improvements that could be made to the program. For example, inactive compounds that are similar to known actives are often incorporated in a structure–activity analysis, and it should be possible to extend the GA to incorporate inactives or to include biological activity. Again, ring closure has recently been incorporated into the algorithm, to allow the conformational analysis of structures with cyclic regions. Here, a cyclic bond is broken and replaced by constraints, thus enabling ring flexibility (Sanderson *et al.*, 1994).

The use of the GA to elucidate possible pharmacophoric patterns has been emphasized in this section. However, other applications of the approach are equally feasible. The overlays may be used as a starting point for investigation of a dataset by 3-D QSAR, which requires an initial alignment of the molecules that are to be analyzed (Cramer *et al.*, 1988). Another possible application is similarity searching in 3-D databases (Dean, 1994). The speed of the GA when performing pairwise superimpositions, which is what is required to match a target structure against each of the structures in a database, is such as to suggest that it might be feasible to consider the use of a modified version of the algorithm for flexible 3-D similarity searching.

CONCLUSIONS

This study has successfully shown that GAs can be applied to a wide variety of complex problems in computer aided drug design. It is worth reviewing the major points raised by each application.

The GA has been shown to be a highly successful algorithm in the field of 3-D flexible searching. One of the great strengths of the GA in an application of this sort is the ease with which new constraints can be added to the existing algorithm. For example, the algorithm described here could be readily modified to incorporate exclusion volumes, torsional and vectorial constraints and NMR NOE constraints (as described by Sanderson *et al.* (1994)).

Likewise the docking GA has proved to be capable of reproducing experimentally observed binding modes. The reason for this success is three-fold. Firstly, the GA is sufficiently powerful an algorithm that it is able successfully to search the important conformational aspects of the protein and ligand. Secondly, the use of a least-squares fitting procedure means that a complete search of molecular rotational and translational space is not required. Lastly, considerable care has been taken in designing an objective function that is capable of swiftly estimating binding affinities. The results and design of the algorithm offer insight into the molecular recognition mechanism, clearly showing the importance of desolvation in hydrogen bond formation. Also of particular interest is the reproduction of known binding modes in the absence of bound water: perhaps bound water is a consequence of binding rather than a contributory factor to the stabilization of the bound ligand.

The final application described a GA for the overlay of flexible molecules. As in the docking GA, the overlay GA benefited from a least squares fitting procedure that made redundant a complete search of molecular rotation and translational space. Likewise, the objective function proved able to reproduce known pharmacophores over a set of diverse data. This is a particularly useful tool for the drug designer, enabling the elucidation of putative pharmacophores with the stated caveats on similarity algorithms. Existing compounds can then be fitted to the pharmacophore using the conformational search GA.

In summary, the GA has shown immense potential in the field of chemical database search and molecular recognition applications. It would appear that the GA paradigm will prove invaluable within the field of computational chemistry.

ACKNOWLEDGEMENTS

We thank the Science and Engineering Research Council and The Wellcome Foundation for funding, Tripos Inc. for hardware and software support and David Clark for assistance with the pharmacophore search experiments. This chapter is a contribution from the Krebs Institute for Biomolecular Research, which is a designated Centre for Biomolecular Sciences of the Biotechnology and Biological Sciences Research Council.

REFERENCES

Beeson, C., Nguyen, P., Shipps, G., and Dix, T.A. (1993). A comprehensive description of the free energy of an intramolecular hydrogen bond as a function of solvation: NMR study. *J. Am. Chem. Soc.* **115**, 6803–6812.

Bernstein, F.C., Koetzle, T.F., Williams, G.J.B., Meyer, F., Bryce, M.D., Rogers, J.R, Kennard, O., Shimanouchi, T., and Tasumi, M. (1977). The Protein Data Bank: A computer-based archival file for macromolecular structures. *J. Mol. Biol.* **112**, 535–542.

Blaney, J.M., Crippen, G.M., Dearing, A., and Dixon, J.S. (1990). *DGEOM: Distance Geometry.* Quantum Chemistry Program Exchange, Department of Chemistry, Indiana University, Indiana, USA.

Blaney, J.M. and Dixon, J.S. (1993). A good ligand is hard to find: Automated docking methods. *Perspect. Drug Discovery Des.* **1**, 301–319.

Blommers, M.J., Lucasius, C.B., Kateman, G., and Kaptein, R. (1992). Conformational analysis of a dinucleotide photodimer with the aid of a genetic algorithm. *Biopolymers* **32**, 45–52.

Bolin, J.T., Filman, D.J., Matthews, D.A., Hamlin, R.C., and Kraut, J. (1982). Crystal structures of *Escherichia coli* and *Lactobacillus casei* dihydrofolate reductase refined at 1.7 Å resolution. *J. Biol. Chem.* **257**, 13650–13662.

Brown, R.D., Jones, G., Willett, P., and Glen, R.C. (1994). Matching two-dimensional chemical graphs using genetic algorithms. *J. Chem. Inf. Comput. Sci.* **34**, 63–70.

Cheung, H.T.A., Birdsall, B., Frienkiel, T.A., Chau, D.D., and Feeney, J. (1993). ^{13}C NMR determination of the tautomeric and ionization states of folate in its complexes with *lactobacillus casei* dihydrofolate reductase. *Biochem.* **32**, 6846–6854.

Clark, M., Cramer III, R.D., and van Opdenbosch, N. (1989). Validation of the general purpose tripos 5.2 force field. *J. Comput. Chem.* **10**, 982–1012.

Clark, D.E., Jones, G., Willett, P., Glen, R.C., and Kenny, P.W. (1994). Pharmacophoric pattern matching in files of three-dimensional chemical structures: Comparison of conformational-searching algorithms for flexible searching. *J. Chem. Inf. Comput. Sci.* **34**, 197–206.

Clark, D.E., Willett, P., and Kenny, P.W. (1992). Pharmacophoric pattern matching in files of three-dimensional chemical structures: Use of bounded distance matrices for the representation and searching of conformationally-flexible molecules. *J. Mol. Graphics* **10**, 194–204.

Clark, R.D., Miller, A.B., Berger, J., Repke, D.B., Weinhardt, K.K., Kowalczyk, B.A., Eglen, R.M., Bonhaus, D.W., Lee, C., Michel, A.D., Smith, W.L., and Wong, E.H.F. (1993). 2-(Quinuclidin-3-yl)pyrido[4,3-b]indol-1-ones and isoquinolin-1-ones. Potent conformationally restricted 5-HT3 receptor antagonists. *J. Med. Chem.* **36**, 2645–2657.

Codding, P.W. and Muir, A.K.S. (1985). Molecular structure of Ro15-1788 and a model for the binding of benzodiazepine receptor ligands. *Mol. Pharmacol.* **28**, 178–184.

Cramer, C.J., Lynch, G.C., Hawkins, G.D., and Truhlar, D.G. (1994). *AMSOL 3.5c User Manual.* Quantum Chemical Program Exchange, Department of Chemistry, University of Indiana, Bloomington, Indiana, USA.

Cramer III, R.D., Patterson, D.E., and Bunce, J.D. (1988). Comparative molecular field analysis (CoMFA). 1. Effect of shape on binding of steroids to carrier proteins. *J. Am. Chem. Soc.* **110**, 5959–5967.

Davis, L. (1991). *Handbook of Genetic Algorithms.* Van Nostrand Reinhold, New York, p. 385.

Davis, J.F., Delcamp, T.J., Prendergast, N.J., Ashford, V.A., Freisheim, J.H., and Kraut, J. (1990). Crystal structures of recombinant human dihydrofolate reductase complexed with folate and 5-deazafolate. *Biochem.* **29**, 9467–9479.

Dean, P.M. (1995). *Molecular Similarity in Drug Design.* Blackie Academic and Professional, Glasgow, p. 342.

DesJarlais, R.L., Sheridan, R., Seibel, G.L., Dixon, J.S., Kuntz, I.D. and Venkataraghavan, R. (1986). Using shape complementarity as an initial screen in designing ligands for a receptor binding site of known three-dimensional structure. *J. Med. Chem.* **31**, 722–729.

Digby, P.G.N. and Kempton, R.A. (1987). *Multivariate Analysis of Ecological Communities.* Chapman and Hall, London.

Goldberg, D.E. (1989). *Genetic Algorithms in Search, Optimization and Machine Learning.* Addison-Wesley Publishing Company, Wokingham, England.

Goodsell, D.S. and Olsen, A.J. (1990). Automated docking of substrates to proteins by simulated annealing. *Proteins: Struct. Funct. Genet.* **8**, 195–202.

Hirschfelder, J.O., Curtiss, C.F., and Bird, R.B. (1964). *Molecular Theory of Gases and Liquids.* John Wiley, New York.

Ho, C.M.W. and Marshall, G.R. (1990). Cavity search: An algorithm for the isolation and display of cavity like binding regions. *J. Comput.-Aided Mol. Des.* **4**, 337–354.

Howard, A.E. and Kollman, P.A. (1988). An analysis of current methodologies for conformational searching of complex molecules. *J. Med. Chem.* **31**, 1669–1675.

Hurst, T. (1994). Flexible 3D searching: The directed tweak technique. *J. Chem. Inf. Comput. Sci.* **34**, 190–196.

Jones, G. (1995). *Genetic Algorithms for Chemical Structure Handling and Molecular Recognition*. Ph.D. thesis, University of Sheffield.

Jones, G., Willett, P., and Glen, R.C. (1995a). Molecular recognition of receptor sites using a genetic algorithm with a description of desolvation. *J. Mol. Biol.* **245**, 43–53.

Jones, G., Willett, P., and Glen, R.C. (1995b). A genetic algorithm for flexible molecular overlay and pharmacophore elucidation. *J. Comput.-Aided Mol. Des.* **9**, 532–549.

Judson, R.S., Jaeger, E.P. and Treasurywala, A.M. (1994). A genetic algorithm based method for docking flexible molecules. *J. Mol. Struct. (Theochem)* **308**, 191–206.

Klebe, G. (1993). Structural alignment of molecules. In, *3D QSAR in Drug Design* (H. Kubinyi, Ed.). ESCOM, Leiden, pp. 173–199.

Kolb, V.M. (1991). Opiate receptors: Search for new drugs. *Progress Drug Res.* **36**, 49–70.

Kuntz, I.D., Blaney, J.M., Oatley, S.J., Langridge, R., and Ferrin, T.E. (1982). A geometric approach to macromolecule-ligand interactions. *J. Mol. Biol.* **161**, 269–288.

Lawrence, M.C. and Davis, P.C. (1992). CLIX – A search algorithm for finding novel ligands capable of binding proteins of known 3-dimensional structure. *Proteins: Struct. Funct. Genet.* **12**, 31–41.

Leach, A.R. (1991). A survey of methods for searching the conformational space of small and medium size molecules. In, *Reviews in Computational Chemistry II* (K.B. Lipkowitz and D.B. Boyd, Eds.). VCH, New York, pp. 1–56.

Leach, A.R. (1994). Ligand docking to proteins with discrete side-chain flexibility. *J. Mol. Biol.* **235**, 345–356.

Leach, A.R. and Kuntz, I.D. (1992). Conformational analysis of flexible ligands in macromolecular receptor sites. *J. Comput. Chem.* **13**, 730–748.

Lucasius, C.B., Blommers, M.J.J., Buydens, L.M.C., and Kateman, G. (1991). A genetic algorithm for conformational analysis of DNA. In, *Handbook of Genetic Algorithms* (L. Davis, Ed.). Van Nostrand Reinhold, New York, pp. 251–281.

Marshall, G.R., Barry, C., Bossard, H.E., Dammkoehler, R.A., and Dunn, D.A. (1979). The conformational parameter in drug design: The active analogue approach. In, *Computer-Assisted Drug Design. ACS Symposium Series 112* (E.C. Olson and R.E Christofferson, Eds.). American Chemical Society, Washington, DC, pp. 205–226.

Moock, T.E., Henry, D.R., Ozkabak, A.G., and Alamgir, M. (1994). Conformational searching in ISIS/3D databases. *J. Chem. Inf. Comput. Sci.* **34**, 184–189.

Newman, W.M. and Sproull, R.F. (1981). *Principles of Interactive Computer Graphics.* McGraw-Hill.

Oshiro, C.M., Kuntz, I.D., and Dixon, J.S. (1995). Flexible docking using a genetic algorithm. *J. Comput.-Aided Mol. Des.* **9**, 113–130.

Payne, A.W.R. and Glen, R.C. (1993). Molecular recognition using a binary genetic search algorithm. *J. Mol. Graphics* **11**, 74–91.

Perkins, T.D.J. and Dean, P.M. (1993). An exploration of a novel strategy for superimposing several flexible molecules. *J. Comput.-Aided Mol. Des.* **7**, 155–172.

Rinaldi, D., Rivail, J.L., and Rguini, N. (1993). Fast geometry optimisation in self-consistent reaction field computations on solvated molecules. *J. Comput. Chem.* **13**, 675–680.

Rusinko III, A., Sheridan, R.P., Nilakantan, R., Haraki, K.S., Bauman, N., and Venkataraghavan, R. (1989). Using CONCORD to construct a large database of three-dimensional coordinates from connection tables. *J. Chem. Inf. Comput. Sci.* **29**, 251–255.

Sanderson, P.N., Glen, R.C., Payne, A.W.R., Hudson, B.D., Heide, C., Tranter, G.E., Doyle, P.M., and Harris, C.J. (1994). Characterisation of the solution conformation of a cyclic RGD peptide analogue by NMR spectroscopy allied with a genetic algorithm approach and constrained molecular dynamics. *Int. J. Peptide Protein Res.* **43**, 588–596.

Shoichet, B.K., Stroud, R.M., Santi, D.V., Kuntz, I.D., and Perry, K.M. (1993). Structure-based discovery of inhibitors of thymidylate synthase. *Science* **259**, 1445–1450.

Smellie, A.S., Crippen, G.M., and Richards, W.G. (1991). Fast drug receptor mapping by site-directed distances: A novel method of predicting new pharmacological leads. *J. Chem. Inf. Comput. Sci.* **31**, 386–392.

Starkweather, T., Whitley, D., and Mathias, K. (1990). Optimisation using distributed genetic algorithms. In, *Parallel Problem Solving From Nature* (H.P. Schwefel and R. Manner, Eds.). Springer-Verlag, Berlin, pp. 176–185.

Stewart, J.J.P. (1990). MOPAC: A semiempirical molecular orbital program. *J. Comput.-Aided Mol. Des.* **4**, 1–105.

Tanese, R. (1989). Distributed genetic algorithms. In, *Proceedings of the Third International Conference on Genetic Algorithms and their Applications* (D. Schaffer, Ed.). Morgan Kaufmann, San Mateo, CA, pp. 434-439.

Tripos Inc. (1994). *SYBYL: Molecular Modelling Software.* 1699 South Hanley Road, Suite 303, St Louis, MO63144, USA.

Vermersch, P.S., Lemon, D.D., Tesmer, J.J., and Quiocho, F.A. (1991). Sugar-binding and crystallographic studies of an arabinose-binding protein mutant ($Met^{108}Leu$) that exhibits enhanced affinity and altered specifity. *Biochem.* **30**, 6861–6866.

von Itzstein, M., Wu, W., Kok, G.B., Pegg, A.M.S., Dyason, J.C., Jin, B., Phan, T.V., Smythe, A.M.L., White, A.H.F., Oliver, A.S.W., Colman, A.P.M., Varghese, J.N., Ryan, D.M., Woods, J.M., Bethell, R.C., Hotham, V.J., Cameron, M., and Penn, C.R. (1993). Rational design of potent sialidase-based inhibitors of influenza virus replication. *Nature* **363**, 418–423.

Zamora, A. (1976). An algorithm for finding the smallest set of smallest rings. *J. Chem. Inf. Comput. Sci.* **16**, 40–43.

10 Genetic Selection of Aromatic Substituents for Designing Test Series

C. PUTAVY, J. DEVILLERS*, and D. DOMINE
CTIS, 21 rue de la Bannière, 69003 Lyon, France

A classical genetic algorithm (GA) was used for the selection of a set of aromatic substituents highly representative of a physicochemical parameter space. The parameters considered were the π constant, the H-bonding acceptor (HBA) and donor (HBD) abilities, the molar refractivity (MR), and the inductive and resonance parameters of Swain and Lupton (F and R). Different fitness functions based on the calculation of correlation coefficients or Euclidean distances were tested. The different test series proposed by the GA were compared from displays on nonlinear maps and calculation of variance coefficients.

KEYWORDS: *genetic algorithm; test series; aromatic substituent constants; nonlinear mapping; Euclidean distances; intraclass correlation coefficient; variance coefficient.*

INTRODUCTION

In quantitative structure–activity relationship (QSAR) studies, to relate the physicochemical properties of aromatics and aliphatics to pharmaceutical and toxicological activities observed *in vivo* or *in vitro*, many parameters describing the hydrophobic, steric, and electronic effects of their substituents have been derived (Taft, 1953; Bowden, 1990; Dearden, 1990; Hansch *et al.*, 1991; Silipo and Vittoria, 1991). Among them the most widely used are the π contribution of Hansch which depicts the lipophilic character of the aromatic substituents (Hansch and Fujita, 1964), the Hammett σ constants which are used to account for electronic processes (Bowden, 1990; Ludwig

* Author to whom all correspondence should be addressed.

In, *Genetic Algorithms in Molecular Modeling* (J. Devillers, Ed.)
Academic Press, London, 1996, pp. 243–269.
ISBN 0-12-213810-4

et al., 1992), the Swain and Lupton F and R parameters derived from the σ constants which separate the inductive and resonance effects of the substituents (Swain and Lupton, 1968), and the molar refractivity (MR) used to describe the steric bulk of substituents (Dunn, 1977). The above substituent constants have shown their efficiency in numerous QSAR models (Karcher and Devillers, 1990; Hansch and Leo, 1995) but also in the selection of test series (Wootton *et al.*, 1975; Dove *et al.*, 1980; Streich *et al.*, 1980; Wootton, 1983; Domine *et al.*, 1994a,b; Devillers, 1995). Indeed, the synthesis of new compounds and the analysis of their biological activity is extremely costly. Each compound tested should therefore give a maximum information to enable rational drug development. The information content of a set of compounds deals with biological effects, in one or more *in vitro* and *in vivo* test systems, and the relationships between their chemical structures and their physicochemical properties.

Thus, the problem of selecting a set of substituents with independence among several parameters has concerned researchers for a long time. The advent of large chemical databases has occasioned the use of computer-based approaches for selecting compounds, and a number of methods have been proposed for selecting test series with low collinearity of molecular descriptors and high data variance. Most of these selection methods are based on the inspection of graphs usually derived from linear and nonlinear multivariate analyses. Among these graphical approaches, we can cite simple 2-D plots (Craig, 1971), 'decision trees' (Topliss, 1972), dendrograms (Hansch *et al.*, 1973), spectral maps (Dove *et al.*, 1980), score plots (Alunni *et al.*, 1983; van de Waterbeemd *et al.*, 1989), nonlinear maps (Goodford *et al.*, 1976; Domine *et al.*, 1994a,b; Devillers, 1995), Kohonen self organizing maps (Simon *et al.*, 1993), and N2M (Domine *et al.*, 1996).

The aim of this study is to show that besides the above graphical methods, genetic algorithms (GAs) (Goldberg, 1989; Davis, 1991; Michalewicz, 1992) can represent an attractive alternative for selecting valuable test series from substituent constants.

MATERIALS AND METHODS

Data matrix of aromatic substituents

The study was performed from a data matrix of 166 aromatic substituents (Table I) described by means of six substituent constants encoding their hydrophobic, steric, and electronic effects. These parameters were the π constant, the H-bonding acceptor (HBA) and donor (HBD) abilities, MR, and the inductive and resonance parameters of Swain and Lupton (1968) F and R. The data were retrieved from the literature (Hansch and Leo, 1979). All inductive and resonance field constants F and R were recalculated from

Table I *Aromatic substituents.*

No.	Substituent	No.	Substituent
1	Br	2	Cl
3	F	4	SO_2F
5	SF_5	6	I
7	IO_2	8	NO
9	NO_2	10	NNN
11	H	12	OH
13	SH	14	$B(OH)_2$
15	NH_2	16	NHOH
17	SO_2NH_2	18	$NHNH_2$
19	5-Cl-1-Tetrazolyl	20	$N{=}CCl_2$
21	CF_3	22	OCF_3
23	SO_2CF_3	24	SCF_3
25	CN	26	NCS
27	SCN	28	CO_2^-
29	1-Tetrazolyl	30	NHCN
31	CHO	32	CO_2H
33	CH_2Br	34	CH_2Cl
35	CH_2I	36	NHCHO
37	$CONH_2$	38	CH=NOH
39	CH_3	40	$NHCONH_2$
41	$NHC{=}S(NH_2)$	42	OCH_3
43	CH_2OH	44	$SOCH_3$
45	SO_2CH_3	46	OSO_2CH_3
47	SCH_3	48	$SeCH_3$
49	$NHCH_3$	50	$NHSO_2CH_3$
51	CF_2CF_3	52	C≡CH
53	$NHCOCF_3$	54	CH_2CN
55	$CH{=}CHNO_2$-(trans)	56	$CH{=}CH_2$
57	$NHC{=}O(CH_2Cl)$	58	$COCH_3$
59	$SCOCH_3$	60	$OCOCH_3$
61	CO_2CH_3	62	$NHCOCH_3$
63	$NHCO_2CH_3$	64	$C{=}O(NHCH_3)$
65	$CH{=}NOCH_3$	66	$NHC{=}S(CH_3)$
67	$CH{=}NNHC{=}S(NH_2)$	68	CH_2CH_3
69	$CH{=}NNHCONHNH_2$	70	CH_2OCH_3
71	OCH_2CH_3	72	SOC_2H_5
73	SC_2H_5	74	SeC_2H_5
75	NHC_2H_5	76	$SO_2C_2H_5$
77	$N(CH_3)_2$	78	$NHSO_2C_2H_5$
79	$P(CH_3)_2$	80	$PO(OCH_3)_2$
81	$C(OH)(CF_3)_2$	82	CH=CHCN
83	Cyclopropyl	84	COC_2H_5
85	$SCOC_2H_5$	86	$CO_2C_2H_5$
87	$OCOC_2H_5$	88	$CH_2CH_2CO_2H$
89	$NHCO_2C_2H_5$	90	$CONHC_2H_5$
91	$NHCOC_2H_5$	92	$CH{=}NOC_2H_5$
93	$NHC{=}S(C_2H_5)$	94	$CH(CH_3)_2$
95	C_3H_7	96	$NHC{=}S(NHC_2H_5)$
97	$OCH(CH_3)_2$	98	OC_3H_7

Table I *continued.*

No.	Substituent	No.	Substituent
99	$CH_2OC_2H_5$	100	SOC_3H_7
101	$SO_2C_3H_7$	102	SC_3H_7
103	SeC_3H_7	104	NHC_3H_7
105	$NHSO_2C_3H_7$	106	$N(CH_3)_3^+$
107	$Si(CH_3)_3$	108	$CH{=}C(CN)_2$
109	1-Pyrryl	110	2-Thienyl
111	3-Thienyl	112	$CH{=}CHCOCH_3$
113	$CH{=}CHCO_2CH_3$	114	COC_3H_7
115	$SCOC_3H_7$	116	$OCOC_3H_7$
117	$CO_2C_3H_7$	118	$(CH_2)_3CO_2H$
119	$CONHC_3H_7$	120	$NHCOC_3H_7$
121	$NHC{=}OCH(CH_3)_2$	122	$NHCO_2C_3H_7$
123	$CH{=}NOC_3H_7$	124	$NHC{=}S(C_3H_7)$
125	C_4H_9	126	$C(CH_3)_3$
127	OC_4H_9	128	$CH_2OC_3H_7$
129	$N(C_2H_5)_2$	130	NHC_4H_9
131	$P(C_2H_5)_2$	132	$PO(OC_2H_5)_2$
133	$CH_2Si(CH_3)_3$	134	$CH{=}CHCOC_2H_5$
135	$CH{=}CHCO_2C_2H_5$	136	$CH{=}NOC_4H_9$
137	C_5H_{11}	138	$CH_2OC_4H_9$
139	C_6H_5	140	$N{=}NC_6H_5$
141	OC_6H_5	142	$SO_2C_6H_5$
143	$OSO_2C_6H_5$	144	NHC_6H_5
145	$NHSO_2C_6H_5$	146	2,5-di-Me-1-pyrryl
147	$CH{=}CHCOC_3H_7$	148	$CH{=}CHCO_2C_3H_7$
149	Cyclohexyl	150	2-Benzthiazolyl
151	COC_6H_5	152	$CO_2C_6H_5$
153	$OCOC_6H_5$	154	$N{=}CHC_6H_5$
155	$CH{=}NC_6H_5$	156	$NHCOC_6H_5$
157	$CH_2C_6H_5$	158	$CH_2OC_6H_5$
159	$C{\equiv}CC_6H_5$	160	$CH{=}NNHCOC_6H_5$
161	$CH_2Si(C_2H_5)_3$	162	$CH{=}CHC_6H_5$-(trans)
163	$CH{=}CHCOC_6H_5$	164	Ferrocenyl
165	$N(C_6H_5)_2$	166	$P{=}O(C_6H_5)_2$

the σ_m and σ_p constants of Hammett with equations of Swain and Lupton (1968). Corrections were made when values obtained were different from those reported in the compilation of substituent constants (Hansch and Leo, 1979). For the GA analysis, data were centered (i.e., zero mean) and reduced (i.e. unit variance).

GA implementation

GAs are computational models exploiting the principles of natural evolution in complicated computer optimization tasks (Goldberg, 1989; Davis, 1991;

Michalewicz, 1992). They work with a population of 'individuals' represented by chromosomes interacting through genetic operators to carry out an optimization *via* a reproduction process. The algorithm consists in successively transforming one generation of individuals into the next using the operations of selection, crossover, and mutation (Goldberg, 1989; Davis, 1991; Michalewicz, 1992). According to Michalewicz (1992), a GA for a particular problem must have the following five components:

- a genetic representation for potential solutions to the problem;
- a way to create an initial population of potential solutions;
- an evaluation function rating solutions in terms of their fitness;
- genetic operators (mutation, crossover) altering the composition of children during reproduction;
- values for various parameters used by the GA (e.g., population size, probabilities of applying genetic operators).

There is no unique GA procedure, since all of the above steps are problem-dependent. In addition, for the same problem several operators can be used. Therefore, for each problem, a new protocol has to be designed. However, from a general point of view, the aim of a GA is to search the best chromosome(s) for a given problem. Therefore, in a first step a chromosome constituted of a string of genes must be defined. In our study, the aim was to find the most representative series of aromatic substituents. As a result, a gene in a chromosome corresponded to a substituent. Genes were encoded by means of integers from 1 to 166 which simply corresponded to the substituent numbers (Table I). Thus, '1' corresponded to bromine, '2' to chlorine, and so on (Table I). Once the chromosome was defined, an initial population of randomly selected chromosomes was generated. Then, the GA proceeded in four steps which were iteratively repeated until the best population was found (Figure 1). These steps were the fitness evaluation, the selection of parents, the crossover, and the mutation. Due to the importance of the selection of the fitness function, this parameter will be described in the next section. There are several different ways for selecting parents (Goldberg, 1989; Lucasius and Kateman, 1994). In our study, a roulette wheel selection procedure was used. Thus, the probability that an individual was selected for mating was proportional to its fitness (Goldberg, 1989; Brown *et al.*, 1994). To perform this step with maximum efficiency, fitness values were scaled using the scaling routines (i.e., prescale, scale, scalepop) of Goldberg (1989, p. 79). In GA, crossovers facilitate the large-scale exploration of the search space. Different procedures such as one-point, two-point and multiple-point crossovers can be used (Eshelman *et al.*, 1989; Goldberg, 1989; Davis, 1991). In the present study, a uniform crossover, with a probability Pcross, was selected. Lastly, in order to facilitate the small-scale exploration of the search space, a mutation operator was used. It was characterized by a probability Pmut.

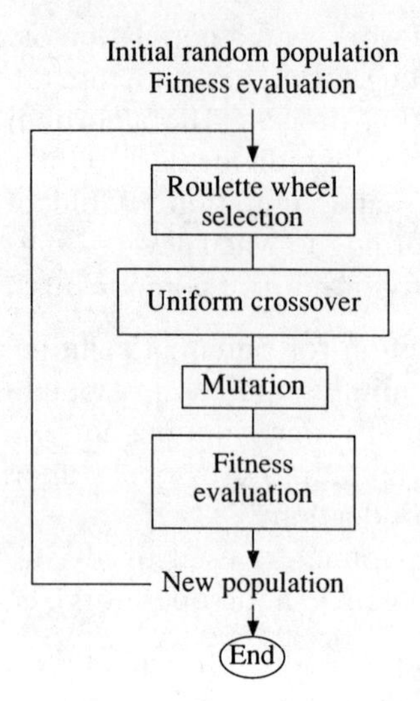

Figure 1 Genetic algorithm flow chart.

There exist several ways for generating a population at each cycle. In our study, the entire population was not regenerated at each generation. We used an elitist strategy in which the best individuals were duplicated. One copy was passed directly onto the next generation without crossover or mutation and the other participated in the reproduction process. This ensured that the best individuals were not lost but continued to be available for mating.

Fitness evaluation

A chromosome can contain substituents with various levels of differences. Thus, in Figure 2, three levels of differences between three couples of genes in a chromosome are represented. The aim of this study was to find sets of different substituents with the largest information content. Each gene (or substituent) in a chromosome should be selected in such a way that it conveyed the maximum amount of information. As a result, the fitness function had to be designed for quantifying the differences between the substituents. For this purpose, basically, correlation coefficients, distance measurements, or variance coefficients can be used. In the present study, different coefficients were used.

The square intraclass correlation coefficient (Dove et al., *1980)*
For two genes (i.e. substituents) 1 and 2, this coefficient is defined as follows:

$$r_{I(1,2)} = \frac{s_B^2 - s_I^2}{s_B^2 + s_I^2} \tag{1}$$

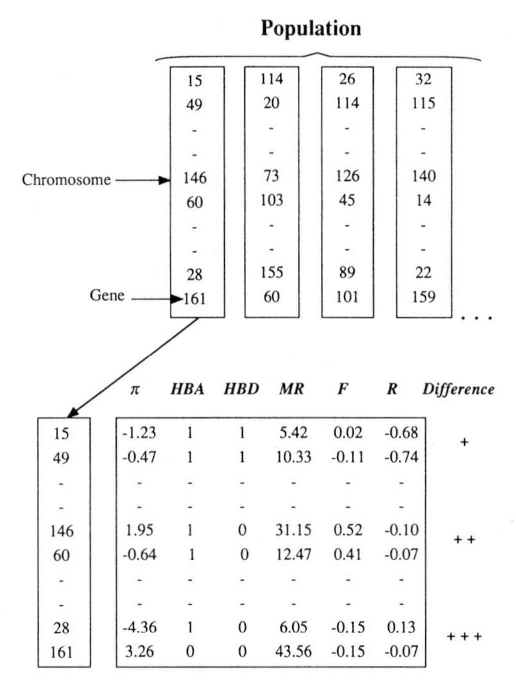

Figure 2 Representation of a population of chromosomes and characterization of three levels of differences between three couples of genes in a chromosome.

where

$$s_B^2 = \frac{1}{m} \sum_{i=1}^{m} \left(\frac{z_{i1} - z_{i2}}{2} - \bar{z} \right)^2 \tag{2}$$

and

$$s_I^2 = \frac{1}{4m} \sum_{i=1}^{m} \left(z_{i1} - z_{i2} \right)^2 \tag{3}$$

with

$$\bar{z} = \frac{1}{2m} \sum_{i=1}^{m} \sum_{j=1}^{2} z_{ij} \tag{4}$$

In the above equations, z_{ij} represents the value of the ith parameter for the jth gene and m represents the number of parameters.

Note that if there is no relationship between the parameter values of two genes 1 and 2, $r_{I(1,2)}$ and $r^2_{I(1,2)}$ become equal to zero (Dove *et al.*, 1980). Therefore, to select optimal test series, the aim was to obtain minimal values for $r^2_{I(1,2)}$. The intraclass correlation coefficient was calculated for each possible couple of genes in a chromosome. The number of combinations (N) for n genes is given by Eq. (5).

$$N = C_n^2 = \frac{n!}{2!(n-2)!} = \frac{n(n-1)}{2} \quad (5)$$

This yields the following fitness function:

$$\text{Fitness} = 1 - \left(\frac{\sum_k^N (r_I^2)_k}{N} \right) \quad (6)$$

The mean Euclidean coefficient (Turner et al., *1995)*
For two genes 1 and 2 in a chromosome, this coefficient is given by the following formula:

$$d_{12} = \frac{\sqrt{\sum_{i=1}^{m} (z_{i1} - z_{i2})^2}}{m} \quad (7)$$

This coefficient was calculated for each possible couple of genes in the chromosome (Eq. (5)). Four different fitness functions were elaborated from Eq. (7). In all cases, the aim was to obtain a maximal d_{ij} value. The first one was performed by means of the following equation:

$$\text{Fitness} = \frac{\sum_{ij} d_{ij}}{N} \quad (8)$$

where the fitness of a chromosome was the mean of the N calculated distances. In the second fitness function a constraint was added. This constraint implied that at least one of the terms of the distance equation had to be superior to a given value. It was defined from a threshold value calculated as follows:

$$C_i = x \times \sqrt{(\max_i - \min_i)^2} \quad (9)$$

In Eq. (9), x is a coefficient which has to be adjusted from a trial and error procedure. '$\max_i$' and '$\min_i$' represent the maximum and minimum values for π, HBA, HBD, MR, F, or R. If for all couples of genes in a chromosome, the constraint is satisfied (Eq. (9)), the chromosome is maintained in the population otherwise it is eliminated. This strategy allowed to remove chromosomes having couples of substituents which were relatively distant but presented only average differences for all the parameters.

The last two fitness functions used the minimum value of the N calculated mean Euclidean distance (Eq. (10)). In both cases, the aim was to maximize this value but in one case, the constraint previously described (i.e., Eq. (9)) was added.

$$\text{Fitness} = \min(d_{ij}) \quad (10)$$

Analysis of the results obtained with the different fitness functions was performed from two different methodologies. The first was based on the use of the nonlinear mapping method (Sammon, 1969) to graphically analyze the results (Domine *et al.*, 1993; Devillers, 1995). The second dealt with the calculation of a variance coefficient (Eq. (11)) often used in this kind of study to evaluate the representativity of a test series (Dove *et al.*, 1980; Streich *et al.*, 1980; Wootton, 1983).

$$V_s = \frac{1}{m}\sum_{i=1}^{m}\frac{S_{is}^2}{S_{ip}^2} \quad (11)$$

In Eq. (11), S_{is}^2 is the variance of the *i*th parameter within the test series and S_{ip}^2 is the variance of the *i*th parameter within the scaled data matrix. The nonlinear mapping and GA analyses were performed with the STATQSAR Package (STATQSAR, 1994). Graphics were obtained from ADE-4 (http://biomserv.univ.lyon.fr/ADE-4.html)

RESULTS AND DISCUSSION

Validity of information contained in the nonlinear map for the evaluation of the GA designed test series

In order to evaluate the interest of the different fitness functions for selecting valuable test series, a nonlinear map of the 166 substituents was used. Indeed, the NLM analysis of the data matrix after a scaling transformation (Domine *et al.*, 1994a) allowed to obtain a 2-dimensional map easily interpretable with a minimal loss of information (Figure 3.1). The statistical significance of the map was verified by inspecting the total mapping error and the goodness of fit of each point (Figure 3.2). Indeed, the total mapping error (E) was low (i.e. E = 6.4 10^{-2}) and the representation of the individual mapping errors (Domine *et al.*, 1993) showed that no points carried a large part of the total mapping error (Figure 3.2). In Figure 3, the three couples of substituents (15/49, 146/60, and 28/161) corresponding to the three levels of difference cited above (Figure 2) are represented. It is interesting to note that the distances separating the substituents of each couple on the map agree with the levels of difference shown in Figure 2. To confirm that the nonlinear map obtained (Figure 3.1) was representative of the physicochemical space studied and was chemically coherent, collections of maps were drawn by representing the physicochemical data used to run the NLM (Figure 4) and structural information (Figure 5) on the nonlinear map (Figure 3). In Figure 4, the larger the squares (positive values), the higher the physicochemical values and the larger the circles (negative values) the lower the values. Figure 4 shows that except for MR (Figure 4.4) the substituents are distributed and

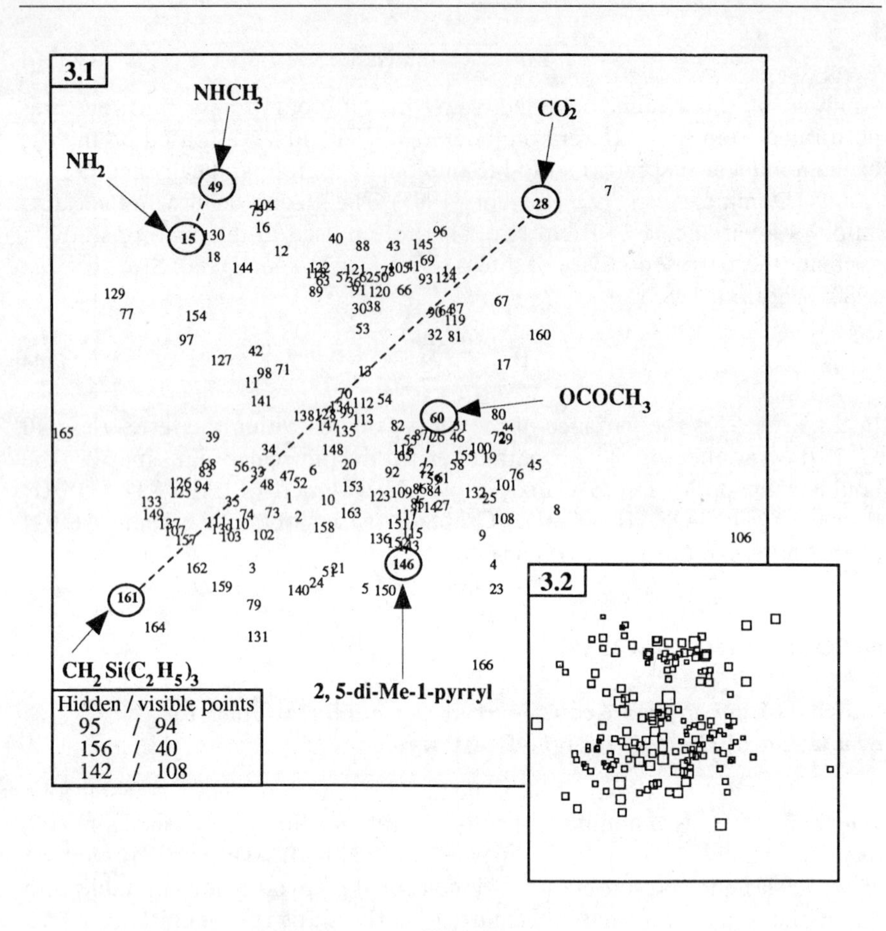

Figure 3
3.1 Nonlinear map of the 166 aromatic substituents described by six substituent constants (i.e., π, HBA, HBD, MR, F and R).
3.2 Plot of the individual mapping errors on each substituent of the nonlinear map. Squares are proportional in size to the magnitude of the errors.

clustered on the nonlinear map according to their substituent constants. Thus, for example, the π values of the substituents generally decrease along an axis running from the bottom left to the top right-hand side of the map. Clusters of substituents presenting HBA and/or HBD abilities can be observed in Figures 4.2 and 4.3, respectively. If the couples of substituents shown in Figure 2 are represented on Figure 4 (in light grey, black, and dark grey for couples 15/49, 146/60, and 28/161, respectively), it is possible to visualize for each property the level of difference between these substituents.

Figure 5 presents a collection of maps drawn by plotting the presence or frequency of some atoms or functional groups on the nonlinear map (Figure 3).

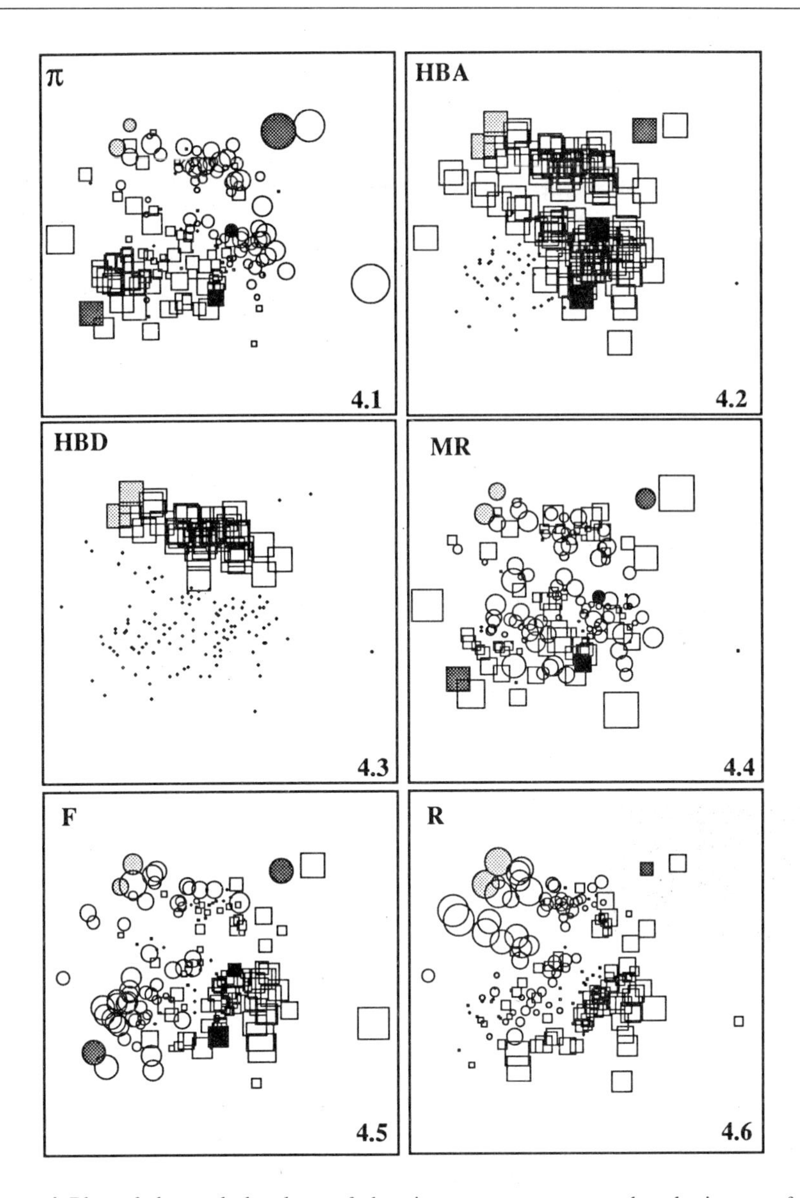

Figure 4 Plot of the scaled values of the six parameters on each substituent of the nonlinear map. Squares (positive values) and circles (negative values) are proportional in size to the magnitude of the parameters. In Figures 4.2 and 4.3, the dots indicate the substituents which do not have the ability to accept and donate H-bonds, respectively.

Figure 5.1 reveals that substituents containing only C and H atoms form a cluster in the bottom left-hand corner of the map. This location of the substituents is generally linked to high values of π (Figure 4.1) and low F values (Figure 4.5). A closer inspection of Figure 5.1 reveals that a gradient linked to the number of carbon atoms running downwards can be drawn.

For the presence of C=O or C=S groups (Figure 5.2), two clusters are formed depending on the ability of the substituents to donate or not H-bonds. The substituents of the top cluster are characterized by HBA and HBD abilities while the others can only accept H bonds (Figures 4.2 and 4.3).

Figure 5.3 shows that substituents containing primary or secondary amine groups cluster at the top of the nonlinear map. This is associated with HBA and HBD abilities (Figures 4.2 and 4.3). These substituents also generally have low π values (Figure 4.1). The cluster formed can be divided into three subclusters. The first located on the left-hand side contains substituents with formula NHR where R = H, OH, NH_2, alkyl, and phenyl. The substituents in which the NH group is bound to a C=O, C=S, CN, CO_2, or a SO_2 group (including their derivatives) and to the substitution site are found in the middle. Subcluster III is constituted of the substituents for which the amine group is not the group bound to the derivatives (e.g., $CONHCH_3$). Comparison with Figures 4.5 and 4.6 shows that this repartition of the three subclusters is due to increased F and R values.

The substituents containing a tertiary amine group are represented in Figure 5.4. These substituents form two clusters which principally differ by their F (Figure 4.5) and R (Figure 4.6) values. The non cyclic substituents are found in the same cluster in the left-hand side of the map ($N(CH_3)_2$, $N(C_2H_5)_2$, and $N(C_6H_5)_2$) while the cyclic ones (5-Cl-1-tetrazolyl, 1-tetrazolyl, 1-pyrryl, and 2,5-di-Me-1-pyrryl) are found in the bottom right-hand cluster. The quaternary amine $N(CH_3)_3^+$ appears as an outlier due to its very low π value and high F value. Another difference is that it cannot accept H-bond unlike tertiary amines.

The presence of SO, SO_2 or SO_3 groups in the substituents is plotted on Figure 5.5. These substituents form two clusters in the right-hand side of the cloud of points displayed on the map, indicating that their π values are generally low (Figure 4.1). The substituents of the first cluster can accept and donate H-bonds while those of the second cluster can only accept H-bonds (Figures 4.2 and 4.3). The substituents constituting the second cluster differ from those of the first cluster by higher F and R values (Figures 4.5 and 4.6). Between these two clusters, SO_2NH_2 occupies an intermediate location with HBA and HBD ability but rather high values for F and R.

Therefore, analysis of Figures 4 and 5 shows that the nonlinear map (Figure 3) summarizes the main part of the information contained in the large substituent constants data table of Hansch and Leo (1979) and it is coherent in terms of chemical information. Other projections of chemical information confirming the above statement can be found in Domine *et al.*

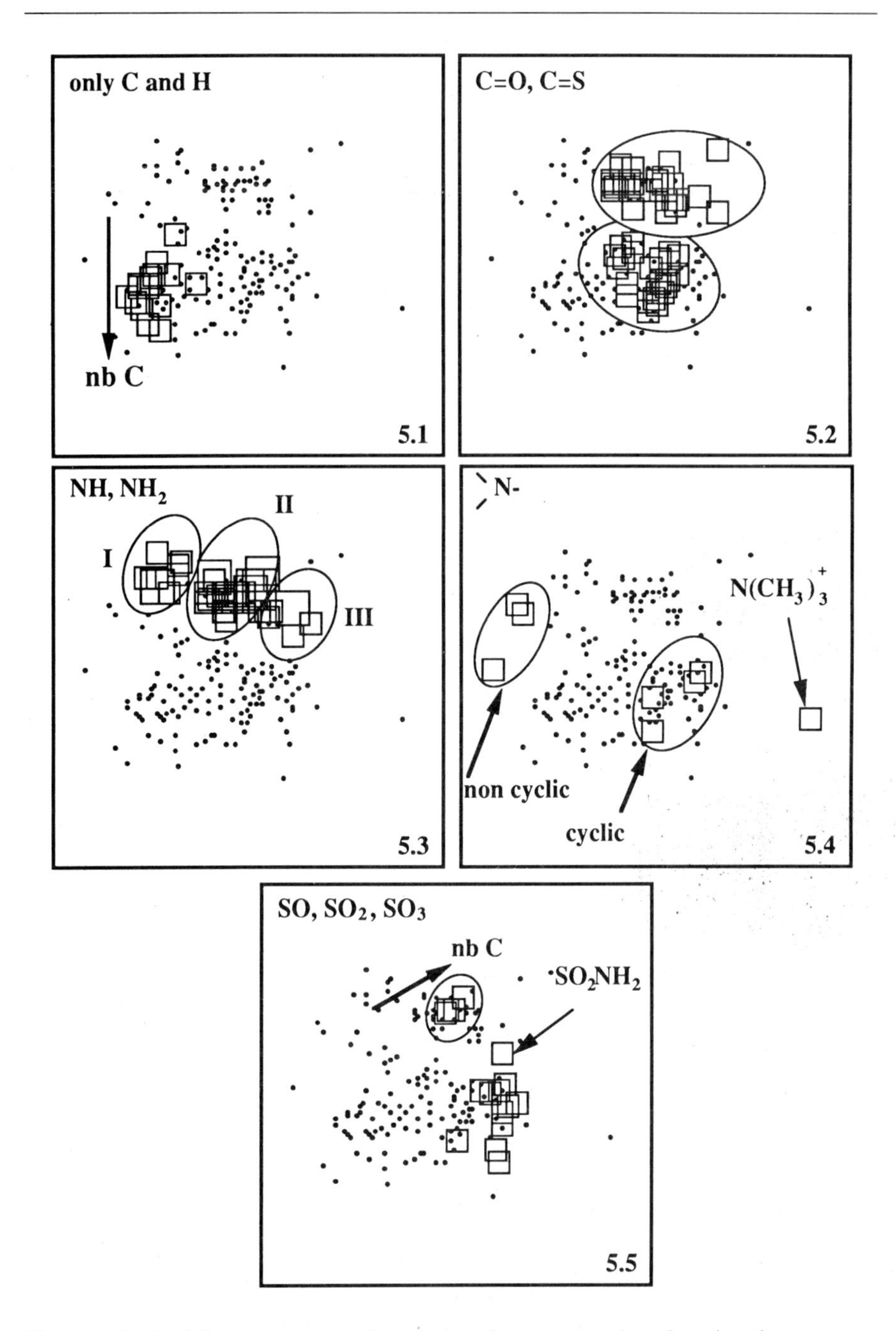

Figure 5 Plot of the presence or frequency of some atoms or functional groups on the nonlinear map (Figure 3). Squares are proportional in size to the number of functional groups. The absence of a descriptor is represented by a dot.

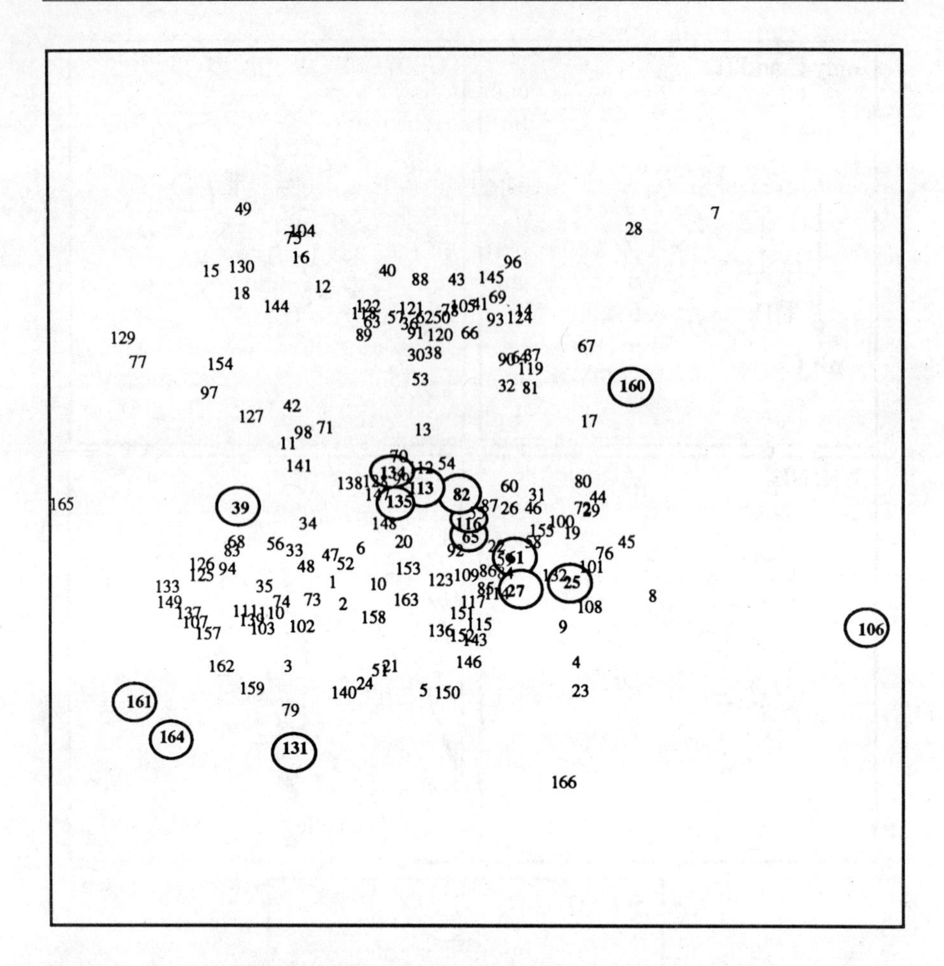

Figure 6 Graphical display on the nonlinear map of the substituents constituting the best chromosome when Eq. (6) is used as fitness function.

(1994a). As a result, we assume that the nonlinear map (Figure 3) can be used for estimating the quality of the selections performed by the GA.

Genetic selection of test series

The parameters used to run the GA were optimized by means of a trial and error procedure from a huge number of runs. A chromosome length of 15 genes (i.e., aromatic substituents) was selected. This represented a rather good dimension for a matrix containing 166 substituents. The population size was fixed to 200. The crossover and mutation probabilities equaled 0.8 and 0.01, respectively. A 75% population permutation was selected. The GA was stopped after 1000 generations or if no improvement of the best chromosome was observed during 300 generations.

The substituents constituting the best chromosome selected by the GA, when Eq. (6) is used as fitness function, are circled in Figure 6. On this figure there is an aggregation of the selected substituents in the middle part of the map (i.e., substituents nos. 25, 27, 61, 65, 82, 113, 116, 134, 135). On the contrary, substituents in other parts of the map are not selected (e.g., substituents nos. 12, 66, 77, 102, and 166). It is also noteworthy that some selected substituents present rather similar structures (e.g., $CH{=}CHCO_2CH_3$, $CH{=}CHCOC_2H_5$, $CH{=}CHCO_2C_2H_5$, substituents nos. 113, 134, and 135, respectively). When the physicochemical data of the 15 selected substituents are represented on the nonlinear map, it appears that some substituents selected bear similar physicochemical properties (Figure 7). For substituents numbers 161 and 164, for example, it can be noted that they possess almost identical physicochemical properties for the six parameters studied. This is also the case for substituents numbers 113, 134, and 135. From the above, it appears that Eq. (6) does not represent a valuable fitness function for designing optimal test series.

The second fitness function (Eq. (8)) using the mean Euclidean coefficient without constraint does not give satisfactory results (Figure 8). As could be expected with this fitness function, the substituents located in the periphery of the cloud of points are selected. Figure 9 shows that even if the selected substituents can present largely different physicochemical properties (large squares and circles), the problem encountered with Eq. (6) (i.e., substituents presenting rather similar structures and/or physicochemical constant values) still exists. Thus, for example, the GA using Eq. (8) selects both NH_2 and $NHCH_3$ (substituents numbers 15 and 49, respectively). In the same way, substituents numbers 161 and 164 which have similar properties are also selected in the same chromosome. In addition, it is interesting to note that most of the selected substituents present an extreme value for at least one parameter. This implies that many can be regarded as outliers, for example $N(CH_3)_3^+$ (substituent number 106) which is atypical due to its positive charge. From a practical point of view, it is obvious that the selection of too many substituents in the periphery of the cloud of points and/or outliers is not ideal. Therefore, Eq. (8) appears to be not suitable for optimal test series selection.

When a constraint is added (Eq. (9) with $x = 0.3$), a slightly better selection is obtained (Figure 10). However, the repartition of the selected points on the nonlinear map is not ideal since many regions in the middle of the map (Figure 10) are not represented and many outliers are still selected. This is confirmed in Figure 11 when the values of the six parameters for the 15 substituents are plotted on the nonlinear map.

Eq. (10) which is based on the minimum Euclidean distance allows the selection of better test series (Figure 12). Indeed, the repartition of the substituents on the nonlinear map is wide so that we can consider that

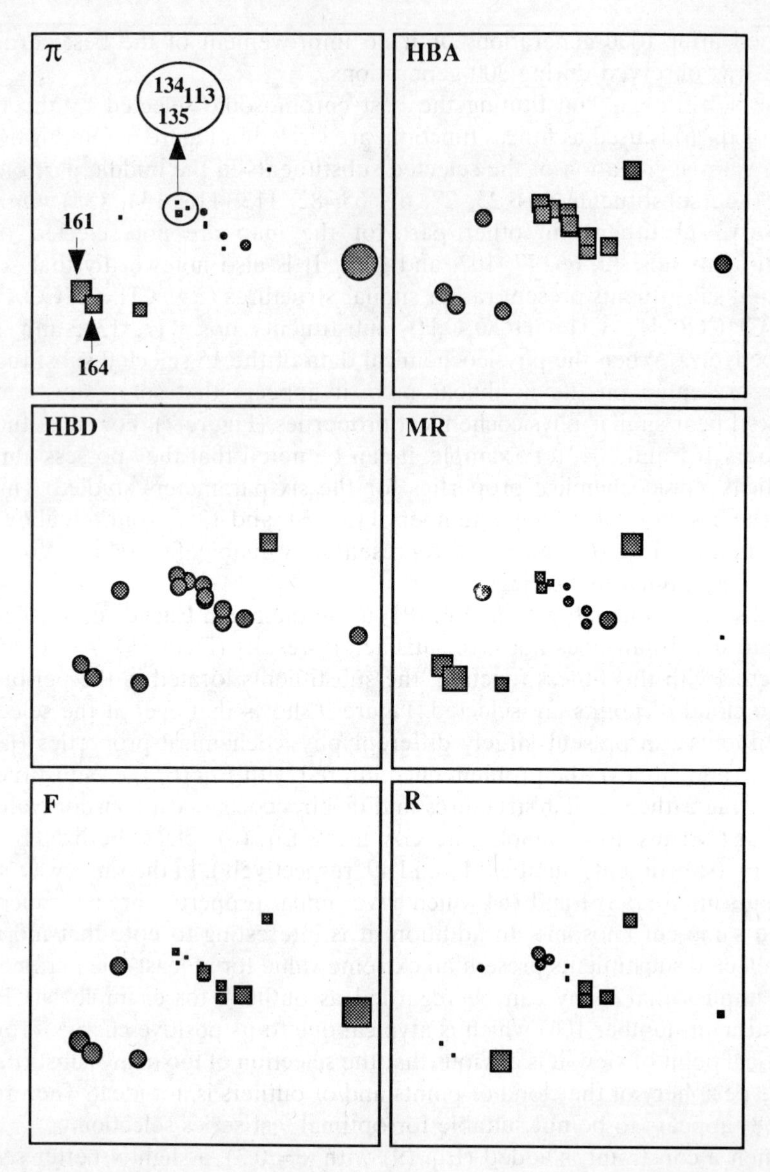

Figure 7 Plot of the scaled values of the six parameters for the 15 substituents selected with Eq. (6) on the nonlinear map. Squares (positive values) and circles (negative values) are proportional in size to the magnitude of the parameters. The other points of the nonlinear map are not represented. For the HBA map, the squares represent HBA substituents while the circles indicate the selected substituents which do not have the ability to accept H-bonds. The same applies to the HBD map.

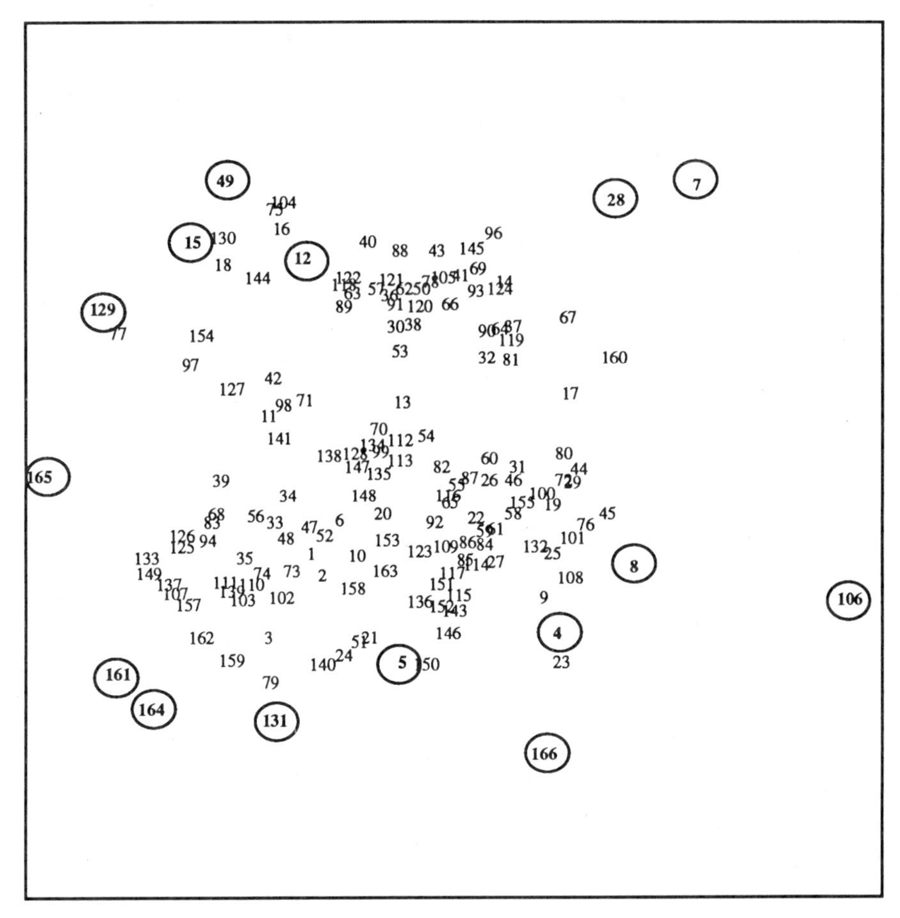

Figure 8 Graphical display on the nonlinear map of the substituents constituting the best chromosome when Eq. (8) is used as fitness function.

the space of physicochemical properties is efficiently represented. The representation of the physicochemical constant data for the 15 selected substituents (Figure 13) shows that for each parameter a wide variation in the values is obtained. Thus, for π, squares and circles of different sizes are found on the map, indicating that this property is well represented. In the same way, for the Swain and Lupton F parameter, the map (Figure 13) contains seven circles and height squares of different sizes. In addition, Eq. (10) prevents the selection of similar substituents. Indeed, even if there are substituents which appear relatively close to each other, inspection of Figure 13 indicates that they present sufficiently different properties to justify their selection by the GA. Thus, for example, if we consider $B(OH)_2$ and $NHC{=}S(NHC_2H_5)$ (i.e., substituents numbers 14 and 96), Figure 13 indicates

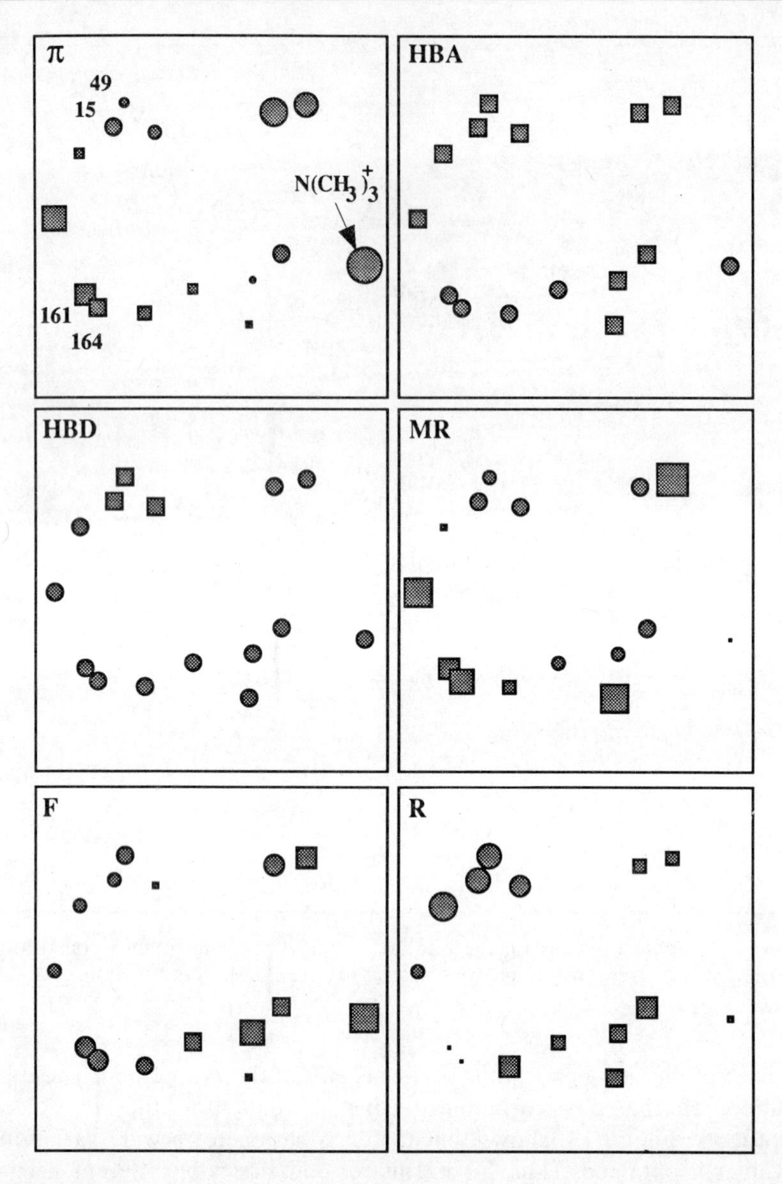

Figure 9 Plot of the scaled values of the six parameters for the 15 substituents selected with Eq. (8) on the nonlinear map. Squares (positive values) and circles (negative values) are proportional in size to the magnitude of the parameters. The other points of the nonlinear map are not represented. For the HBA map, the squares represent HBA substituents while the circles indicate the selected substituents which do not have the ability to accept H-bonds. The same applies to the HBD map.

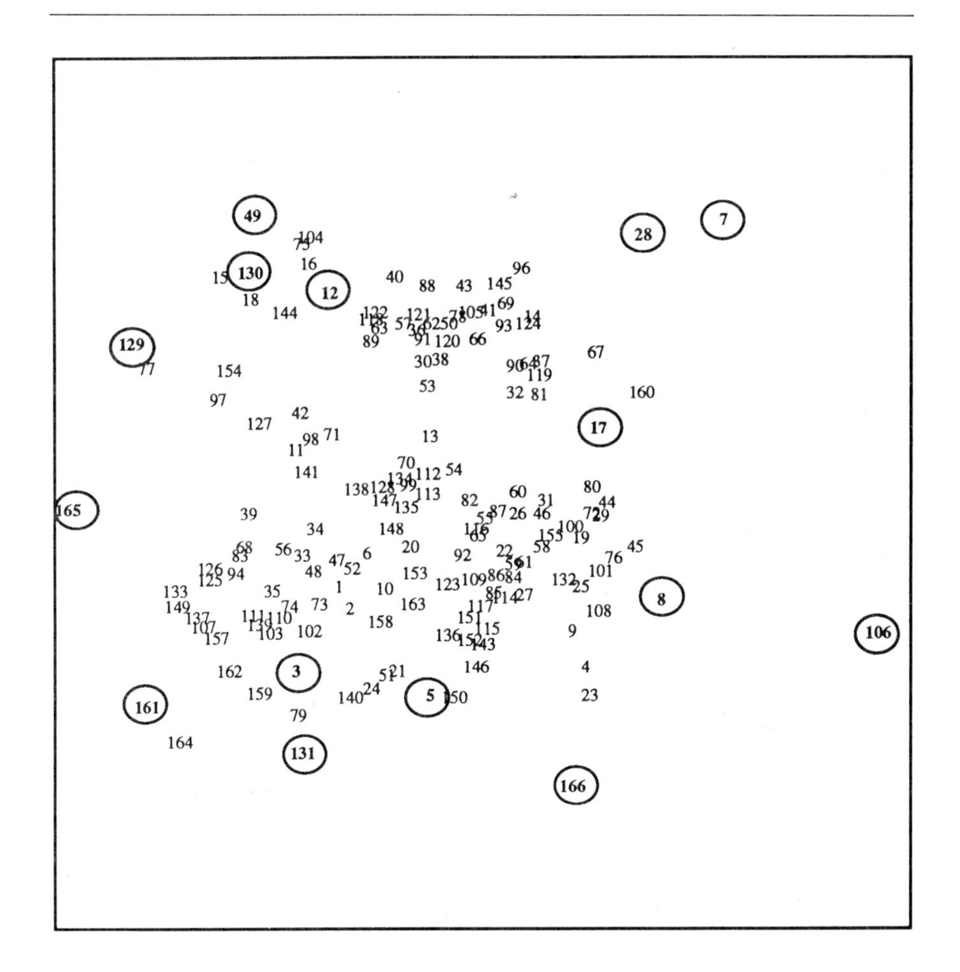

Figure 10 Graphical display on the nonlinear map of the substituents constituting the best chromosome when Eq. (8) with a constraint (Eq. (9)) is used as fitness function.

that their MR, F, and R values are significantly different.

The results obtained when a constraint (Eq. (9) with $x = 0.2$) is added to Eq. (10), are also interesting (Figure 14). Projection of the physicochemical properties (Figure 15) allows us to draw similar conclusions as with Eq. (10) used alone.

However, even if the best results have been obtained with the last two fitness functions, it is obvious that it could still be optimized since some regions of the nonlinear map still remain not well represented.

In order to assess the results obtained with the different fitness functions, a variance coefficient (Eq. (11)) was also calculated (Table II). The values

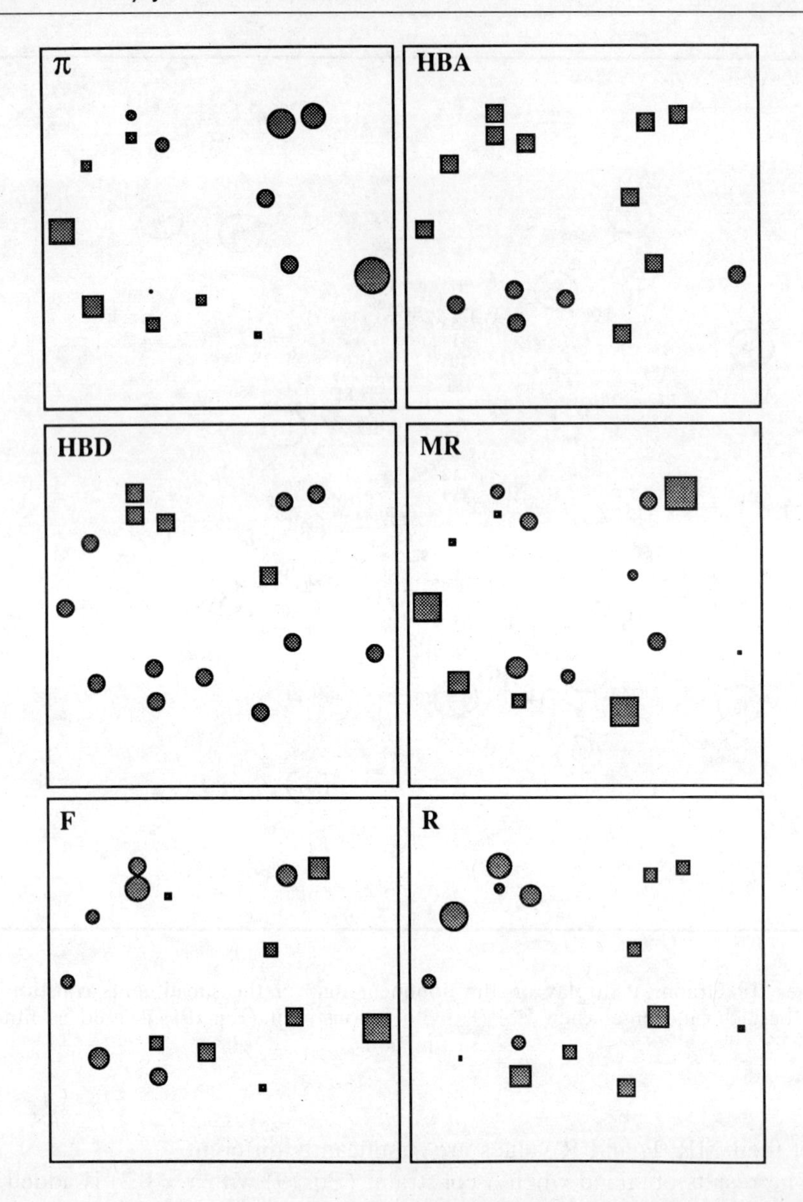

Figure 11 Plot of the scaled values of the six parameters for the 15 substituents selected with Eqs. (8) and (9) on the nonlinear map. Squares (positive values) and circles (negative values) are proportional in size to the magnitude of the parameters. The other points of the nonlinear map are not represented. For the HBA map, the squares represent HBA substituents while the circles indicate the selected substituents which do not have the ability to accept H-bonds. The same applies to the HBD map.

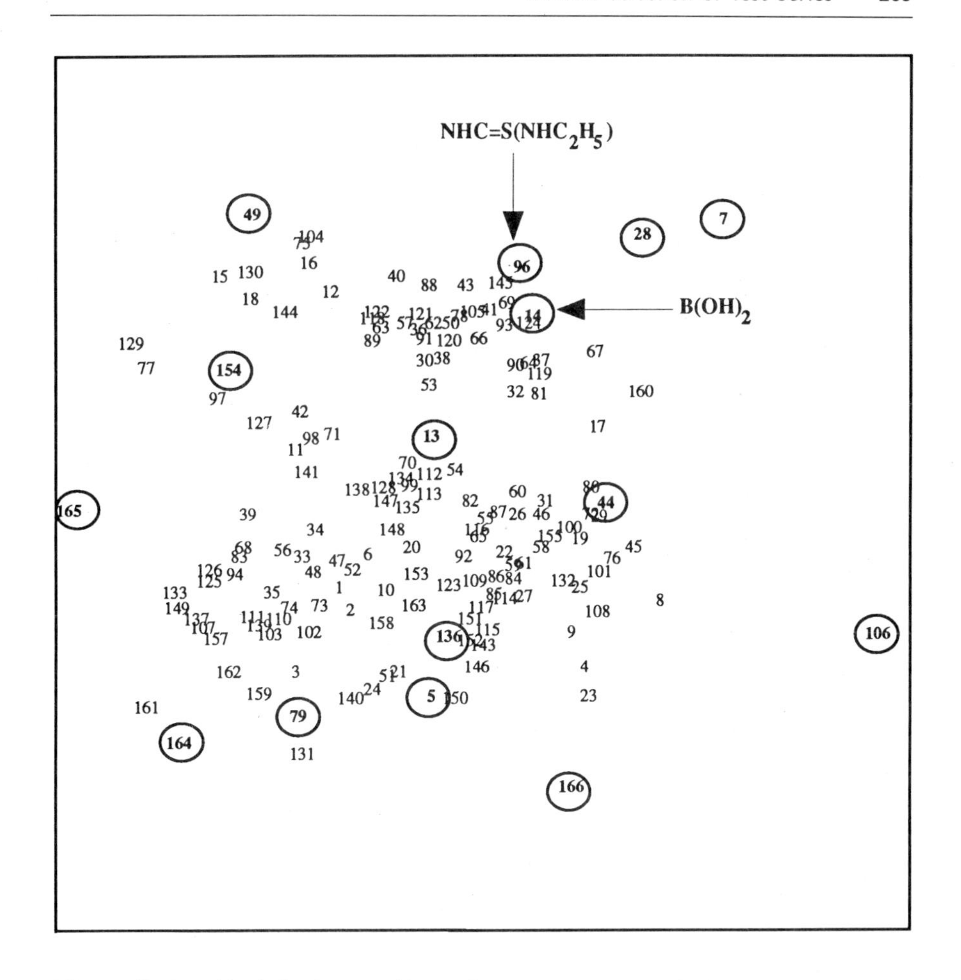

Figure 12 Graphical display on the nonlinear map of the substituents constituting the best chromosome when Eq. (10) is used as fitness function.

Table II *Variance coefficient for the best chromosome obtained from the different fitness functions.*

Fitness function	Variance coefficient
Eq. (6)	1.223
Eq. (8)	2.717
Eqs. (8)(9)	2.575
Eq. (10)	2.187
Eqs. (10)(9)	2.147

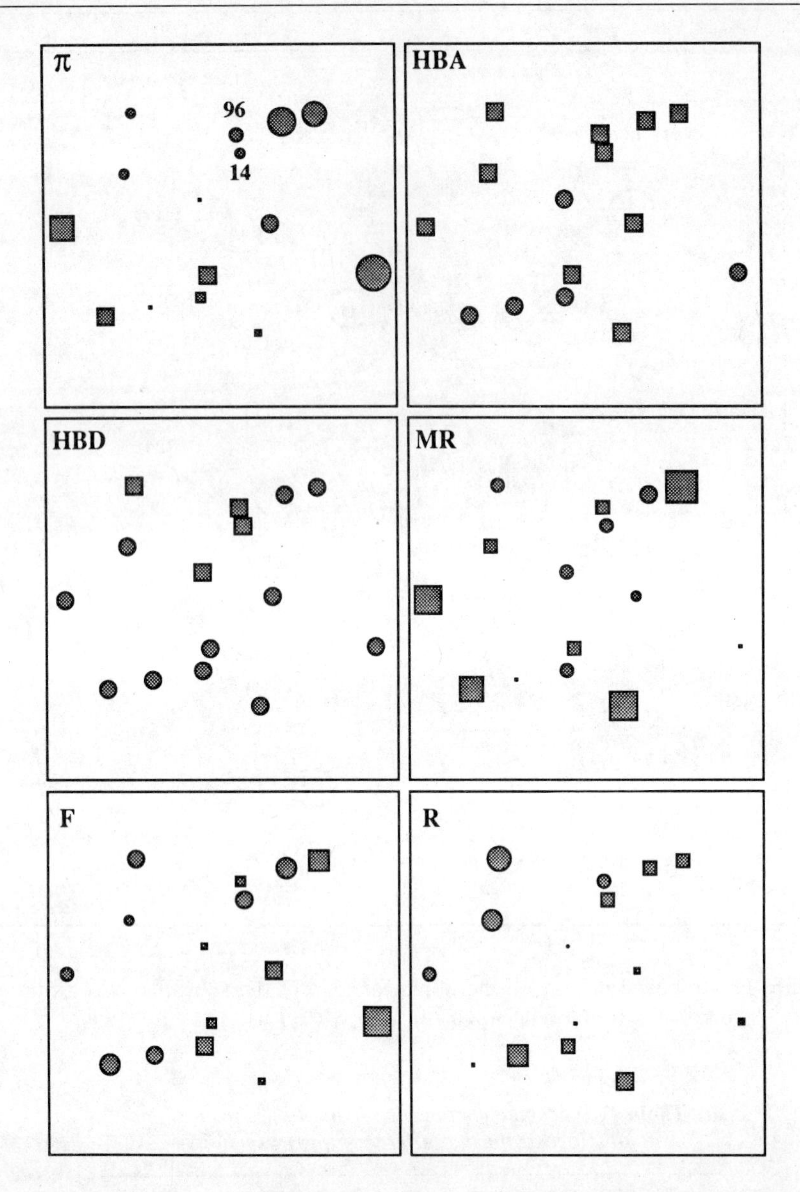

Figure 13 Plot of the scaled values of the six parameters for the 15 substituents selected with Eq. (10) on the nonlinear map. Squares (positive values) and circles (negative values) are proportional in size to the magnitude of the parameters. The other points of the nonlinear map are not represented. For the HBA map, the squares represent HBA substituents while the circles indicate the selected substituents which do not have the ability to accept H-bonds. The same applies to the HBD map.

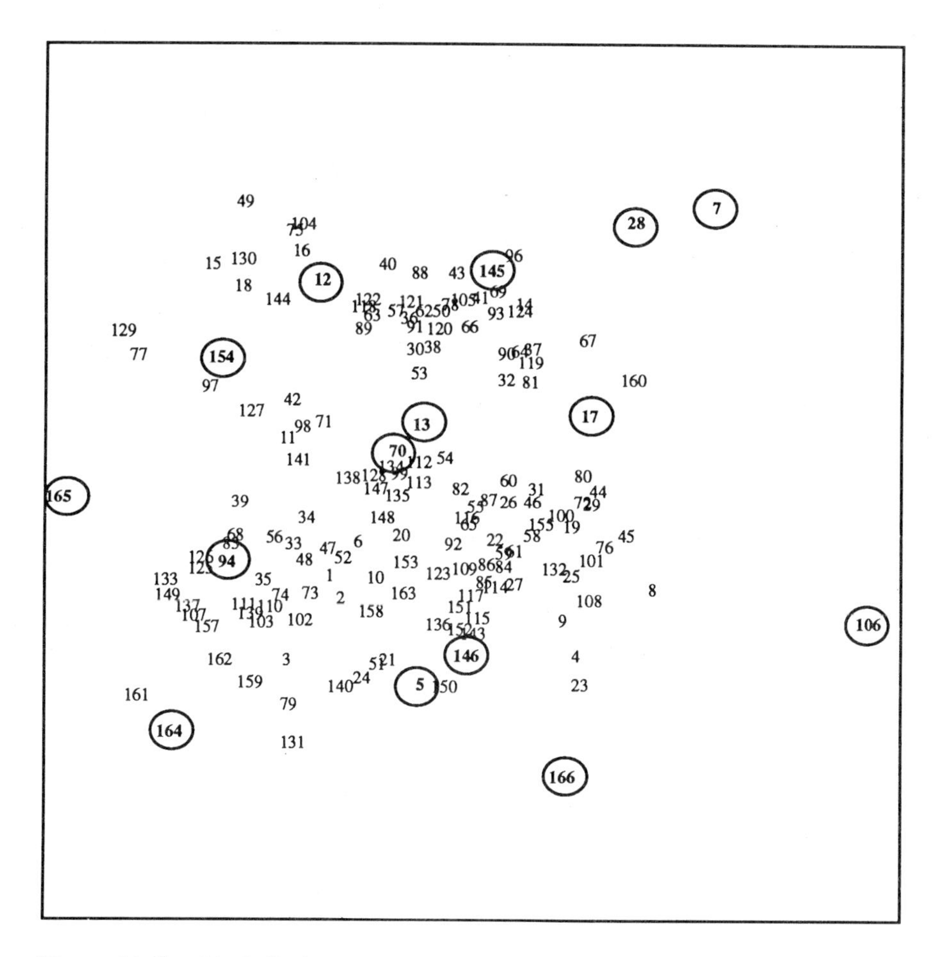

Figure 14 Graphical display on the nonlinear map of the substituents constituting the best chromosome when Eq. (10) with a constraint (Eq. (9)) is used as fitness function.

reported in Table II indicate that the lowest variance value is obtained with Eq. (6). It corresponds to large aggregations of points on the nonlinear map (Figure 6). The highest variance value is obtained with Eq. (8). However, this equation yields the selection of substituents only located in the periphery of the cloud of points on the nonlinear map (Figure 8). Eq. (8) with a constraint (Eq. (9) yields a slightly lower value. When the selection is plotted on the nonlinear map and the structures and properties of the selected substituents are compared (Figures 10 and 11), it appears that slightly better results are obtained. Lower variance coefficients are obtained with Eq. (10) (alone) and Eq. (10) with a constraint (Eq. (9)). On the map, this results in

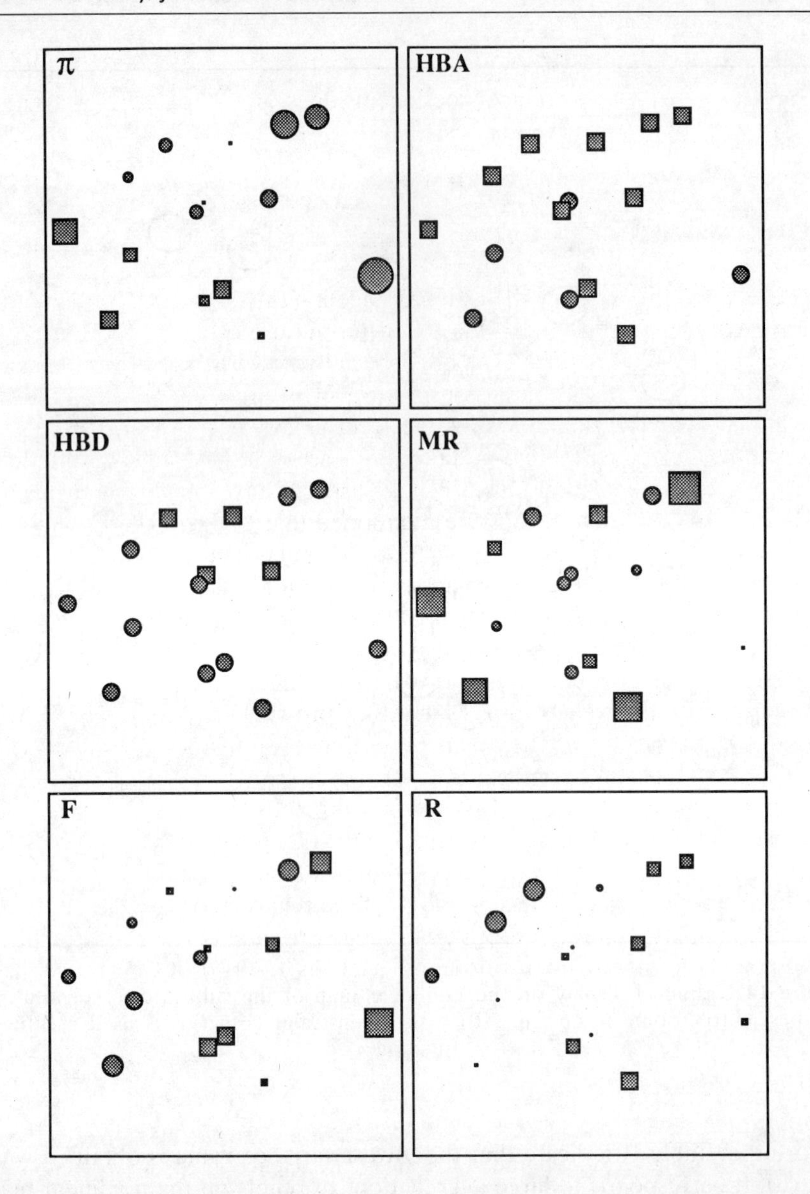

Figure 15 Plot of the scaled values of the six parameters for the 15 substituents selected with Eqs. (10) and (9) on the nonlinear map. Squares (positive values) and circles (negative values) are proportional in size to the magnitude of the parameters. The other points of the nonlinear map are not represented. For the HBA map, the squares represent HBA substituents while the circles indicate the selected substituents which do not have the ability to accept H-bonds. The same applies to the HBD map.

a better representation of the physicochemical space. Under these conditions, it appears that the variance coefficient does not represent an optimal parameter for the selection of test series.

CONCLUSION

In this study, a method for the rational selection of test series with high information content is presented. It is based on the use of a GA. Our approach is particularly suitable since GAs are especially powerful for exploring spaces with a large number of dimensions. However, it must be kept in mind that the aim in such a study is not only to select the very best set. Indeed, in some cases, other constraints must be taken into account, such as synthetic feasibility, cost, reactivity and so on. These aspects have major importance in practical drug research. It must be mentioned that to direct the GA towards the solution optimizing all the possible constraints such as those cited above, it is possible to include in the fitness equation terms encoding them. In the procedure based on the GA method, these aspects can be considered at the beginning of the optimization process. Also, in this study, we have only represented the best chromosome for each fitness function. However, the advantage of genetic algorithms, is that they provide a population of valuable chromosomes and all of them can be useful to optimize a selection taking into account the other constraints.

Our preliminary results are very promising and convincing. Simple calculations provide representative series of substituents and we do not only end up with one best series but with a population which allows synthetic chemists or those who perform the tests some freedom regarding the selection. Lastly, it must be pointed out that the methodology presented here is open and can be applied to many other problems, such as multisubstitution problems, protein sequences, priority setting, and so on.

REFERENCES

Alunni, S., Clementi, S., Edlund, U., Johnels, D., Hellberg, S., Sjöström, M., and Wold, S. (1983). Multivariate data analysis for substituent descriptors. *Acta Chem. Scand. B* **37**, 47–53.

Bowden, K. (1990). Electronic effects in drugs. In, *Comprehensive Medicinal Chemistry* (C.A. Ramsden, Ed.). Pergamon Press, Oxford, Vol. 4, pp. 205–239.

Brown, R.D., Jones, G., Willett, P., and Glen, R.C. (1994). Matching two-dimensional chemical graphs using genetic algorithms. *J. Chem. Inf. Comput. Sci.* **34**, 63–70.

Craig, P.N. (1971). Interdependence between physical parameters and selection of substituent groups for correlation studies. *J. Med. Chem.* **14**, 680–684.

Davis, L. (1991). *Handbook of Genetic Algorithms.* Van Nostrand Reinhold, New York, p. 385.

Dearden, J.C. (1990). Physico-chemical descriptors. In, *Practical Applications of Quantitative Structure–Activity Relationships (QSAR) in Environmental Chemistry and Toxicology* (W. Karcher and J. Devillers, Eds.). Kluwer Academic Publishers, Dordrecht, pp. 25–59.

Devillers, J. (1995). Display of multivariate data using non-linear mapping. In, *Chemometric Methods in Molecular Design* (H. van de Waterbeemd, Ed.). VCH, Weinheim, pp. 255–263.

Domine, D., Devillers, J., and Chastrette, M. (1994a). A nonlinear map of substituent constants for selecting test series and deriving structure–activity relationships. 1. Aromatic series. *J. Med. Chem.* **37**, 973–980.

Domine, D., Devillers, J., and Chastrette, M. (1994b). A nonlinear map of substituent constants for selecting test series and deriving structure–activity relationships. 2. Aliphatic series. *J. Med. Chem.* **37**, 981–987.

Domine, D., Devillers, J., Chastrette, M., and Karcher, W. (1993). Non-linear mapping for structure–activity and structure–property modelling. *J. Chemom.* **7**, 227–242.

Domine, D., Wienke, D., Devillers, J., and Buydens, L. (1996). A new nonlinear neural mapping technique for visual exploration of QSAR data. In, *Neural Networks in QSAR and Drug Design* (J. Devillers, Ed.). Academic Press, London (in press).

Dove, S., Streich, W.J., and Franke, R. (1980). On the rational selection of test series. 2. Two-dimensional mapping of intraclass correlation matrices. *J. Med. Chem.* **23**, 1456–1459.

Dunn, W.J. (1977). Molar refractivity as an independent variable in quantitative structure–activity studies. *Eur. J. Med. Chem. – Chim. Ther.* **12**, 109–112.

Eshelman, L.J., Caruana, R.A., and Schaffer, J.D. (1989). Biases in the crossover landscape. In, *Proceedings of the Third International Conference on Genetic Algorithms* (J.D. Schaffer, Ed.). Morgan Kaufmann Publishers, San Mateo, pp. 10–19.

Goldberg, D.E. (1989). *Genetic Algorithms in Search, Optimization and Machine Learning.* Addison-Wesley Publishing Company, Reading, p. 412.

Goodford, P.J., Hudson, A.T., Sheppey, G.C., Wootton, R., Black, M.H., Sutherland, G.J., and Wickham, J.C. (1976). Physicochemical-activity relationships in asymmetrical analogues of methoxychlor. *J. Med. Chem.* **19**, 1239–1247.

Hansch, C. and Fujita, T. (1964). ρ-σ-π Analysis. A method for the correlation of biological activity and chemical structure. *J. Am. Chem. Soc.* **86**, 1616–1626.

Hansch, C. and Leo, A. (1979). *Substituent Constants for Correlation Analysis in Chemistry and Biology.* John Wiley & Sons, New York.

Hansch, C. and Leo, A. (1995). *Exploring QSAR. Fundamentals and Applications in Chemistry and Biology.* American Chemical Society, Washington, DC, p. 557.

Hansch, C., Leo, A., and Taft, R.W. (1991). A survey of Hammett substituent constants and resonance and field parameters. *Chem. Rev.* **91**, 165–195.

Hansch, C., Unger, S.H., and Forsythe, A.B. (1973). Strategy in drug design. Cluster analysis as an aid in the selection of substituents. *J. Med. Chem.* **16**, 1217–1222.

Karcher, W. and Devillers, J. (1990). *Practical Applications of Quantitative Structure–Activity Relationships (QSAR) in Environmental Chemistry and Toxicology*. Kluwer Academic Publishers, Dordrecht, p. 475.

Lucasius, C.B. and Kateman, G. (1994). Understanding and using genetic algorithms. Part 2. Representation, configuration and hybridization. *Chemom. Intell. Lab. Syst.* **25**, 99–145.

Ludwig, M., Wold, S., and Exner, O. (1992). The role of *meta* and *para* benzene derivatives in the evaluation of substituent effects: A multivariate data analysis. *Acta Chem. Scand.* **46**, 549–554.

Michalewicz, Z. (1992). *Genetic Algorithms + Data Structures = Evolution Programs*. Springer-Verlag, Berlin, p. 250.

Sammon, J.W. (1969). A nonlinear mapping for data structure analysis. *IEEE Trans. Comput.* **C18**, 401-409.

Silipo, C. and Vittoria, A. (1991). *QSAR: Rational Approaches to the Design of Bioactive Compounds*. Elsevier, Amsterdam, p. 575.

Simon, V., Gasteiger, J., and Zupan, J. (1993). A combined application of two different neural network types for the prediction of chemical reactivity. *J. Am. Chem. Soc.* **115**, 9148–9159.

STATQSAR (1994). CTIS, Lyon, France.

Streich, W.J., Dove, S., and Franke, R. (1980). On the rational selection of test series. 1. Principal component method combined with multidimensional mapping. *J. Med. Chem.* **23**, 1452–1456.

Swain, C.G. and Lupton, E.C. (1968). Field and resonance components of substituent effects. *J. Am. Chem. Soc.* **90**, 4328–4337.

Taft, R.W. (1953). The general nature of the proportionality of polar effects of substituent groups in organic chemistry. *J. Am. Chem. Soc.* **75**, 4231–4238.

Topliss, J.G. (1972). Utilization of operational schemes for analog synthesis in drug design. *J. Med. Chem.* **15**, 1006–1011.

Turner, D.B., Willett, P., Ferguson, A.M., and Heritage, T.W. (1995). Similarity searching in files of three-dimensional structures: Evaluation of similarity coefficients and standardisation methods for field-based similarity searching. *SAR QSAR Environ. Res.* **3**, 101–130.

van de Waterbeemd, H., El Tayar, N., Carrupt, P.A., and Testa, B. (1989). Pattern recognition study of QSAR substituent descriptors. *J. Comput.-Aided Mol. Design* **3**, 111–132.

Wootton, R. (1983). Selection of test series by a modified multidimensional mapping method. *J. Med. Chem.* **26**, 275–277.

Wootton, R., Cranfield, R., Sheppey, G.C., and Goodford, P.J. (1975). Physicochemical-activity relationships in practice. 2. Rational selection of benzenoid substituents. *J. Med. Chem.* **18**, 607–613.

11 Computer-Aided Molecular Design Using Neural Networks and Genetic Algorithms

V. VENKATASUBRAMANIAN*, A. SUNDARAM,
K. CHAN, and J.M. CARUTHERS
Laboratory for Intelligent Process Systems, School of Chemical Engineering, Purdue University, West Lafayette, IN 47907, USA

The design of materials possessing desired physical, chemical and biological properties is a challenging problem in the chemical, petrochemical and pharmaceutical industry. This involves modeling important interactions between basic structural units for property prediction as well as efficiently locating viable structures that can yield desired performance on synthesis. This chapter describes a neural network (NN) based approach for solving the forward problem of property prediction based on the structural characteristics of the molecular sub-units, and a genetic algorithm (GA) based approach for the inverse problem of constructing a molecular structure given a set of desired macroscopic properties. The neural network approach was found to develop nonlinear structure–property correlations more accurately and easily in comparison with other conventional approaches. Similarly, the genetic algorithm was found to be very effective in locating globally optimal targets for the design of fairly complex polymer molecules.

To gain a better understanding of the nature of the search space and the mechanics of GA's search, we explored the search space in detail in an attempt to analyze and characterize its features. A pseudo-Hamming distance analysis using 'super-groups' suggested that the average fitness and the maximum fitness of the molecules increased as they approached the target, a feature which the genetic algorithm exploits through its 'survival of the fittest' principle to carry out an effective search. Since real-life CAMD problems are so difficult and complex, a successful approach would require a combination of several

* Author to whom all correspondence should be addressed.

In, *Genetic Algorithms in Molecular Modeling* (J. Devillers, Ed.)
Academic Press, London, 1996, pp. 271–302.
ISBN 0-12-213810-4

techniques integrated into a unified, interactive, framework. We present one such framework with the GA-based design engine integrated with expert systems and math programming methodologies to improve the effectiveness of the overall design process. We also envision such a system as an interactive one, where the molecular designer actively participates in the design process.

KEYWORDS: *computer aided molecular design; neural networks; genetic algorithms; QSAR; polymer molecules.*

INTRODUCTION

Computer aided molecular design: background

The process of designing new molecules possessing desired physical, chemical and biological properties is an important endeavor in chemical, material and

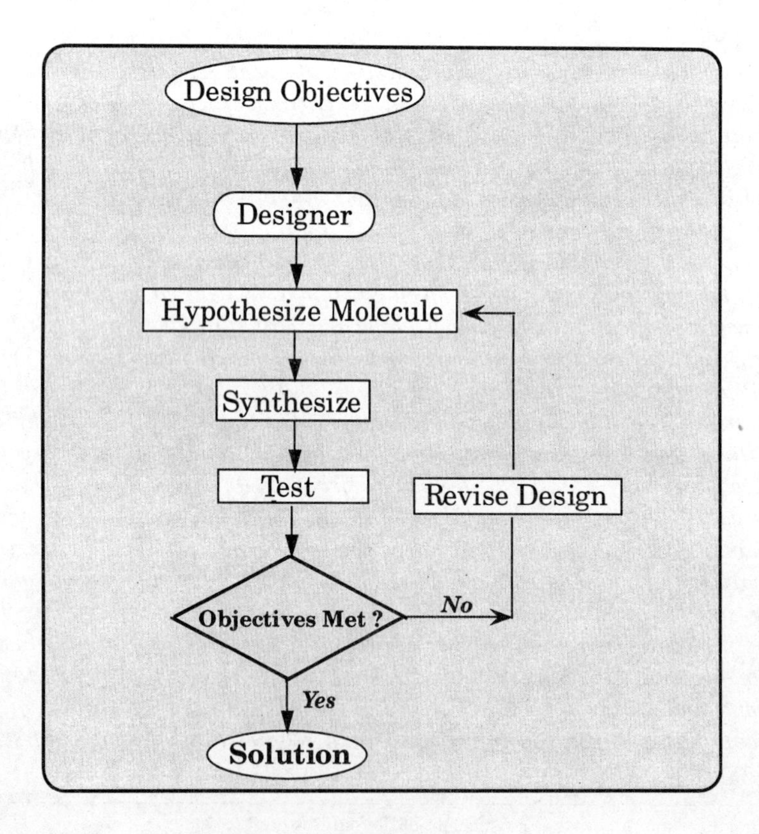

Figure 1 The molecular design problem.

pharmaceutical industries. Industrial applications include designing composites and blends, drugs, agricultural chemicals such as pesticides or herbicides, refrigerants, solvents, paints and varnishes, etc. With recent developments such as stricter penalties on environmentally unfriendly products and emphasis on value-added products and designer molecules, the search for novel materials has become an essential part of research in the above fields. The traditional approach for formulating new molecules requires the designer to hypothesize a compound, synthesize the material, evaluate to see if it meets the desired design targets, and to reformulate the design if the desired properties are not achieved. Figure 1 shows a schematic of this iterative design process. This process is usually lengthy and expensive, involving the preliminary screening of dozens or hundreds of candidates. Computer tools such as molecular graphics, molecular modeling, reaction pathway synthesis, and structure–property prediction systems have somewhat alleviated the time and expense involved in this process. However, there are still major problems that remain to be addressed as discussed below.

Forward problem in CAMD

The explosion in computational power of modern computers as well as their inexpensive availability has prompted the development of computer-assisted procedures for designing new materials to ease the protracted design, synthesis and evaluation cycle. Computational molecular design systems require the solution of two problems: the 'forward' problem which predicts physical, chemical and biological properties from the molecular structure, while the 'inverse' problem requires the identification of the appropriate molecular structure given the desired macroscopic properties (Figure 2).

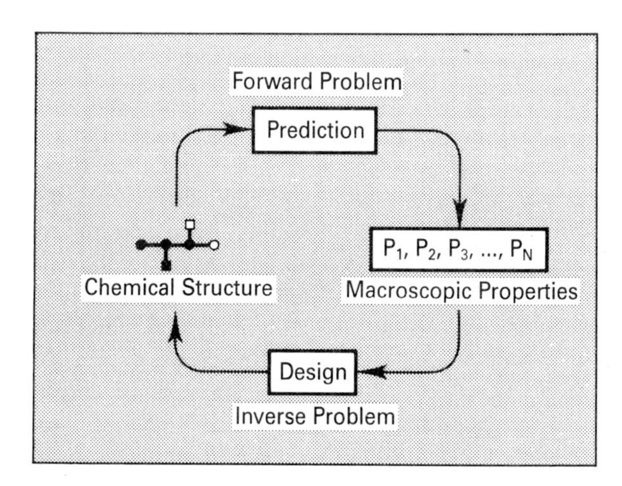

Figure 2 Iterative molecular design process.

In the last decade, significant advances have been made in a variety of methods for structure–property predictions including molecular and empirical models, group contribution methods, quantitative structure–activity relationships (QSARs) and other correlation methods, etc. (Reid *et al.*, 1977, 1987; Joback and Reid, 1987). However, most of these approaches usually assume that physicochemical properties depend on chemical structure in a relatively simple manner. Hence, they are generally less successful for complex molecules with nonlinear structure–property relationships.

Inverse problem in CAMD

The inverse problem has similarly been approached by a variety of methods. The earliest work of Gani and Brignole (1983) and Brignole and coworkers (Brignole *et al.*, 1986) presented an enumeration approach for solvent design for liquid–liquid extraction. The enumeration approach was also used by Derringer and Markham (1985) for polymer design. Following this, a heuristic-guided enumeration technique to combat combinatorial explosion was proposed by Joback and Stephanopoulos (1989) for polymer, refrigerant, and solvent design. Subsequently, Macchietto and coworkers (Macchietto *et al.*, 1990) applied a mixed integer nonlinear programming (MINLP) formulation to the design of solvents for liquid–liquid extraction and gas absorption. Knowledge-based approaches for drug design (Bolis *et al.*, 1991), polymer design as well as refrigerant and solvent design have been reported (Gani *et al.*, 1991). Zefirov and coworkers (Skvortsova *et al.*, 1993) and Kier and coworkers (Kier *et al.*, 1993) developed graph reconstruction methods in which molecular structures can be derived from a QSAR based on topological indices.

While all these methods have some appeal, they suffer from drawbacks due to combinatorial complexity of the search space, design knowledge acquisition difficulties, nonlinear structure–property correlations and problems in incorporating higher-level chemical and biological knowledge. The combinatorial complexity associated with CAMD makes random, exhaustive or heuristic enumeration procedures less effective for large-scale molecular design problems. Mathematical programming methods consider molecular design as an optimization problem where the objective is to minimize the error between the desired target values and the values attained by the current design. The solutions to MINLP formulations are susceptible to local minima traps for problems with nonlinear constraints, as is the case with most structure–property relations. The solutions to these problems are also computationally very expensive, especially for highly nonlinear systems. The knowledge-based systems approach assumes that expert rules exist for manipulating chemical structures to achieve the desired physical properties. However, many nonlinear structure–property relationships can not be easily expressed as rules, especially when designing for multiple design objectives. Furthermore, extraction of such design expertise from experts on molecular design is not easy, making the knowledge acquisition problem a very difficult one. Lastly, graph reconstruction methods require expressing all structure–property relations in terms of topological indices. This may not be appropriate or feasible for all properties. Furthermore, topological indices are not unique and there currently does not exist a general graph reconstruction method for all molecular indices. In addition, since one often deals with a number of design criteria to be satisfied and not just one or two properties, this approach may not be feasible in general.

Using genetic algorithms and neural networks

Recently, Venkatasubramanian and coworkers (Venkatasubramanian *et al.*, 1994, 1995) proposed an evolutionary molecular design approach using genetic algorithms (GAs) to address the drawbacks of the other methods for the inverse problem. GAs are general purpose, stochastic, evolutionary search and optimization strategies based on the Darwinian model of natural selection and evolution. Their study showed that the genetic design (GD) approach was able to locate globally optimal designs for many target molecules with multiple specifications.

In this paper, we extend this approach further by describing a CAMD framework that uses both neural networks and genetic algorithms to address the difficulties in both the forward and inverse problems. The neural network (NN) based property prediction methodology addresses the forward problem, while the genetic algorithmic component tackles the inverse problem. This paper also discusses some recent results on the characterization and analysis of complex search spaces for molecular design to gain further insight into the operation of GAs. The paper also describes an interactive CAMD system that is under development.

THE FORWARD PROBLEM USING NEURAL NETWORKS

Approaches to the forward problem

The forward problem refers to the prediction of macroscopic properties of a product given an abstraction of its basic structure. Depending on the kind of final product desired, the basic structure can be as detailed as a three-dimensional atomic/molecular structure or a formulation of the basic components of the final product. Since this paper is focused on molecular design, the basic structure of interest to us is the molecule. Forward methods or property estimation methods can be broadly classified into three categories: (1) group contribution, (2) equation oriented, and (3) model-based techniques. The model-based approaches are further subdivided into: (a) topological, (b) pattern recognition, (c) molecular modeling methods.

The group contribution or additivity approach assumes that a physical property is equal to the sum of the individual contributions of the molecule

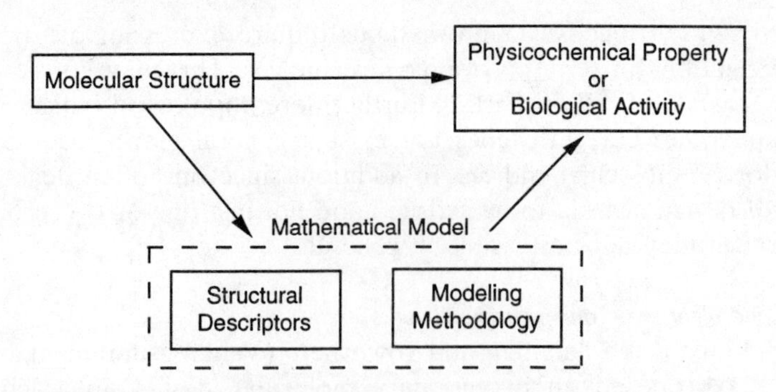

Figure 3 Schematic diagram of model-based approaches for structure-property/activity studies.

fragments or groups to that property. It is developed by utilizing experimental data for a large number of compounds in a regression scheme to estimate the contributions of each structural fragment. Additivity schemes for estimating thermodynamic physicochemical properties of small molecules are well developed and widely used (Reid *et al.*, 1977, 1987). Van Krevelen and Hoftyzer (1972, 1976, 1992), presented group methods as well as semi-empirical relationships utilizing group additive parameters for thermal, mechanical, optical, electrical, and rheological properties of polymers. Equation oriented approaches assume the dependence of a final physico-chemical property on a basic set of other physico-chemical properties. This approach is essentially empirical in nature and is of limited utility.

Model-based approaches represent the molecular structure by a set of descriptors in order to develop a mathematical model which relates descriptors to physicochemical properties or biological activity. These approaches are in general known as structure–property or structure–activity methods (QSARs) as shown in Figure 3. These are further classified based on the nature of descriptors as discussed below.

Topological techniques

Topological techniques are based on graph theory. They consider molecular structure as planar graphs where atoms are represented by vertices and chemical bonds by edges. The matrix representation of the chemical constitutional graph is transformed to an invariant which is normally an index. These non-empirical descriptors are sensitive to constitutional features such as size, shape, symmetry, and atom heterogeneity. Structure activity models have been reviewed in detail by Kier and Hall (1976, 1986).

Pattern recognition

The basis of pattern recognition techniques is that a collection of structural descriptors representing points in high-dimensional space will assemble in localized regions of common physicochemical property. Thus, the assumption is similar chemical structures exhibit similar physicochemical properties. The objective is to find discriminant functions or decision regions which separate the data into subsets (i.e. classes) with similar properties. There has been a considerable amount of work in applying pattern recognition techniques to biological, chemical, environmental, and pharmaceutical problems (Jurs and Isenhour, 1975; Strouf, 1986).

Molecular modeling techniques

Molecular modeling techniques are separated into three branches: (1) *ab initio* or Gaussian methods, (2) semi-empirical molecular orbital methods, and (3) molecular mechanics methods. Quantum mechanical methods are based on computing (*ab initio* methods) or approximating analytically (semi-empirical molecular orbital methods) the wave function from first principles, namely from the Schrödinger equation, without resorting to any kind of regression with experimental data. On the other hand, molecular mechanics is a computational realization of the ball-and-stick model (Burkert and Allinger, 1982). A molecule is considered to be a series of masses (atoms) attached by springs (bonds), where the interactions between the nuclei are treated according to the laws of classical mechanics. These force laws and its constants constitute the internal energy of a molecule or the force-field, V, which is then approximated by a sum of different types of energy contributions such as bond stretching, angle bending, bond rotations, non-bonded interactions, hydrogen bonding, etc. The molecular model is then parameterized by assigning the force constants and natural values for angles and lengths which allow for the reproduction of experimental data.

Comparison of the approaches to the forward problem

The property prediction methods may be evaluated based on their classification as empirical, semi-empirical, theoretical, and hybrid approaches. The empirical methods usually require extensive data collection and result in linear or simple nonlinear structure–property relations. Computations are very rapid at the expense of prediction accuracy. In addition, these methods require a specific functional form which may not always be available and the parameters determined by regression from the data. At the opposite end of the spectrum are theoretical approaches which require an extensive understanding of the first-principles to develop the models. They are also computationally expensive, but provide excellent property estimations. Most approaches settle for the middle ground by utilizing simplifying assumptions as those found in semi-empirical methods and hybrid approaches. These methods provide the best compromise between model development effort, computational time and property prediction accuracy. In this regard, NN based methods offer advantages of ease of development and implementation, and execution speed, while maintaining a high degree of accuracy of predictions. NN based models are relatively model free, in the sense that the underlying functional form is not as rigorous as in the traditional model based methods. This adds to the generality of these methods.

Neural networks for the forward problem

NN applications to chemistry have been explored in recent years. These include systems for secondary protein structure prediction (Qian and Sejnowski, 1988), QSAR parameters to determine biological activities (Aoyama *et al.*, 1990), functional group identification from mass and infrared spectra (Robb and Munk, 1990), connection tables for predicting chemical composition of aromatic substitution reactions (Elrod *et al.*, 1990) and a host of other applications which are summarized by Zupan and Gasteiger (1991). Sumpter and coworkers (Sumpter *et al.*, 1994) give an excellent review on the recent advances in neural network applications in chemistry ranging from molecular spectroscopy (Wythoff *et al.*, 1990) to applications to QSAR and quantitative structure–property relationships (QSPR) for polymers (Sumpter and Noid, 1994) and pharmaceuticals (Rose *et al.*, 1991).

NNs in general can be viewed as nonlinear regression models, where the functional structure is some combination of activation functions that operate on linear (the weighted sum of) inputs. Different types of activation functions (sigmoidal, radial, etc.), also known as basis functions, can be used and this adds to the generality of the NN structure. Specifying a basis function fixes the most basic functional unit of the NN. The complex interactions between the inputs are captured by the NN by adaptive alteration of its connection weights while matching the training output patterns. However, NNs are not completely parameter free since one still needs to specify the number of layers and the number of nodes in the hidden layers in the architecture and these drastically affect the performance of the NN. Cybenko (1988) has proved the capability of two layer NNs as universal function approximators, which helps with the number of layers issue. Yet, the choice of the number of nodes to use for a particular problem is not an easy one and is often determined by trial and error.

The NN-based methodology used in this paper treats polymer property prediction as one of function approximation. The function to be approximated is the one correlating a basic set of structural descriptors to a desired set of properties. In other words, we wish to approximate, using a NN, a function F such that:

$$\text{Property} = F(\text{structural descriptors})$$

The key element in the above approach is the selection of an appropriate set of structural descriptors. The chosen set of descriptors should adequately capture the phenomena underlying the macroscopic behavior or property of the material. Depending upon the property sought after and the nature of the material, the structural descriptors could be of any form. It is also important that these descriptors are obtained without a lot of computational effort since they have to be computed for every molecule whose property needs to be predicted, not to mention the ones used for training the NN. Once these structural descriptors are identified the NN is trained to accurately correlate the property values to the values of the descriptors. In this regard, traditional NN based function approximators use the backpropagation algorithm to arrive at the optimal set of connection weights. However, these weights may retain some redundancy due to the particular choice of descriptors and hence lead to poor generalization performance of the NN.

Neural network architecture for property prediction: polymers case studies

Figure 4 shows the schematic of the NN architecture for structure–property predictions. The polymer sub-structural sequence (at the bottom of the figure) represents a chemical structure to structural descriptor encoding. Molecular mechanics (MM) techniques were chosen over the less rigorous topological and physicochemical methods used to quantify chemical structure. The computational speed coupled with the accuracy with which molecular mechanics methods calculate the ground state energies and geometries make them more suitable than the much slower *ab initio* and semi-empirical methods.

Chemical structural descriptors are obtained from MM conformational analysis. Two monomer units were used in present MM simulations since it has been demonstrated that the conformational energy rapidly converges with increasing monomer units beyond three (Orchard *et al.*, 1987). However,

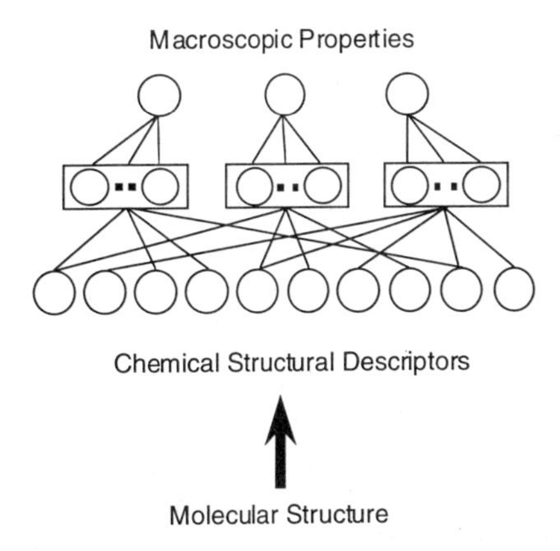

Figure 4 Neural network architecture for structure-property prediction.

for polymers with many flexible units, the high dimensionality of the dihedral angles or residuals to the contribution of the total energy prevent ready conceptualization of the features of the conformational energy surface. Further simplification is effected by representing the total energy, V, as a sum of contributions $V_i(f_i)$ from each of the several residuals of the polymer. This approximation of independent, first-neighbor interactions allows conceptualization of the conformational energy of a polymer of any size to a consideration of dimeric polymer segments. Thus, a judicious selection of dihedral angles representing the major relaxation processes is required.

Polymer configurations are another important consideration in the MM conformational analysis. Consequently, head-to-head, head-to-tail, and, tail-to-tail as well as syndiotactic, isotactic, and atactic forms must be considered. This leads to multiple MM simulations for each polymer. Average structural descriptors were then obtained from Boltzmann statistics. The configurational complexities encountered with copolymers, however, are not considered in this study.

The NN architecture consists of a single hidden layer and the number of nodes used in this layer is varied to estimate the improvement in the performance obtained. However, increasing the number of nodes in this layer can eventually lead to memorization of the training sets and poor generalization characteristics. The number of input nodes is fixed by the number of structural descriptors used and there is only one output node corresponding to the property being predicted. Sigmoidal activation functions were used as the basis functions.

The properties considered in the case studies were density and glass transition temperature. The objective of the case studies was to implement a NN based polymer property predictor and evaluate its performance in comparison to existing techniques. The case study also explores alternative architectures for the NN in an effort to study its influence on performance. The structural parameters for density and T_g were taken from van Krevelen and Hoftyzer (1990) and from Hopfinger and coworkers (1988). The polymer data for network training were taken from Hopfinger and coworkers (Hopfinger *et al.*, 1988) and are listed in Table I. Hopfinger and coworkers (1988) and Koehler and Hopfinger (1989) combined molecular mechanics, group additive methods, and multivariate regression analysis for developing molecular models for estimating the glass transition (T_g) and crystal-melt transition temperatures (T_m) of linear polymers. The intermolecular energetic contribution was later added to the previous intramolecular flexibility parameters. The final set of descriptors that are assumed to reflect the dominant type of polymer intermolecular interactions were, $\langle E_D \rangle$ the non-bonded/dispersion interactions, $\langle E_+ \rangle$ and $\langle E_- \rangle$ which are the electrostatic interactions. The trained network was then tested on a different polymer set to evaluate its generalization characteristics. Table II lists the polymer set used for the testing/generalization phase.

Table I *Polymer set used for NN training.*

1) Poly(methylene oxide)	18) Poly(hexadecyl 1-methacrylate)
2) Poly(ethylene oxide)	19) Poly(isopropyl methacrylate)
3) Poly(trimethylene oxide)	20) Poly(benzyl methacrylate)
4) Poly(tetramethylene oxide)	21) Poly(isobutylene)
5) Polyethylene	22) Poly(vinyl alcohol)
6) Polypropylene	23) Poly(1-butene)
7) Polystyrene	24) Poly(1-pentene)
8) Poly(vinyl fluoride)	25) Poly(vinylmethyl ether)
9) Poly(vinylidene fluoride)	26) Poly(vinylethyl ether)
10) Poly(1,2-difluoroethylene)	27) PMA
11) Poly(vinyl chloride)	28) PEA
12) PTFCE	29) Polyacrylate
13) PET	30) Poly(butyl acrylate)
14) Poly(metaphenylene isophthalate)	31) Poly(decamethylene terephthalate)
15) Poly(4,4' isopropylidene diphenylene carbonate)	32) Poly(metaphenylene isophthalate)
	33) Poly(hexamethylene adipamide)
16) PMMA	34) Poly(1-octene)
17) PEMA	35) Poly(vinylhexyl ether)

Table II *Polymer set used for NN prediction.*

1) Poly(propyl acrylate)
2) PPMA
3) PBMA
4) Poly(phenyl methacrylate)
5) Poly(1-hexane)

Case study 1: Density

Density can be directly calculated as the ratio of monomer mass to the molar volume of the unit. The NN input descriptors employed in this case study consist of the monomer mass, molar volume, as well as Hopfinger's intermolecular parameters ($\langle E_+ \rangle$ and $\langle E_- \rangle$). The error measure used is the root-mean-square (RMS) error, E, for all exemplars, P, defined as:

$$E = \sum_{p=1}^{P} \frac{E_p}{P} \tag{1}$$

and

$$E_p = \left[\sum_{i=1}^{N} \frac{(d_i - o_i)^2}{N} \right]^{1/2} \tag{2}$$

where E_p is the root-mean-square error for all output nodes, N, and d_i , o_i are the desired and actual outputs of node i, respectively. The input and output data are discretized between 0.2 and 0.8. Several network architectures

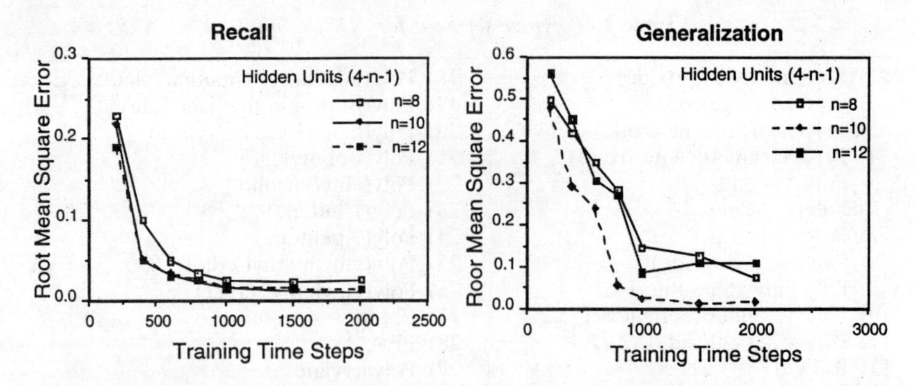

Figure 5 Error during the recall and generalization phases for density case-study.

were considered starting with the (2–6–1) and (4–10–1) networks. The network architecture specification within the parenthesis should be interpreted, from left to right, as the number of input units, hidden units, and output units, respectively. Initially, twice the number of hidden neurons as the number of inputs were considered as in the case of the (2–6–1) and (4–10–1) networks. The selection of these architectures was based on prior experience with this set of structural descriptors and a general rule of thumb based on the performance of different architectures. These networks were evaluated as starting points only and both smaller and larger architectures (based on the number of hidden neurons) were analyzed to evaluate the best performance.

As seen from Figure 5, for each architecture of hidden units, the recall error decreases monotonically with an increase in the learning iterations. Furthermore, hidden units greater than eight in the (4–n–1) networks weakly influenced the recall error. Generalization results, also shown in Figure 5, however, suggest a degradation of property prediction accuracy for hidden units less than 8 and greater than 12. Similar results are seen in the (2–n–1) networks. Thus, the property estimation accuracy is strongly dependent on the NN architectures and is an important research issue. A graphical summary of the van Krevelen and the NN recall results for the two networks is presented in Figure 6. Table III lists the NN prediction and van Krevelen group additive results for density. Consequently, the NN recall and generalization results demonstrate a significant improvement over the van Krevelen results.

Case study 2: Glass transition temperature

The glass transition temperature is dependent on a number of factors, including the bulkiness and flexibility of movable units, intermolecular interactions, and the free volume fraction. The T_g case study employed Hopfinger's descriptors, such as conformational entropy, mass moments,

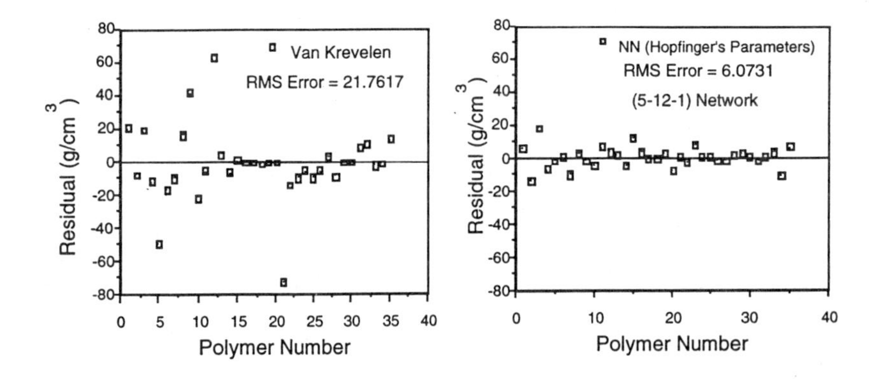

Figure 6 Comparison of NN recall results to van Krevelen's group contribution: density.

and the intermolecular descriptors, $<E_D>$, $<E_+>$, and $<E_->$, for NN structure–property correlations. The NN architectures considered were (5–10–1), (5–12–1), and (5–14–1) networks. Figure 7 illustrates the recall and generalization errors for each architecture of hidden units as a function of NN iterations, respectively. A graphical summary of the network's recall and generalization results is illustrated in Figure 8. Table IV lists the NN prediction, Hopfinger's results, and van Krevelen group additive results for T_g. Once again, the NN recall and generalization results are a considerable improvement over both van Krevelen and Hopfinger's results. Moreover, the limitation of linear regression analysis is evident as Hopfinger's MM-based (RMS error = 21.07) descriptors yield only a slight improvement over van Krevelen's group methods (RMS error = 21.76) while the NN correlation methodology using the same Hopfinger parameters result in an improvement in both recall (RMS error = 6.07) and generalization results (RMS error = 17.93).

Table III *NN prediction results compared to van Krevelen's predictions: density(gm/cm^3).*

Polymer	Actual	V. Krevelen	NN(4–10–1)
1) Poly(propyl acrylate)	1.08	1.08	1.08
2) PPMA	1.08	1.03	1.07
3) PBMA	1.05	1.01	1.04
4) Poly(phenyl methacrylate)	1.21	1.15	1.21
5) Poly(1-hexane)	0.86	0.85	0.86
RMS error		0.0395	0.0063

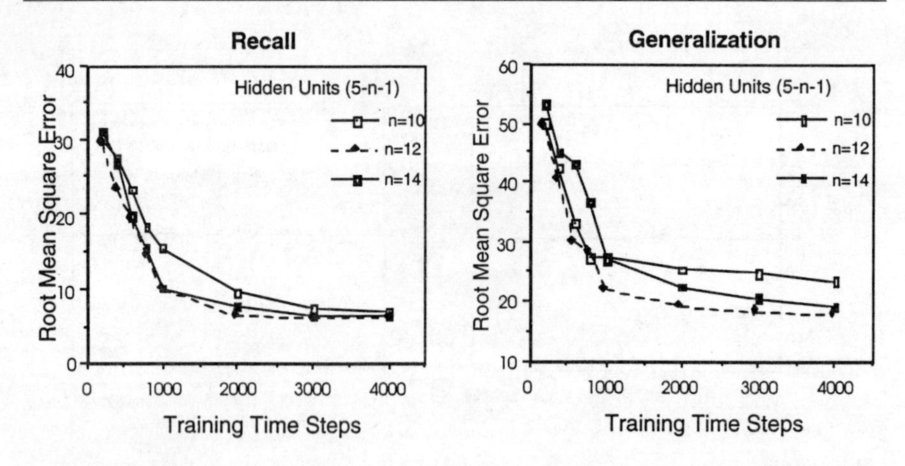

Figure 7 Error during the recall and generalization phases for T_g case study.

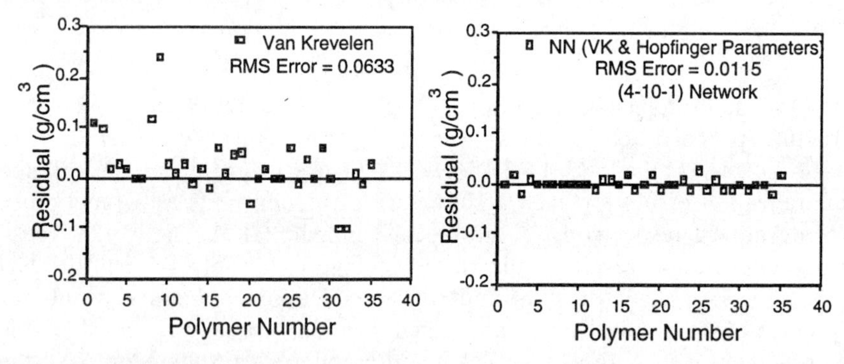

Figure 8 Comparison of NN recall results to van Krevelen's group contribution: T_g.

Table IV *NN prediction results compared to van Krevelen and Hopfinger's methods: T_g (K).*

Polymer	Actual	V. Krevelen	Hopfinger	NN(5–12–1)
1) Poly(propyl acrylate)	229	210	249	242
2) PPMA	308	243	350	289
3) PBMA	249	238	240	231
4) Poly(phenyl methacrylate)	378	313	400	363
5) Poly(1-hexane)	223	254	273	200
RMS error		44.48	32.34	17.93

As demonstrated, NNs were able to acquire structure–property relations from examples of structural descriptors and properties for complex polymer molecules. The NN recall accuracy to trained properties was near perfect, with the best RMS error of 0.115 and 6.0731 for the density case study and T_g case study, respectively. The best RMS values for density and T_g prediction (i.e. generalization) were 0.0063 and 17.93, respectively. These results are graphically illustrated in Figure 9 and represent a significant improvement over traditional methods. These investigations strongly encourage the use of NNs for the forward problem for complex nonlinear molecules which cannot be adequately modeled using traditional methods.

While NNs have proven to be quite effective, it is important to note some weaknesses of this approach. First, it is a 'black-box' model, which means that no insights or explanations can be provided for the nonlinear correlation that has been developed. Second, the neural net model will only be as good as the data that was used to train the network. Hence, it is important to have a substantial amount of good data for developing the network model. Third, the NN models are generally more reliable for interpolation than extrapolation of data. That is, if one tries to use the network to predict the properties of molecules that differ substantially from the molecules used in the training set, the predictions are likely to be suspect. For instance, in the case studies performed, the emphasis was on improving the ability of the NN for recall without compromising the generalization performance. Hence, the NN trained as above would probably perform very well in predicting properties for molecules that are structurally similar to those used in training. A large connectivity for the NN as used in the case studies is motivated towards developing very good interpolation characteristics. However, any reasonable prediction as to the performance of the network for molecules with structural characteristics outside that of the training set can be made only with extensive testing. Finally, as noted earlier, the effectiveness of the

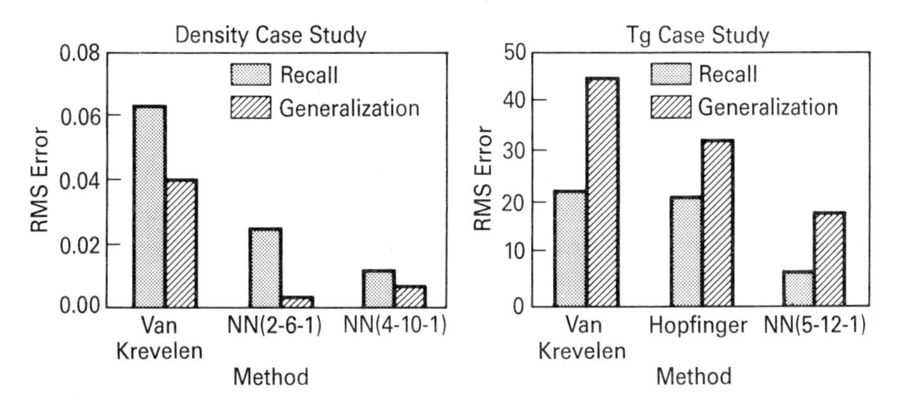

Figure 9 NN error during recall and generalization for density and T_g case studies.

neural net model depends critically on the choice of the input descriptors which characterize the molecule's structure. Some of these shortcomings can be circumvented by using hybrid NN models. In the standard 'black-box' neural net approach, one does not use *a priori* information about well-established functional relationships between the property and the structural descriptors. In a hybrid approach, one incorporates such *a priori* knowledge and use these networks to approximate only unmodeled phenomena. Such hybrid NNs have been shown to perform better than the 'black-box' models, particularly for generalization problems (Psichogios and Ungar, 1992; Aoyama and Venkatasubramanian, 1995).

GENETIC ALGORITHMS FOR THE INVERSE PROBLEM

Overview

GAs, originally developed by Holland (1975), are stochastic optimization procedures based on natural selection and Darwin's theory of survival of the fittest. They have been demonstrated to provide globally optimal solutions for a variety of problems and their effectiveness has been explored in different fields including engineering (Goldberg, 1989) and chemistry (Hibbert, 1993). Under a GA framework, a population of solutions competes for survival based on their closeness to a predefined measure of the optimum (or a characteristic of the optimum). The closeness to the optimum is captured by a normalized value between 0 and 1 called the 'fitness' of the solution, where a fitness of 1 implies the achievement of the optimum. Typically, the solution candidates are represented in the form of strings which are manipulated in a manner similar to that of genes in nature. The surviving members from a population are then allowed to propagate their genetic materials, which in this case, are components of their underlying structure (components of the string) to the next generation. However, the nature of this propagation is dependent on the use of genetic operators such as crossover, mutation, etc. which are random in nature. This evolutionary process continues until convergence is obtained, i.e. no significant improvement is obtained in the solution over generations. The real advantage of using a GA is its ability to provide several near optimal solutions even when convergence is lacking. This characteristic of the GA is highly desirable especially with regard to the problem of CAMD as the designer is finally left with a set of candidates which can then be synthesized and evaluated instead of a single sub-optimal solution.

GAs for CAMD

To circumvent the difficulties faced by most CAMD approaches as discussed earlier, Venkatasubramanian and coworkers (Venkatasubramanian *et al.*,

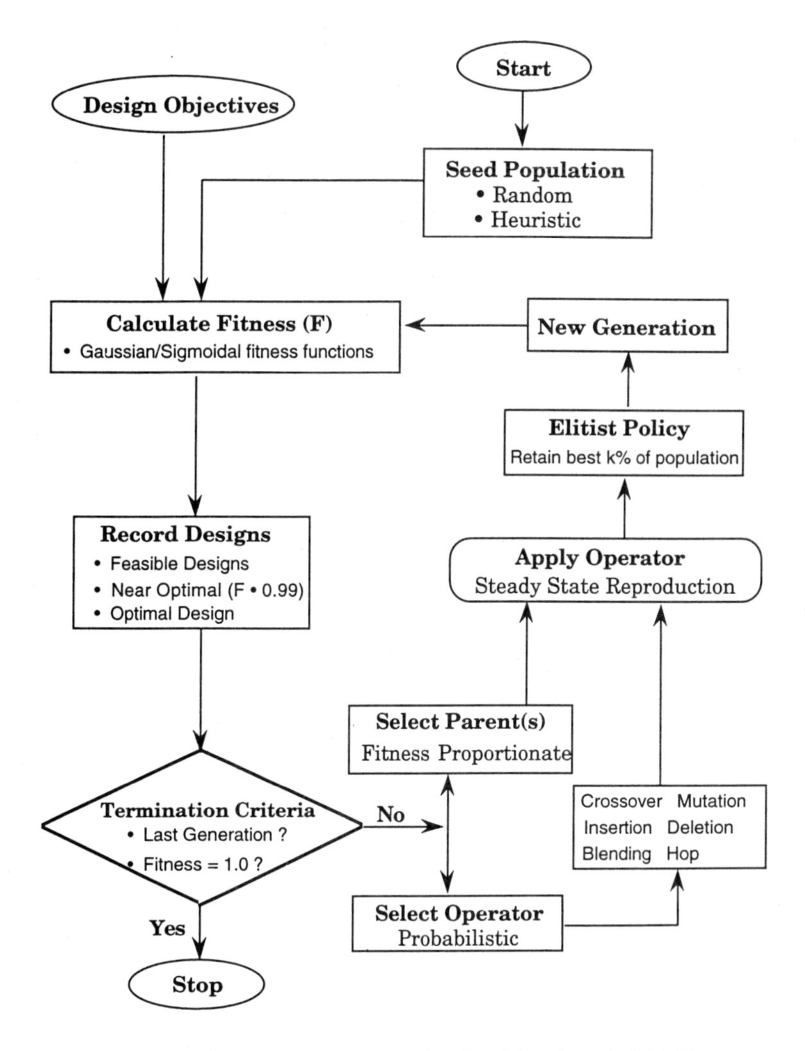

Figure 10 Flowchart for genetic algorithm based CAMD.

1994, 1995) proposed a GA-based framework, a schematic of which has been shown in Figure 10. The basic steps of this procedure are: initialization of a population of solutions, evaluation of the population, survival of the fittest, probabilistic application of genetic operators and elitist propagation of solutions to create the next generation and finally application of a test for convergence. The GA-based CAMD makes use of several genetic operators such as: crossover (main-chain and side-chain), mutation, hopping, blending, insertion and deletion. The representation of the solutions as well as the details of the operators and the evaluation procedures are discussed at length elsewhere (Venkatasubramanian *et al.*, 1994, 1995). Venkatasubramanian and coworkers demonstrated that this approach was quite effective in locating globally optimal designs for significantly complex polymer molecules. Since the details can be found elsewhere, only a short summary of the results will be provided here. The focus then shifts to the problem of characterizing and analyzing the search space for molecular design to gain insights into the mechanics of genetic search.

Design case studies

The effectiveness of GA as an optimization procedure lies in its handling of different forms of the objective function while including constraints on its variables. This was explored in the first case study performed by Venkatasubramanian and coworkers (Venkatasubramanian *et al.*, 1994), where open-ended constraints were used on the target properties and the fitness function was modified to reflect this. The GA-based CAMD approach was applied to the semi-conductor encapsulation case-study performed by Joback (1989) and it was very successful in locating several feasible solutions which satisfied the property constraints. This case study was performed with base-groups that occupy positions along the main-chain only and no independent manipulations of the side-chains was allowed.

The second case study was performed to evaluate the performance of the GA in locating an exact target. For this purpose, three target polymers were chosen (Polyethylene terephthalate, Polyvinylidene copolymer, Poly carbonate) of varying degrees of complexity in structure. A base-group set of four main-chain and four side-chain groups were chosen. This set is reproduced from Venkatasubramanian and coworkers (Venkatasubramanian *et al.*, 1994) in Table V. The target properties were now exact values with a tolerance specification, and the constraints were closed-ended and hence tighter. The GA-based CAMD was allowed to perform manipulations along the main-chain as well as the side-chain. The fitness function was modified to reflect the closed nature of the target constraints. A statistically significant number of runs were performed and the GA-based CAMD was found to be

Table V *Base groups considered for structure characterization.*

• 4 mainchain groups, 4 sidechain groups			
Mainchain groups			
>C<	–⟨phenylene⟩–	–O–C(=O)–	–O–
Sidechain groups			
–H	$-CH_3$	–F	–Cl

100% successful in locating the target molecules when the initial population was a random combination of the base groups. An extension to this case study included a larger set of target polymers and a comparison with the performance of random search on the same problems. The GA-based CAMD outperformed the random search with more than twice the success rate and consistently located some target molecules that the random search failed to identify in any of the runs. The results of this have been previously published (Venkatasubramanian *et al.*, 1994). These case studies have demonstrated the ability of the GA to perform an efficient search in a combinatorially explosive space of solutions with several local optima.

The advantage of using a GA-based approach, especially for CAMD, is its ability to provide several near-optimal solutions even when it fails to converge. From a design point of view, this implies that one has a set of alternative structures to synthesize and evaluate which are expected to perform close to their specifications. In fact, some of these alternative structures could be more attractive for synthesis from an economic perspective. In addition to this advantage, any GA-based method has the flexibility of including complex criteria in its objective function and hence use additional knowledge about the desired designs. These two aspects, and the effectiveness of GAs for the design of significantly complex molecules, were demonstrated on a large case study involving 17 main-chain groups and 15 side-chain groups (Venkatasubramanian *et al.*, 1995). The case study presented considers nine target polymers and a set of five target properties. Additional chemical knowledge in the form of restrictions on the bulkiness of the molecules as well as their stability was incorporated into the objective function. We clearly

Table VI *Four target polymers and their properties.*

Target polymer	Density $r\ [=]\ \frac{gm}{cm^3}$	Glass trans. temp. $T_g\ [=]\ K$	Thermal expansion coeff. $a\ [=]\ K^{-1}$	Heat capacity $Cp\ [=]\ \frac{J}{Kg.K}$	Bulk modulus $K\ [=]\ \frac{n}{m^2}$
TP1: [–O–C(=O)–⟨ring⟩–C(=O)–O–CH₂–CH₂–]n	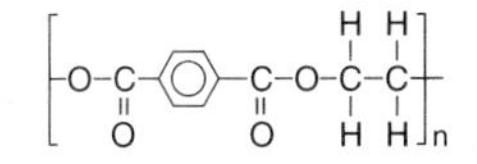 1.34	350.75	2.96×10^{-4}	1152.67	5.18×10^{9}
TP2: [–⟨ring⟩–S–⟨ring⟩–O–⟨ring⟩–C(CH₃)(CH₃)–⟨ring⟩–O–]n	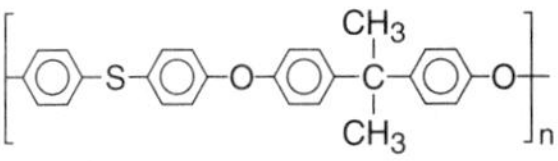 1.19	406.83	2.90×10^{-4}	1073.96	5.39×10^{9}
TP3: [–O–C(=O)–⟨ring⟩–C(=O)–O–CH₂–CH₂–]n	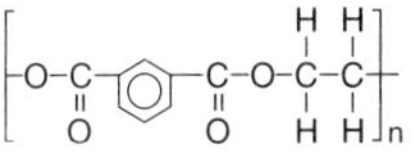 1.27	322.12	2.81×10^{-4}	1152.67	3.42×10^{9}
TP4: [–⟨ring⟩–SO₂–⟨ring⟩–O–⟨ring⟩–⟨ring⟩–O–]n	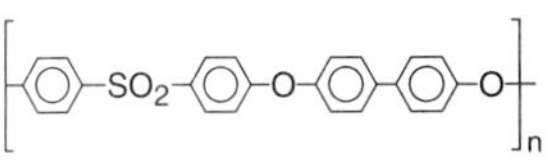 1.28	472.00	2.89×10^{-4}	995.95	5.31×10^{9}

Table VII *Summary of GA runs for four target polymers.*

Target polymer	Part	Standard GA random MC, SC	Feasible MC random MC, SC	Feasible MC penalize complex random MC, SC
[–O–C(=O)–⬡–C(=O)–O–CH₂–CH₂–]ₙ	a)	12% (3)	28% (7)	64% (16)
	b)	184	233	428
	c)	282	281	166
[–⬡–SO₂–⬡–O–⬡–⬡–O–]ₙ	a)	32% (8)	48% (12)	92% (23)
	b)	400	317	420
	c)	175	197	99
[–O–C(=O)–⬡–C(=O)–O–CH₂–CH₂–]ₙ	a)	68% (17)	76% (19)	96% (24)
	b)	210	147	81
	c)	162	158	125
[–⬡–S–⬡–O–⬡–C(CH₃)₂–⬡–O–]ₙ	a)	0% (0)	0% (0)	0% (0)
	b)	–	–	–
	c)	861	910	570

Legend: a) Success in locating target percentage (number of times out of 25 runs).
b) Average generation number when target was located.
c) Total number of distinct polymers with fitness greater than 0.99.

demonstrate the ability of the GA-based CAMD to locate several near-optimal solutions of fitness greater than 0.99 in some cases. The results also established an improvement in the performance of the approach when including additional knowledge. A summary of some of the results of the case study (Venkatasubramanian *et al.*, 1995) is provided in Tables VI, VII and VIII. Table VI gives the target properties to be designed for by the GA, corresponding to four target polymers. Table VII summarizes the results of the GA design procedure performed repeatedly from a random initial population over 25 runs. Table VII also compares the effect of knowledge augmentation to include complexity and stability criteria on the success rate with that of the standard GA (each of these results based on 25 different runs). The results establish an improvement in performance of the approach when including additional knowledge. It is also clear that the heuristic procedure

Table VIII *Near optimal designs for target polymer #4.*

Polymer design	Overall error	Fitness
Target polymer: #4		
$[-Ph-S-Ph-O-Ph-C(CH_3)_2-Ph-O-]_n$	0%	1.0
Case 1: standard GA		
$[-Ph-Ph-Ph-CH_2-O-C(=O)-Ph-CH_2-]_n$	0.74%	0.995
$[-Ph-Ph-C(C_2H_5)(CN)-Ph-Ph-Ph-O-O-O-C(=O)-O-]_n$	1.18%	0.991
Case 2: modified GA, stability		
$[-Ph-Ph-CH(OH)-CH(CH_3)-Ph-O-C(=O)-Ph-Ph-O-]_n$	0.25%	0.999
$[-C(=O)-O-Ph-CH(C_2H_5)-S-Ph-Ph-Ph-Ph-]_n$	1.10%	0.991
Case 3: modified GA, stability and complexity		
$[-CH_2-O-C(=O)-Ph-Ph-Ph-]_n$	0.21%	0.999
$[-C(CH_3)_2-Ph-Ph-O-C(=O)-Ph-O-C(=O)-]_n$	0.83%	0.995

offers no guarantee of convergence to the target as is evident from the results for target polymer TP4 in Table VII. However, Table VIII shows some of the very high fitness alternatives to the same set of target properties, as identified by the GA-based CAMD during the different runs. In fact there are around 500 such alternatives that are usually located by the GA for this target property set. The number of such high fitness alternatives (fitness $\geq$ 0.99) is shown against the (c) entry in Table VII for the different polymers. This feature of locating several near optimal solutions, even when the GA fails to converge, is one of the important advantages of the GA based CAMD.

The GA approach suffers from two main drawbacks. One is the heuristic nature of the search and that there is no guarantee of finding the target solution. But then, this criticism would be applicable to the other heuristic approaches as well. The other drawback is that the selection of the genetic design parameter values would require some experimentation.

While the GA-based design framework performed quite well under difficult circumstances, it failed to locate the target in a few instances. To gain a better understanding of the nature of the search space and the mechanics of the GA search which would shed some light on the success and failure of GA, we recently carried out certain investigations to characterize and analyze the search space for molecular design. The rest of this chapter discusses these results.

CHARACTERIZATION OF THE SEARCH SPACE

The key to understanding the characteristics of the search space is to understand the correlation between the structural similarity of molecules and the closeness in their fitness values. Towards this end, a target molecule (polyvinylidene copolymer) was chosen (Venkatasubramanian *et al.*, 1994) along with the set of four main-chain and side chain groups. The GA-based CAMD located this target in all the 25 runs as previously reported (Venkatasubramanian *et al.*, 1994). The properties of this target are invariant to the ordering of the base groups. Every possible feasible combination of the base groups (both main-chain and side-chain) within a certain maximum length (7 in this case) of the monomer unit was generated and its fitness evaluated. Figure 11 shows the distribution of fitness over a portion of the search space. It is clear from this figure that the search space is hostile with several local optima. Figure 12 and Table IX show the histogram and the breakdown of the distribution in fitness respectively.

Super-groups representation and pseudo-Hamming distance

To structurally characterize the search space, each molecule was identified by a coordinate in the space spanned by the base groups and its distance from the target molecule was determined. The coordinate structure of a molecule is generated based on a set of 'super-groups'. These super-groups are constructed from the set of base groups chosen for the design. The idea behind the generation of these base groups is to map the essentially 2-dimensional structure of a molecule, i.e. along the main-chain and along the side-chain, into a one-dimensional structure that contains the super-groups along the main-chain only. This simplifies the analysis and interpretation of the search space characteristics. The super-grouping is achieved by generating all possible combinations of every main-chain group that can take side-chains, with the side-chain group set to yield a set of super-groups that only occur along the main-chain. Combinations of these super-groups create the different molecules. Table V shows the set of base groups used for the case study, reproduced from a recent study (Venkatasubramanian *et al.*, 1994). It is clear, from the valency of the main-chain groups that only >C< can

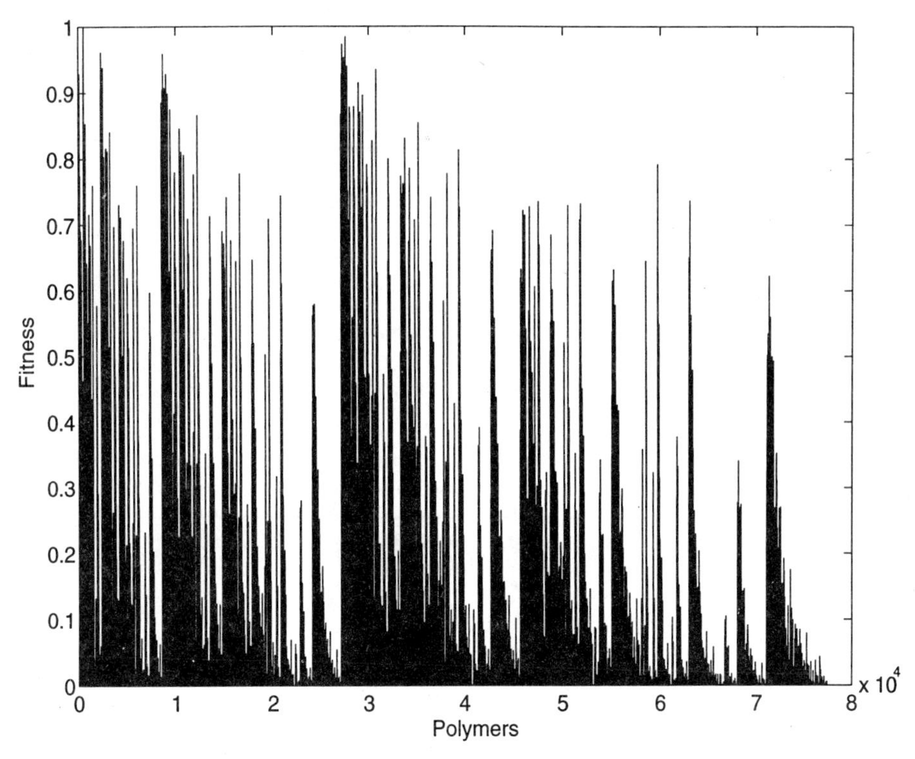

Figure 11 Distribution of fitness in combinatorial search space.

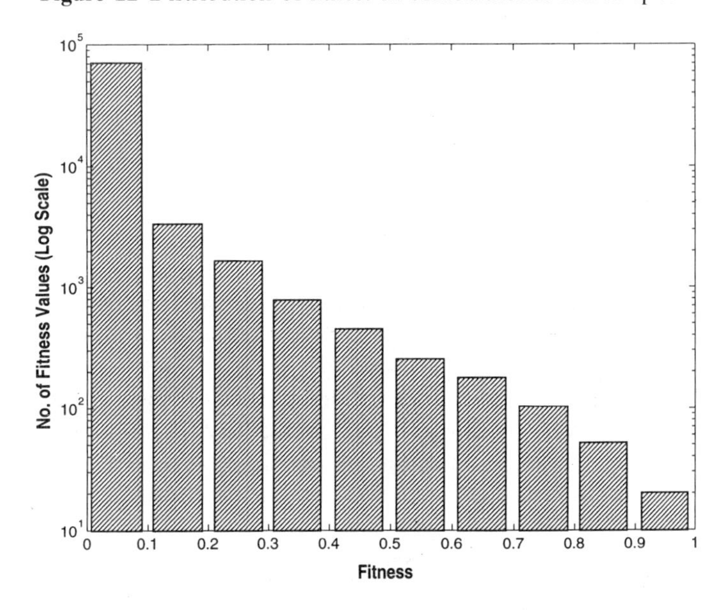

Figure 12 Histogram of fitness distribution.

Table IX *Distribution of distinct molecules over fitness ranges.*

No. of molecules	Fitness (F)
70686	$0.00 \leq F < 0.10$
3217	$0.10 \leq F < 0.20$
1632	$0.20 \leq F < 0.30$
824	$0.30 \leq F < 0.40$
458	$0.40 \leq F < 0.50$
263	$0.50 \leq F < 0.60$
180	$0.60 \leq F < 0.70$
106	$0.70 \leq F < 0.80$
54	$0.80 \leq F < 0.90$
21	$0.90 \leq F < 1.00$

Table X *Super-groups generated from base-groups of Table V.*

Number	Super-group
1	$-CH_2-$
2	$-CH(CH_3)-$
3	$-CHF-$
4	$-CHCl-$
5	$-C(CH_3)_2-$
6	$-C(CH_3)F-$
7	$-C(CH_3)Cl-$
8	$-CF_2-$
9	$-CFCl-$
10	$-CCl_2-$
11	$-C6-$
12	$-COO-$
13	$-O-$

accommodate side-chains. Hence, the set of super-groups includes the three main-chain groups C6, COO, O and all combinations of >C< with the four side-chain groups. The set of super-groups is listed in Table X with a number assigned to each. The molecular coordinate vector has 13 dimensions corresponding to the 13 super-groups and the value of each coordinate of the vector is the number of occurrences of that particular super-group in a molecule. Figure 13 gives an illustrative example for the construction of the coordinate. Once these coordinates are constructed the distance (d^M) of a molecule M from the target molecule T is calculated as:

$$d = \sum_{i=1}^{13} \left| D_i^M - D_i^T \right| \tag{3}$$

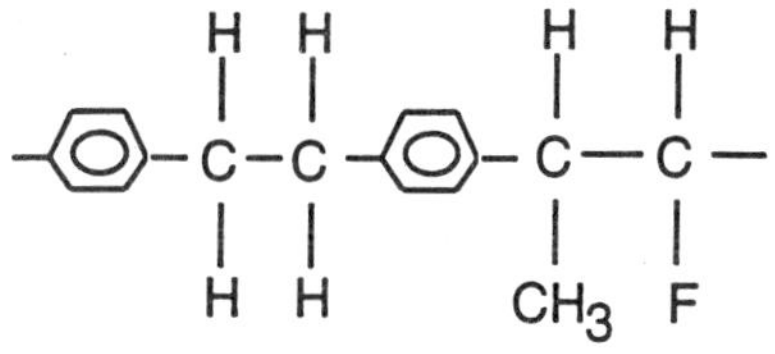

Super-group Notation : -C6-CH2-CH2-C6-CHCH3-CHF-

Super-group (Super-group #, No. of Occurences): C6(11,2);CH2(1,2);CHCH3(2,1);CHF (3,1)

Coordinate Vector : (2 1 1 0 0 0 0 0 0 0 2 0 0)

Figure 13 Illustrative example for coordinate vector construction.

where, D_i^M is the ith coordinate of a molecule M and D_i^T is the ith coordinate of the target molecule.

In this manner the entire search space is characterized in terms of these 'pseudo-Hamming' distances from the target molecule. The distance (d) of a particular molecule M from the target is the number of flips of the super-groups including additions and deletions, required to get to the target molecule from the structure of molecule M. Hence, molecules that are structurally similar will have a small value of d and those that are structurally dissimilar will have larger values. Once the distances of the molecules from the target were obtained, the average and maximum fitness of a group of molecules at a certain distance (d) away from the target was plotted as a function of d. Figure 14 shows the average and maximum fitness of a cluster d units away from the target, the distance d plotted on the X-axis. In this particular case, the property values of the target molecule (and hence the overall fitness) are independent of the ordering of the groups in the molecule and hence the coordinate vector constructed in the above manner uniquely identifies every distinct solution generated. Thus, for order-invariant molecules, for any given fitness value there might exist several molecular configurations that have the same groups but with different group orderings. This means that the number of configurations investigated by the GA will be much more than the number of different fitness points seen in the search space by the GA. For instance in this case study, there are about 78000 different fitness points whereas the search space contains about 10 million different molecular configurations.

It is clear from Figure 12 and Table IX that the search space is hostile with only a small fraction of the molecules having fitness greater than 0.90 (21 molecules out of about 78000) and these are randomly dispersed within the

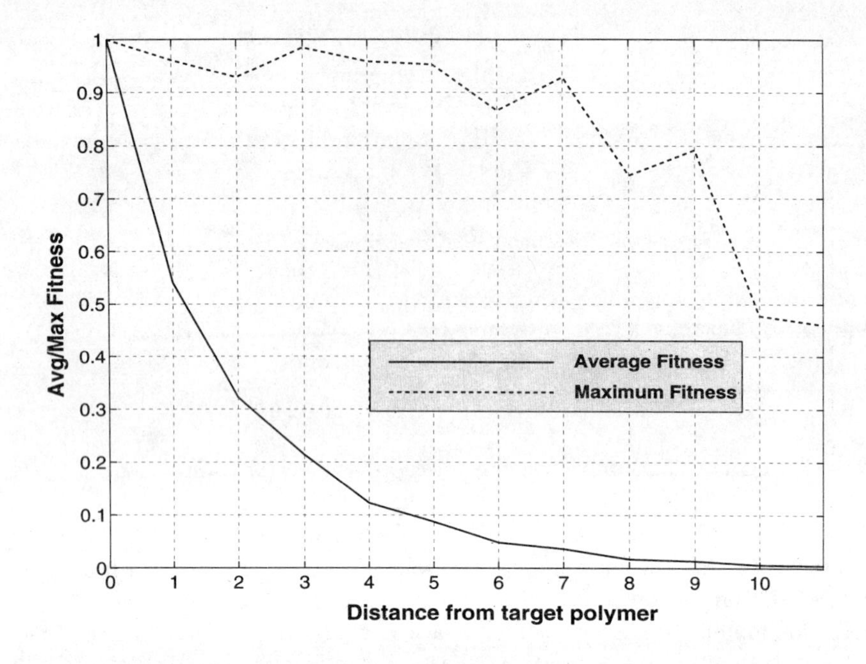

Figure 14 Fitness as a function of distance from target molecule.

search space. However, Figure 14 clearly shows that the gradation in the fitness follows a very clear pattern with respect to the distance from the target. Cast in the coordinate space generated by the super-groups, the average fitness follows a concave path to the global optimum. Similarly, the maximum fitness values also show a decreasing trend with respect to distance though there are small local deviations from this trend. This implies that molecules that are structurally similar to the target in the coordinate space are also of very high average fitness. Thus, the identification of highly fit regions of the search space is very likely to lead to molecules that are structurally similar to the target. Hence, the natural process of exploitation of the solution characteristics by the GA combined with the survival of the fittest paradigm may be hypothesized as leading to the convergence of the solution to the target. It is clear from this exercise that there is a need to characterize the structure of the search space in order to gain a deeper understanding of the convergence properties of the GA. In addition, knowledge of the search landscape would help to enhance the performance of the GA-based CAMD by a judicious selection and use of various operators, their probabilities, etc. Even though this investigation has shed some light on the nature of exploitation of the search space by the GA, it is still far from clear and further explorations are needed to gain a deeper understanding.

AN INTERACTIVE FRAMEWORK FOR EVOLUTIONARY DESIGN

Designing complex molecules with desired properties is a very challenging problem for industrial-scale molecular design applications. Most of the previous approaches for the inverse problem are not satisfactory as they get bogged down by the reasons discussed earlier. Similarly, the forward problem also poses challenges for complex molecules with nonlinear structure–activity relationships. In our past studies on using GA summarized above, we had used group contribution methods for the forward problem as they were found to be adequate for our needs. For more complex design situations, we advocate the use of a NN approach for the forward problem, while using the GA for the inverse problem. This proposed model of CAMD is shown in Figure 15.

While our investigations have shown that both NNs and GAs show great promise in tackling the difficult molecular design problems, we realize that the real-life CAMD problems are so difficult and complex that a successful approach would require a combination of several techniques integrated into a unified, interactive, framework. For instance, the basic genetic design approach can be augmented with additional knowledge. A wealth of information may be available for the CAMD process as such, in terms of the groups to be changed and the combinations to be avoided or favored while constructing solutions for a particular set of target properties. This knowledge is acquired from prior experience with similar problems, and such knowledge may also reside with the expert user. Better moves could be constructed into more promising regions of the search space based on prior experience about the local structure of the search space. Experiential knowledge-based focus of the search procedure could be obtained by incorporating a knowledge-base of rules for the manipulation of structures and even suggestions for operator probabilities. On the other hand, incorporation of user knowledge

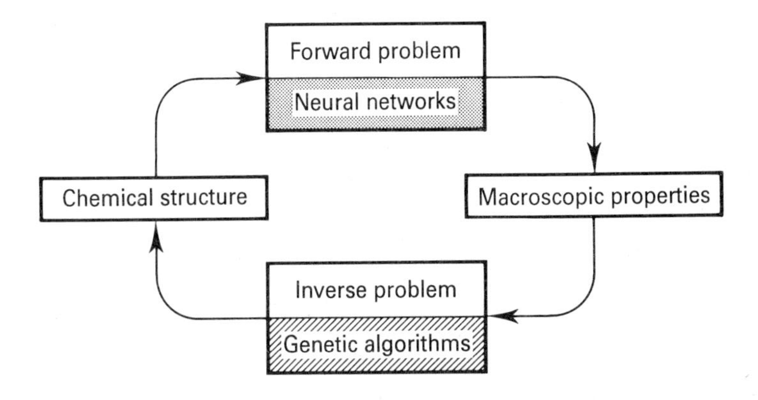

Figure 15 Proposed framework for CAMD.

could be accomplished by an interactive design process which communicates with the user and receives suggestions and directions regarding the structures to delete or add to the current population, as well as the kind of changes desired in terms of the operators. The input from the user is used to guide the search to the more optimal solutions. The speed of the GA-based CAMD could be enhanced by augmenting it with other methodologies such as math programming methods. As mentioned earlier, the GAs by virtue of their global search characteristics can identify regions in the solution space which may contain globally optimal designs. However, lack of use of local information might prolong the convergence. Math programming methods on the other hand, by virtue of their greater focus during local search could converge in on the exact solution based on the local gradient information and using the GA generated designs as their starting points.

Figure 16 shows an interactive design framework that is under development. In this object-oriented framework, the GA-based design engine would be coupled with expert systems as well as nonlinear programming methodologies to improve the effectiveness of the overall design process. We also envision such a system as an interactive one, where the molecular designer actively participates in the design process.

CONCLUSIONS

Computer-aided design of molecules with desired properties is a challenging problem due to a number features such as combinatorial complexity and nonlinearity of the search space, difficulties in knowledge acquisition and incorporation of high-level chemistry-related constraints and so on. To circumvent these difficulties, we propose a CAMD framework that uses NNs for the properties prediction problem and GAs for the inverse problem. We discuss the merits and demerits of these approaches along with results on some case studies.

While the GA-based design framework performed quite well under difficult circumstances, it failed to locate the target in a few instances. To gain a better understanding of the nature of the search space and the mechanics of the GA search which would shed some light on the success and failure of GAs, we explored the search space in detail for one of the molecular case studies in an attempt to analyze and characterize its features. This analysis showed that the search space is indeed quite unfriendly with numerous local optima which will trap many of the standard methods such as gradient-ascent. It was found to exhibit sharp discontinuities, and that only a very small fraction of the search space contained molecules of high fitness. A pseudo-Hamming distance analysis using ‘super-groups’ suggested that the average fitness and the maximum fitness of the molecules increased as they approached the target, a feature which the GA exploits through its

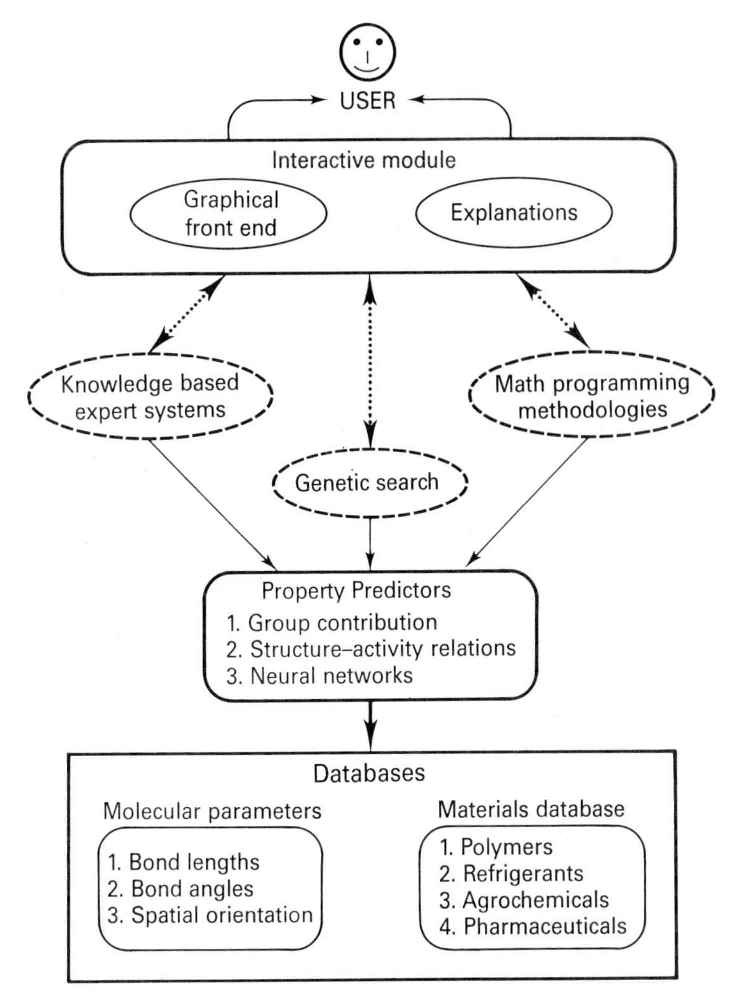

Figure 16 Interactive framework for evolutionary molecular design.

'survival of the fittest' principle to carry out an effective search. Even though these investigations have demonstrated the promise of GAs for CAMD and have shed some light on the nature of its exploitation of the search space, many aspects still remain less clear and further explorations are needed to gain a deeper understanding.

We realize that real-life CAMD problems are so difficult and complex that a successful approach would require a combination of several techniques integrated into a unified, interactive, framework. We present one such framework with the GA-based design engine integrated with expert systems and math programming methodologies to improve the effectiveness of the overall design process. We also envision such a system as an interactive one, where the molecular designer actively participates in the design process.

REFERENCES

Aoyama, A. and Venkatasubramanian, V. (1995). Internal model control framework using neural networks for the modeling and control of a bio-reactor. *Engng. Applic. Artif. Intell.* in press.

Aoyama, T., Suzuki, Y., and Ichikawa, H. (1990). Neural networks applied to quantitative structure–activity relationship. *J. Med. Chem.* **33**, 905–908.

Bolis, G., Pace, L.D., and Fabrocini, F. (1991). A machine learning approach to computer-aided molecular design. *J. Comput.-Aided Mol. Des.* **5**, 617–628.

Brignole, E.A., Bottlini, S., and Gani, R. (1986). A strategy for design and selection of solvents for separation processes. *Fluid Phase Equilibria* **29**, 125.

Burkert, U. and Allinger, N.L. (1982). *Molecular Mechanics.* ACS Monograph 177, American Chemical Society, Washington, DC.

Cybenko, G. (1988). Approximation by superpositions of sigmoidal functions. *Math. Cont. Syst. Sig.* **2**, 303–314.

Derringer, G.C. and Markham, R.L. (1985). A computer-based methodology for matching polymer structure with required properties. *J. Appl. Polym. Sci.* **30**, 4609–4617.

Elrod, D.W., Maggiora, G.M., and Trenary, R.G. (1990). Applications of neural networks in chemistry. 1. Prediction of electrophilic aromatic substitution reactions. *J. Chem. Inf. Comput. Sci.* **30**, 477–484.

Gani, R. and Brignole, E.A. (1983). Molecular design of solvents for liquid extraction based on UNIFAC. *Fluid Phase Equilibria* **13**, 331–340.

Gani, R., Nielsen, B., and Fredenslund, A. (1991). A Group contribution approach to computer-aided molecular design. *AIChE Journal* **37**, 1318–1332.

Goldberg, D.E. (1989). *Genetic Algorithms in Search, Optimization, and Machine Learning*. Addison-Wesley, p. 412.

Hibbert, D.B. (1993). Genetic algorithm in chemistry. *Chemom. Intell. Lab. Syst.* **19**, 277–293.

Holland, J.H. (1975). *Adaptation in Natural and Artificial Systems.* The University of Michigan Press, Ann Arbor.

Hopfinger, A.J., Koehler, M.G., and Pearlstein, R.A. (1988). Molecular modeling of polymers. IV. Estimation of glass transition temperatures. *J. Polym. Sci.: Part B: Polym. Phys.* **26**, 2007–2028.

Joback, K.G. (1989). Designing molecules possessing desired physical property values, Volumes 1 & 2. Ph.D. Thesis in Chemical Engineering, MIT.

Joback, K.G. and Reid, R.C. (1987). Estimation of pure-component properties from group contributions. *Chem. Eng. Comm.* **57**, 233–243.

Joback, K.G. and Stephanopoulos, G. (1989). Designing molecules possessing desired physical property values. In, *Proceedings of FOCAPD '89.* Snowmass, CO, pp. 363–387.

Jurs, P.C. and Isenhour, T.L. (1975). *Chemical Applications of Pattern Recognition.* Wiley & Sons, New York.

Kier, L.B. and Hall, L.H. (1976). *Molecular Connectivity in Chemistry and Drug Research.* Academic Press, New York.

Kier, L.B. and Hall, L.H. (1986). *Molecular Connectivity in Structure–Activity Analysis.* Research Studies Press, Letchworth, Hertfordshire, U.K.

Kier, L.B., Lowell, H.H., and Frazer, J.F. (1993). Design of molecules from quantitative structure–activity relationship models. 1. Information transfer between path and vertex degree counts. *J. Chem. Inf. Comput. Sci.* **33**, 142–147.

Koehler, M.G. and Hopfinger, A.J. (1989). Molecular modeling of polymers: 5. Inclusion of intermolecular energetics in estimating glass and crystal-melt transition temperatures. *Polymer* **30**, 116–126.

Macchietto, S., Odele, O., and Omatsone, O. (1990). Design of optimal solvents for liquid–liquid extraction and gas absorption processes. *Trans. Ind. ChemE.* **68**, Part A, 429–433.

Orchard, B.J., Tripathy, S.K., Pearlstein, R.A., and Hopfinger, A.J. (1987). Molecular modeling of polymer: I. Corrected and efficient enumeration of the intrachain conformational energetics. *J. Comput. Chem.* **8**, 28–38.

Psichogios, D.C. and Ungar, L.H. (1992). A hybrid neural network – First principles approach to process modeling. *AIChE Journal* **38**, 1499–1512.

Qian, N. and Sejnowski, T.J. (1988). Predicting the secondary structure of globular proteins using neural network models. *J. Mol. Biol.* **202**, 865–884.

Reid, RC., Prausnitz, J.M., and Sherwood, T.K. (1977). *The Properties of Gases and Liquids.* McGraw-Hill, New York.

Reid, R.C., Prausnitz, J.M., and Poling, B.E. (1987). *The Properties of Gases and Liquids.* McGraw-Hill, New York.

Rose, V.S., Macfie, H.J.H., and Croall, I.F. (1991). Kohonen topology-preserving mapping: An unsupervised artificial neural-network method for use in QSAR analysis. In, *QSAR: Rational Approaches to the Design of Bioactive Compounds* (C. Silipo and A. Vittoria, Eds.), Pharmacochem. Lib., vol. 16, Elsevier Science, Amsterdam, pp. 213–216.

Robb, E.W. and Munk, M.E. (1990). A neural network approach to infrared spectrum interpretation. *Mikrochim Acta* 31–155.

Skvortsova, M.I., Baskin, I.I., Slovokhotova, O.L., Palyulin, V.A., and Zefirov, N.S. (1993). Inverse problem in QSAR/QSPR studies for the case of topological indices characterizing molecular shape (Kier indices). *J. Chem. Inf. Comput. Sci.* **33**, 630–634.

Sumpter, B.G., Gettino, C., and Noid, D.W. (1994). Theory and applications of neural computing in chemical science. *Annu. Rev. Phys. Chem.* **45**, 439–481

Sumpter, B.G. and Noid, D.W. (1994). Neural networks and graph theory as computational tools for predicting polymer properties. *Makromol. Chem. Theory Simul.* **3**, 363–378.

Strouf, O. (1986). *Chemical Pattern Recognition*. Research Studies Press, Letchworth.

van Krevelen, D.W. and Hoftyzer, P.J. (1972). *Properties of Polymers, their Estimation and Correlation with Chemical Structure*. First Edition. Elsevier Scientific.

van Krevelen, D.W. and Hoftyzer, P.J. (1976). *Properties of Polymers, their Estimation and Correlation with Chemical Structure*. Second Edition. Elsevier Scientific.

van Krevelen, D.W. and Hoftyzer, P.J. (1990). *Properties of Polymers, their Estimation and Correlation with Chemical Structure*. Third Edition. Elsevier Scientific.

Venkatasubramanian, V., Chan, K., and Caruthers, J.M. (1994). Computer-aided molecular design using genetic algorithms. *Comput. Chem. Engng.* **18**, 833–844.

Venkatasubramanian, V., Chan, K., and Caruthers, J.M. (1995). Evolutionary design of molecules with desired properties using the genetic algorithm. *J. Chem. Inf. Comput. Sci.* **35**, 188–195

Wythoff, B.J., Levine, S.P., and Tomelline, S.A. (1990). Spectral peak verification and recognition using a multi-layered neural network. *Anal. Chem.* **62**, 2702–2709.

Zupan, J. and Gasteiger, J. (1991). Neural networks: A new method for solving chemical problems or just a passing phase? *Technical Report*, TU München, D-8046, Garching, Organisch-Chemisches Institute Germany.

12 Designing Biodegradable Molecules from the Combined Use of a Backpropagation Neural Network and a Genetic Algorithm

J. DEVILLERS* and C. PUTAVY
CTIS, 21 rue de la Bannière, 69003 Lyon, France

A hybrid system constituted of a backpropagation neural network (BNN) and a genetic algorithm (GA) was elaborated for designing organic molecules presenting a specific biodegradability. The BNN model was derived from a training set which consisted of the collective judgments of 22 experts as to the approximate time that might be required for aerobic ultimate degradation in receiving waters (AERUD) of 38 highly diverse molecules. Chemicals were described by means of 14 molecular descriptors. The selected 14/2/1 BNN (α = 0.9 and η = 0.4) model correctly classified 45 of 49 chemicals (91.8%) in the testing set. The configuration of the GA was optimized for selecting candidate molecules having low AERUD values. Different constraints were added in the GA in order to test the limits of the hybrid system for proposing biodegradable molecules presenting specific structural features.

KEYWORDS: *backpropagation neural network; genetic algorithm; hybrid system; biodegradation; AERUD.*

INTRODUCTION

Given the broad array of chemicals of environmental concern, attempts have been made to develop rational approaches allowing to relate the structure of organic chemicals to their biodegradability. These methods, termed (Quantitative) Structure–Biodegradation Relationships ((Q)SBRs)

* Author to whom all correspondence should be addressed.

In, *Genetic Algorithms in Molecular Modeling* (J. Devillers, Ed.)
Academic Press, London, 1996, pp. 303–314.
ISBN 0-12-213810-4

are needed for the estimation of a predictive biodegradation capacity for previously untested substances (Kuenemann *et al.*, 1990; Vasseur *et al.*, 1993; Alexander, 1994). Indeed, such a predictive ability is firstly important in industry since chemicals may be synthesized for a specific purpose but with the particularity to possess short lives in the environment. This is also important for regulatory agencies that have to decide whether a new chemical will be persistent or not before it is introduced on the market.

Recently, the backpropagation neural network (BNN) (Eberhart and Dobbins, 1990; de Saint Laumer *et al.*, 1991) has been presented as a powerful statistical tool for biodegradation modeling, due to its ability to learn and generalize from complex and noisy biodegradation data (Cambon and Devillers, 1993; Devillers, 1993; Devillers *et al.*, 1996). The goal of the present study was to show that the powerful modeling potency of a BNN can be used in conjunction with that of a genetic algorithm (GA) (Goldberg, 1989) in order to produce a hybrid system (Goonatilake and Khebbal, 1995) allowing proposal of molecular structures for the design of molecules presenting various degrees of biodegradability.

BACKGROUND

The hybrid system (Figure 1) was basically constituted of a GA for searching combinations of structural fragments among a set of descriptors in order to facilitate the design of biodegradable molecules. The biodegradability of the candidate molecules was estimated from a three-layer BNN model.

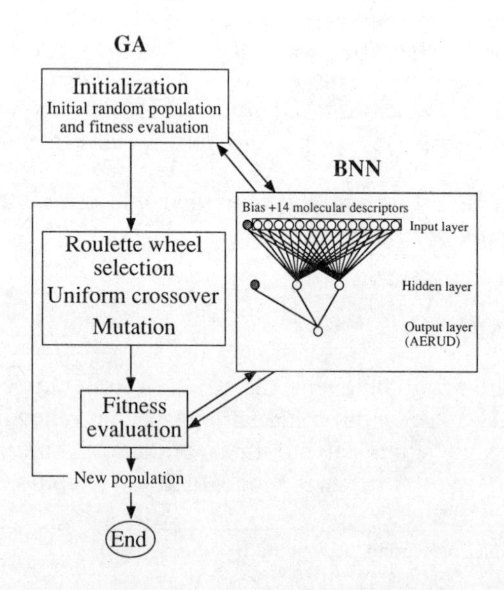

Figure 1 Hybrid system for designing biodegradable molecules.

Table I *Training set.*

No.	Compound name	Observed AERUD	Calculated AERUD	Res.
1	phthalylsulfathiazole	2.64	2.80	–0.16
2	phthalic anhydride	1.76	1.80	–0.04
3	maleic anhydride	1.39	1.41	–0.02
4	camphane	2.82	2.75	0.07
5	2,4-hexadien-1-ol	1.73	1.95	–0.22
6	2-methyl-2-nitro-*n*-butane	2.83	2.83	0.00
7	4-benzamido-2-chloro-5-methylbenzene-diazonium chloride	2.81	2.80	0.01
8	*N*-ethylbenzamide	1.71	1.60	0.11
9	*N,N'*-diphenyl-*p*-phenylenediamine	3.00	2.74	0.26
10	*n*-butylbenzene	1.83	1.85	–0.02
11	hydron yellow	3.50	3.56	–0.06
12	carbazole vat dye	3.04	2.94	0.10
13	epichlorohydrin	2.09	2.16	–0.07
14	copper chelate complex	2.56	2.69	–0.13
15	C.I. direct red 28	3.09	3.12	–0.03
16	vinyl acetate	1.64	1.67	–0.03
17	*o*-benzoylbenzoic acid	2.05	2.17	–0.12
18	2-methyl-3-hexanone	1.92	1.85	0.07
19	resol component of resin	2.36	2.41	–0.05
20	*p-tert*-butylphenol	2.42	2.23	0.19
21	3,3-dimethyl-1,2-epoxybutane	2.68	2.63	0.05
22	2-pyrrolidone-5-carboxylic acid	1.89	1.94	–0.05
23	3-pyrrolidine-1,2-propanediol	2.11	2.14	–0.03
24	di-*n*-octyl ether	2.08	2.12	–0.04
25	tripropylene glycol diacrylate	2.29	2.13	0.16
26	methyl 3-amino-5,6-dichloro-2-pyrazine carboxylate	2.95	2.96	–0.01
27	durene-α_1,α_2-dithiol	2.59	2.54	0.05
28	4-butyl-*N*-(4-ethoxybenzylidene)aniline	2.54	2.63	–0.09
29	butonate	2.71	2.71	0.00
30	4-amino-3,5,6-trichloropicolinic acid	3.13	3.12	0.01
31	chlorendic anhydride	3.80	3.80	0.00
32	acetylpyrazine	2.55	2.54	0.01
33	acid blue 29	3.10	3.10	0.00
34	4-decylaniline	2.32	2.45	–0.13
35	2,4-dichloro-6-ethoxy-1,3,5-triazine	3.33	3.18	0.15
36	(dimethylamino)phenethyl alcohol	2.21	2.17	0.04
37	(4-carboxybutyl)triphenylphosphonium bromide	3.05	3.03	0.02
38	hexatriacontane	2.88	2.90	–0.02

The BNN model was derived from a training set of 38 molecules presenting a high degree of structural heterogeneity (Table I). The aerobic ultimate degradation in receiving waters (AERUD) of these molecules was used as endpoint to estimate their biodegradability (Boethling *et al.*, 1989). The

Table II *Molecular descriptors.*

No.	Descriptor
1	heterocycle N
2	C=O
3	> 2 chlorine atoms
4	fused rings
5	only C, H, O, N
6	NO_2
7	≥ 2 cycles
8	epoxide
9	primary or aromatic OH
10	molecular weight (g/mol)
	< 100
	100–200
	200–300
	≥ 300
11	C–O: ether + OH II and III
12	amines + X=N (X = C or N)
13	conjugated C=O
14	Z=Z except NO_2 (Z = any heteroatom)

AERUD values were calculated from the responses gathered in a survey of 22 biodegradation experts (Boethling *et al.*, 1989). The chemicals under study (Table I) were described by means of 13 structural descriptors (Table II) known to influence the environmental fate of chemicals (Niemi *et al.*, 1987; Howard *et al.*, 1991, 1992; Cambon and Devillers, 1993; Domine *et al.*, 1993; Boethling *et al.*, 1994). They were Boolean descriptors. In addition, the following four classes of molecular weight were defined: < 100, 100–200, 200–300, and ≥ 300 g mol^{-1}.

Molecular descriptors (except no. 10) and the AERUD values were scaled by means of Eq. (1).

$$x'_i = [a(x_i - x_{min}) / (x_{max} - x_{min})] + b \quad (1)$$

In Eq. (1), x'_i and x_i were the scaled and original values, respectively. For the molecular descriptors, x_{min} and x_{max} were the minimum and maximum values found in the different columns, respectively. For descriptor no. 10, the four classes were encoded 0.2, 0.4, 0.6, and 0.8, respectively. For the outputs, x_{min} and x_{max} corresponded to the minimum and maximum AERUD values (i.e. 1 and 4) which could be calculated from the scoring procedure proposed by Boethling *et al.* (1989). In order to have scaled data ranging between 0.05 and 0.95, the values of *a* and *b* equaled 0.9 and 0.05, respectively. Numerous trials were performed to determine the optimal set of parameters for the BNN (i.e., number of hidden neurons, learning rate (η), momentum term (α), number of learning cycles). For this purpose, the STATQSAR package was used

(STATQSAR, 1994). The program automatically tests several times all the possible combinations of the values of the above parameters. The generalization ability of the BNN model was estimated from a testing set of 49 molecules (Table III) designed from the results of a previous study (Cambon and Devillers, 1993). To avoid problems of compatibility among data of different origins, biodegradability was encoded in a qualitative way. Thus, weakly and highly biodegradable chemicals were encoded by 0 and 1, respectively. Allocation of a simulated activity to class 0 or 1 was made by adopting an AERUD cut-off value of 2.5 as recommended by Boethling and Sabljic (1989).

Genetic algorithms (GAs) are powerful search techniques for rapidly finding optimal solutions to very high-dimensional problems for which systematic solutions are not practical. The basic requirements for applying a GA are that a possible solution to the problem can be encoded and that a given solution can be evaluated quantitatively by means of a fitness function that plays the role of the environment. Thus, initially, a random population of individuals is created. Each individual is represented by a chromosome consisting of a string of genes encoding the studied information. Evaluation of the fitness of each individual allows the selection of candidates for mating. Thus, if the fitness of an offspring is sufficiently high, it replaces a less fit member of the population and becomes a parent. Genetic manipulation of the chromosomes (i.e. crossover, mutation) and more generally the reproduction process allow increase of the overall fitness of the population and therefore to obtain an optimal solution for the problem at hand.

In our study, each chromosome was constituted of 14 genes corresponding to the selected molecular descriptors (Table II). The fitness function was based on the estimation of an AERUD value. However, the objective of the search was to minimize the AERUD values rather than to maximize them, since our goal was to select structural fragments increasing the biodegradability of the molecules. As a result, the following transformation was performed (Goldberg, 1989):

$$\mathrm{f}(x) = C_{\mathrm{max}} - g(x) \tag{2}$$

In Eq. (2), $\mathrm{f}(x)$ corresponds to the fitness function, $g(x)$ is the biodegradation activity calculated by the BNN model. As the outputs of the BNN were scaled between 0.05 and 0.95 by means of Eq. (1), C_{max} in Eq. (2) was fixed to 1. The GA operators used during the reproduction-selection process are given in Figure 1.

It is obvious that due to the different nature of the molecular descriptors (Table II), the reproduction-selection process of the GA could lead to impossible combinations. To overcome this problem, different constraints were added by means of a penalty method (Goldberg, 1989). Thus, a penalty was assigned to the fitness for designed molecules which violated the defined constraints. As a result, they were removed from the mating population. Some examples of constraints are given below.

Table III *Testing set.*

No.	Compound name	Observed biodegradability	Calculated biodegradability*
1	*p*-cresol	1	1
2	*p*-nitrophenol	1	1
3	methyl parathion	1	0
4	*p*-chlorophenol	1	1
5	di-*n*-butyl phthalate	1	1
6	hexadecane	1	1
7	naphthalene	1	0
8	octadecane	1	1
9	2-methylnaphthalene	0	0
10	isopropylphenyl diphenyl phosphate	0	0
11	di-2-ethylhexyl phthalate	0	1
12	chlorobenzene	0	0
13	*tert*-butylphenyl diphenyl phosphate	0	0
14	phenanthrene	0	0
15	chlorobenzilate	0	0
16	2,4,5-trichlorophenoxyacetic acid	0	0
17	2,4,5-trichlorophenol	0	0
18	pyrene	0	0
19	benzo[*a*]pyrene	0	0
20	3-methylcholanthrene	0	0
21	hexachlorophene	0	0
22	*p,p*'-DDE	0	0
23	hexachlorobenzene	0	0
24	*n*-butylamine	1	1
25	*p*-methoxyaniline	1	1
26	*o*-cresol	1	1
27	*m*-hydroxybenzoic acid	1	1
28	ethyl *o*-hydroxybenzoate	1	1
29	3,4-dihydroxybenzoic acid	1	1
30	isodecanol	1	1
31	nicotinic acid	1	0
32	stearic acid	1	1
33	4-aminopyridine	0	0
34	2-nitroaniline	0	0
35	3-nitroaniline	0	0
36	*o*-chloronitrobenzene	0	0
37	benz[*a*]anthracene	0	0
38	2,3,6-trichlorobenzoic acid	0	0
39	bromodichloromethane	0	0
40	pentachlorobenzene	0	0
41	decalin	0	0
42	*trans*-2-butene	1	1
43	trichloroethylene	0	0
44	*N,N*'-dimethylpropanamide	1	1
45	diazo amino dye	0	0
46	methyl methacrylate	1	1
47	dichlorobenzalkonium chloride	0	0
48	diasone	0	0

Table III *continued.*

No.	Compound name	Observed biodegradability	Calculated biodegradability*
49	cyclopropyl phenyl sulfide	0	0

*AERUD < 2.50 = 1 (high biodegradability); AERUD ≥ 2.50 = 0 (low biodegradability).

- When a molecule contains more than two chlorine atoms (descriptor no. 3), it cannot be only constituted of C, H, O, and N (descriptor no. 5). Thus, if #3 = 1 => #5 = 0 and if #5 = 1 => #3 = 0.
- When a molecule contains fused rings (descriptor no. 4), it obligatorily contains at least two cycles (descriptor no. 7). Thus, if #4 = 1 => #7 = 1 (the converse is not true).
- When a molecule possesses one or several of the studied structural descriptors, numerous constraints regarding the molecular weight are required. Thus, for example, when more than two chlorine atoms are present in a molecule (descriptor no. 3), its molecular weight is obligatorily superior to 100. Thus, if #3 = 1 => #10 > 0 (the converse is not true).
- When a molecule contains a conjugated C=O double bond (descriptor no. 13), it obligatorily contains a C=O double bond (descriptor no. 2). Thus, if #13 = 1 => #2 = 1 (the converse is not true).

GA analysis was performed with the STATQSAR package (STATQSAR, 1994).

RESULTS AND DISCUSSION

With a 14/2/1 BNN ($\alpha = 0.9$ and $\eta = 0.4$), we succeeded in correctly modeling all the chemicals belonging to the training set. Indeed, all the residuals (Table I) are inferior to 0.3 (absolute value). The convergence was obtained within 2500 cycles. The high quality of the BNN model is clearly illustrated in Figure 2 which represents the observed versus calculated AERUD data for the 38 chemicals belonging to the training set. In addition, inspection of Table III shows that the selected BNN model is able to correctly predict the biodegradability of 91.8% (i.e., 45/49) of the test chemicals. Due to the structural heterogeneity of the chemicals belonging to the training and testing sets, we assume that the selected BNN model can be used to estimate the AERUD values corresponding to the candidate molecules generated by the GA.

As regards the GA configuration, by a trial and error procedure, we found that using a population of 100 was a reasonable compromise between having enough sampling over combinatorial space and not taking too much computer

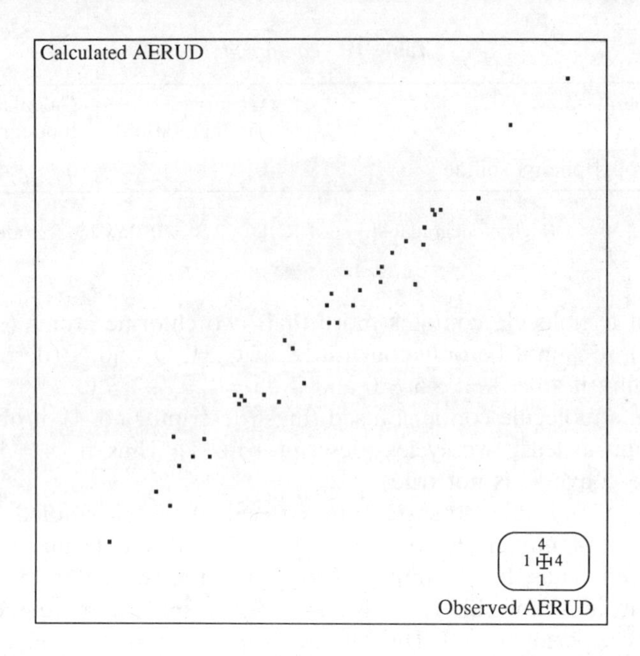

Figure 2 Observed versus calculated AERUD values for the training set (Table I).

time. The crossover and mutation probabilities equaled 0.8 and 0.1, respectively. A 50% population permutation was selected (elitist strategy). Selection was optimized from the scaling routines (i.e., prescale, scale, scalepop) of Goldberg (1989, p. 79). The GA was stopped after 300 generations.

With the hybrid system, it has been possible to propose candidate molecules having low AERUD values. These molecules present particular structures which are summarized on Figure 3. Thus, on the basis of numerous runs, among the 50 best candidates, it is interesting to note, for example, that biodegradable molecules preferentially contain a C=O group and/or a primary or aromatic OH (descriptors no. 2 and 9, respectively). At the opposite, these molecules do not include a nitrogen heterocycle and/or more than two chlorine atoms (descriptors no. 1 and 3, respectively). It must be pointed out that these results are not surprising. Indeed, for example, it is well known that the presence of OH, CHO, and COOH groups on an aromatic ring facilitates biodegradation, as compared with the NO_2 group and halogens. In addition, the rate of biodegradation usually decreases with increasing number of substituents on the aromatic ring. The order of degradability can also be affected by the position of substituents on the aromatic ring (Pitter and Chudoba, 1990). Our results confirm that the proposed hybrid system can be used for designing molecules with a particular biodegradability since, on the basis of this simple exercise, it is able to propose logical solutions.

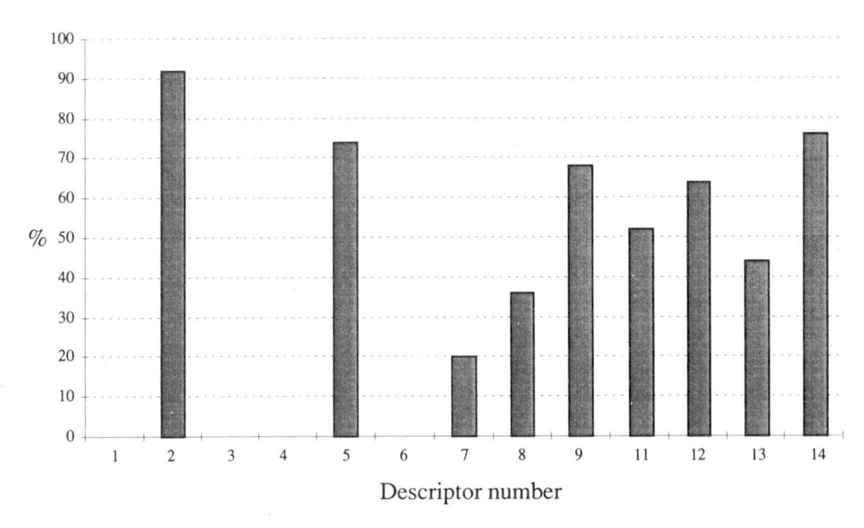

Figure 3 Influence of the 13 Boolean descriptors on the biodegradability of the molecules. See Table II for correspondence between the numbers and the descriptors. Due to the fact that the molecular weight was encoded by means of four classes, it was not represented on this figure.

In the chemical industry, the synthesis of molecules is often under the dependence of structural constraints. Therefore, it could be interesting to test the limits of the hybrid system when structural conditions are voluntarily introduced. In practice, this type of constraint requires prior modification of the initialization process of the population since the selected descriptor(s) is(are) not randomly chosen. Also it is necessary to control the operations of mutations and crossovers in order to avoid the loss of the selected descriptor(s).

In a first series of experiments, the presence of structural fragments decreasing the biodegradability of the molecules (i.e. 1, 3, 4, 6, 7) was imposed. The selection of one of these fragments in the molecule cannot prevent the design of biodegradable molecules by the hybrid system. Indeed, in all cases, it is possible to propose candidate molecules having AERUD values < 2.5 (Table IV). In the same way, the absence of structural features increasing the biodegradability of the molecules (i.e. 2, 5, 9, 12, 14) does not also prevent the proposal of biodegradable candidates. When the presence of two different fragments is imposed, different results can be obtained (Table IV). Thus, the presence of two of the most unfavorable structural fragments (i.e. 1, 3, 4, 6) hardly prevents the design of biodegradable molecules. However, the combination of the presence of one of these unfavorable descriptors with the absence of one of the structural features increasing the biodegradability of the molecules (i.e. 2, 5, 9, 12, 14) can give interesting

Table IV *Number of biodegradable candidates proposed by the GA under constraints.*

Constrained descriptor	Number of candidates with AERUD < 2.5
1, 3, 4, 6, or 7	50 / 50
2, 5, 9, 12, or 14	50 / 50
8, 11, or 13	50 / 50*
1, 3	0 / 50
1, 4	1 / 50
1, 6	1 / 50
3, 4	1 / 50
3, 6	1 / 50
4, 6	5 / 50
1, 2	13 / 50
2, 3	9 / 50
2, 4	16 / 50
2, 6	21 / 50

* Same results are obtained when the presence or absence of these descriptors is imposed.

results. This is exemplified with descriptor no. 2 in Table IV. The observed variability opens interesting perspectives in the design of new compounds.

CONCLUSION

Even if this study has to be considered as a preliminary investigation, it is obvious that the results obtained are very promising. They support the idea that a hybrid system based on the combined use of a GA and a BNN can be employed to propose candidate molecules having a specific biodegradability.

Further investigations are required to increase the efficiency of the proposed hybrid system. Thus, for example, it is obvious that by using the molecular weight as molecular descriptor, predictions are theoretically possible for all organic chemicals. However, additional fragmental descriptors are required to correctly encode all the organic molecules susceptible to contaminate the environment.

From these results, it also appears that the synergy between a BNN and a GA offers a rich field for further explorations in QSAR and drug design.

REFERENCES

Alexander, M. (1994). *Biodegradation and Bioremediation.* Academic Press, San Diego, p. 302.

Boethling, R.S., Gregg, B., Frederick, R., Gabel, N.W., Campbell, S.E., and Sabljic, A. (1989). Expert systems survey on biodegradation of xenobiotic chemicals. *Ecotoxicol. Environ. Safety* **18**, 252–267.

Boethling, R.S., Howard, P.H., Meylan, W., Stiteler, W., Beauman, J., and Tirado, N. (1994). Group contribution method for predicting probability and rate of aerobic biodegradation. *Environ. Sci. Technol.* **28**, 459–465.

Boethling, R.S. and Sabljic, A. (1989). Screening-level model for aerobic biodegradability based on a survey of expert knowledge. *Environ. Sci. Technol.* **23**, 672–679.

Cambon, B. and Devillers, J. (1993). New trends in structure–biodegradability relationships. *Quant. Struct.–Act. Relat.* **12**, 49–56.

de Saint Laumer, J.Y., Chastrette, M., and Devillers, J. (1991). Multilayer neural networks applied to structure-activity relationships. In, *Applied Multivariate Analysis in SAR and Environmental Studies* (J. Devillers and W. Karcher, Eds.). Kluwer Academic Publishers, Dordrecht, pp. 479–521.

Devillers, J. (1993). Neural modelling of the biodegradability of benzene derivatives. *SAR QSAR Environ. Res.* **1**, 161–167.

Devillers, J., Domine, D., and Boethling, R.S. (1996). Use of neural networks and autocorrelation descriptors for predicting the biodegradation of organic chemicals. In, *Neural Networks in QSAR and Drug Design* (J. Devillers, Ed.). Academic Press, London (in press).

Domine, D., Devillers, J., Chastrette, M., and Karcher, W. (1993). Estimating pesticide field half-lives from a backpropagation neural network. *SAR QSAR Environ. Res.* **1**, 211–219.

Eberhart, R.C. and Dobbins, R.W. (1990). *Neural Network PC Tools: A Practical Guide*. Academic Press, San Diego, p. 414.

Goldberg, D.E. (1989). *Genetic Algorithms in Search, Optimization & Machine Learning*. Addison-Wesley Publishing Company, Reading, p. 412.

Goonatilake, S. and Khebbal, S. (1995). *Intelligent Hybrid Systems*. John Wiley & Sons, Chichester, p. 325.

Howard, P.H., Boethling, R.S., Stiteler, W., Meylan, W., and Beauman, J. (1991). Development of a predictive model for biodegradability based on BIODEG, the evaluated biodegradation data base. *Sci. Total Environ.* **109/110**, 635–641.

Howard, P.H., Boethling, R.S., Stiteler, W.M., Meylan, W.M., Hueber, A.E., Beauman, J.A., and Larosche, M.E. (1992). Predictive model for aerobic biodegradability developed from a file of evaluated biodegradation data. *Environ. Toxicol. Chem.* **11**, 593–603.

Kuenemann, P., Vasseur, P., and Devillers, J. (1990). Structure–biodegradability relationships. In, *Practical Applications of Quantitative Structure–Activity Relationships (QSAR) in Environmental Chemistry and Toxicology* (W. Karcher and J. Devillers, Eds.). Kluwer Academic Publishers, Dordrecht, pp. 343–370.

Niemi, G.J., Veith, G.D., Regal, R.R., and Vaishnav, D.D. (1987). Structural features associated with degradable and persistent chemicals. *Environ. Toxicol. Chem.* **6**, 515–527.

Pitter, P. and Chudoba, J. (1990). *Biodegradability of Organic Substances in the Aquatic Environment.* CRC Press, Boca Raton, p. 306.

STATQSAR Package (1994). CTIS, Lyon, France.

Vasseur, P., Kuenemann, P., and Devillers, J. (1993). Quantitative structure–biodegradability relationships for predictive purposes. In, *Chemical Exposure Predictions* (D. Calamari, Ed.). Lewis Publishers, Boca Raton, pp. 47–61.

Annexe

The data set published by Selwood and coworkers (Selwood *et al.*, 1990) dealing with the antifilarial activity of antimycin analogs has often been studied in the QSAR literature using various statistical techniques. It is a classical QSAR data matrix for testing new chemometrical techniques and it is used in different chapters of this book. It is therefore presented herein and the full data set is provided.

Antimycin A1 is a potent inhibitor of mammalian electron-transport systems. Several filarial species possess functional electron-transport systems. Under these conditions, antimycin analogs presenting a high potency on filarial species and reduced mammalian toxicity are developed.

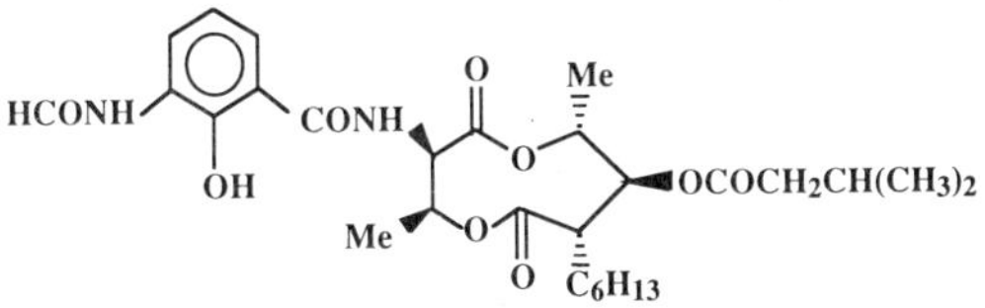

Figure 1 Antimycin A1.

Antimycin A1 (Figure 1) is an amide of 3-formamido-2-hydroxysalicylic acid with the amido portion being a complex dilactone moiety. The Selwood data set deals with a series of 31 antifilarial antimycin analogs where the dilactone moiety of the molecules was replaced by simpler lipophilic groups and the substituents on the aromatic ring were varied. The general formula of these analogs is shown in Figure 2. Table I lists the substituents investigated and Table II gives the *in vitro* and *in vivo* biological activities of these

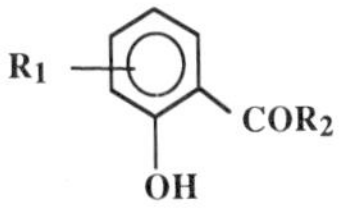

Figure 2 Antimycin analogs.

Table I *Structures of compounds 1–31.*

No.	R_1	R_2
1	3-NHCHO	$NHC_{14}H_{29}$
2	3-NHCHO	NH-3-Cl-4-(4-$ClC_6H_4O)C_6H_3$
3	5-NO_2	NH-3-Cl-4-(4-$ClC_6H_4O)C_6H_3$
4	5-SCH_3	NH-3-Cl-4-(4-$ClC_6H_4O)C_6H_3$
5	5-$SOCH_3$	NH-3-Cl-4-(4-$ClC_6H_4O)C_6H_3$
6	3-NO_2	NH-3-Cl-4-(4-$ClC_6H_4O)C_6H_3$
7	5-CN	NH-3-Cl-4-(4-$ClC_6H_4O)C_6H_3$
8	5-NO_2	NH-4-(4-$CF_3C_6H_4O)C_6H_4$
9	3-SCH_3	NH-3-Cl-4-(4-$ClC_6H_4O)C_6H_3$
10	5-SO_2CH_3	NH-3-Cl-4-(4-$ClC_6H_4O)C_6H_3$
11	5-NO_2	NH-4-$(C_6H_5O)C_6H_4$
12	5-NO_2	NH-3-Cl-4-(4-$ClC_6H_4CO)C_6H_3$
13	5-NO_2	NH-4-(2-Cl-4-$NO_2C_6H_3O)C_6H_4$
14	5-NO_2	NH-3-Cl-4-(4-$CH_3OC_6H_4O)C_6H_3$
15	3-SO_2CH_3	NH-3-Cl-4-(4-$ClC_6H_4O)C_6H_3$
16	5-NO_2	NH-3-Cl-4-(4-$ClC_6H_4S)C_6H_3$
17	3-NHCHO	NHC_6H_{13}
18	3-NHCHO	NHC_8H_{17}
19	3-$NHCOCH_3$	$NHC_{14}H_{29}$
20	5-NO_2	$NHC_{14}H_{29}$
21	3-NO_2	$NHC_{14}H_{29}$
22	3-NO_2-5-Cl	$NHC_{14}H_{29}$
23	5-NO_2	NH-4-$C(CH_3)_3C_6H_4$
24	5-NO_2	$NHC_{12}H_{25}$
25	3-NO_2	$NHC_{16}H_{33}$
26	5-NO_2	NH-3-Cl-4-(4-$ClC_6H_4NH)C_6H_3$
27	5-NO_2	NH-4-(3-$CF_3C_6H_4O)C_6H_4$
28	5-NO_2	NH-3-Cl-4-(4-$SCF_3C_6H_4O)C_6H_3$
29	5-NO_2	NH-3-Cl-4-(3-$CF_3C_6H_4O)C_6H_3$
30	5-NO_2	NH-4-$(C_6H_5CHOH)C_6H_4$
31	5-NO_2	4-ClC_6H_4

chemicals. For the derivation of QSAR, Selwood and coworkers retrieved from the literature and calculated a set of 53 physicochemical descriptors which are listed in Table III and whose values are given in Table IV.

Thanks are due to Dr D.J. Livingstone (ChemQuest, UK) for kindly providing and validating the values of the 53 physicochemical descriptors reported in Table IV and the activity categories given in Table II.

Table II *Biological activity of compounds 1–31.*

No.	*In vitro* activity: EC_{50}*(μM)	– log *in vitro* activity	Category[†]	*In vivo* activity: % worm reduction[‡]
1	7.0	–0.85	1	89
2	2.4	–0.38	1	toxic
3	0.04	1.40	3	80.6
4	0.48	0.32	2	58
5	7.5	–0.88	1	84
6	0.15	0.82	2	toxic
7	0.0145	1.84	3	17
8	0.095	1.02	2	72
9	0.38	0.42	2	52
10	1.0	0.00	1	28
11	0.8	0.10	2	89
12	0.074	1.13	3	14
13	0.12	0.92	2	70
14	0.17	0.77	2	61
15	0.5	0.30	2	42
16	0.044	1.36	3	70
17	>10	–1.0	1	NT
18	2.6	–0.41	1	NT
19	8.0	–0.9	1	NT
20	0.128	0.89	2	NT
21	0.152	0.82	2	NT
22	0.0435	1.36	3	NT
23	0.59	0.23	2	44
24	0.039	1.41	3	0
25	1.1	–0.04	1	NT
26	0.37	0.43	2	48
27	0.094	1.03	3	80
28	0.028	1.55	3	85; some toxicity
29	0.085	1.07	3	74
30	>10	–1.0	1	39
31	0.33	0.48	2	61

* EC_{50} = effective concentration at which 50% of the adenine taken up was released into the medium.
[†] 1: Inactive; 2: Intermediate; 3: Active.
[‡] NT = not tested.

Table III *Physicochemical descriptors.*

1–10	Partial atomic charges (ATCH) for atoms 1–10
11–13	Vectors (X, Y, and Z) of the dipole moment (DIPV)
14	Dipole moment (DIPMOM)
15–24	Electrophilic superdelocalizabilities (ESDL) for atoms 1–10
25–34	Nucleophilic superdelocalizabilities (NSDL) for atoms 1–10
35	van der Waal's volume (VDWVOL)

Table III *continued.*

36	Surface area (SURF_A)
37–39	Moments of inertia (X, Y, and Z) (MOFI)
40–42	Principal ellipsoid axes (X, Y, and Z) (PEAX)
43	Molecular weight (MOL_WT)
44–49	Parameters describing substituent dimensions in the X, Y, and Z directions (S8_1D) and the coordinates of the center of the substituent (S8_1C)
50	Calculated log P (LOGP)
51	Melting point (M_PNT)
52–53	Sums of the F and R substituent constants (SUM)

Table IV *Values of the 53 physicochemical parameters for compounds 1–31.*

No.	ATCH1	ATCH2	ATCH3	ATCH4	ATCH5	ATCH6	ATCH7
1	0.1687	0.0391	–0.0092	–0.1005	0.0027	–0.2403	–0.2521
2	0.1750	0.0408	–0.0055	–0.0980	0.0009	–0.2499	–0.2521
3	0.2609	–0.1463	0.0923	–0.1616	0.1116	–0.2897	–0.2131
4	0.2325	–0.1373	0.0525	–0.2566	0.0753	–0.2834	–0.2310
5	0.2473	–0.1403	0.0876	–0.4688	0.1117	–0.2873	–0.2236
6	0.2834	–0.1642	0.1018	–0.1469	0.0833	–0.2946	–0.1885
7	0.2373	–0.1369	0.0619	–0.0760	0.0738	–0.2813	–0.2237
8	0.2601	–0.1466	0.0920	–0.1618	0.1112	–0.2894	–0.2138
9	0.2323	–0.2545	0.0650	–0.1308	0.0401	–0.2751	–0.2319
10	0.2621	–0.1435	0.1007	–0.5146	0.1454	–0.2904	–0.2170
11	0.2593	–0.1462	0.0916	–0.1611	0.1105	–0.2882	–0.2125
12	0.2605	–0.1453	0.0925	–0.1603	0.1104	–0.2901	–0.2105
13	0.2602	–0.1461	0.0928	–0.1610	0.1113	–0.2898	–0.2132
14	0.2604	–0.1469	0.0917	–0.1623	0.1116	–0.2889	–0.2132
15	0.2913	–0.5230	0.1274	–0.1442	0.0806	–0.2874	–0.2260
16	0.2600	–0.1458	0.0923	–0.1610	0.1103	–0.2888	–0.2113
17	0.1685	0.0386	–0.0084	–0.1010	0.0035	–0.2410	–0.2524
18	0.1686	0.0391	–0.0089	–0.1006	0.0028	–0.2401	–0.2521
19	0.1683	0.0395	–0.0095	–0.1004	0.0020	–0.2405	–0.2520
20	0.2562	–0.1471	0.0895	–0.1628	0.1108	–0.2808	–0.2125
21	0.2791	–0.1652	0.0970	–0.1480	0.0840	–0.2854	–0.1878
22	0.2838	–0.1612	0.1127	–0.0938	0.0995	–0.2808	–0.1818
23	0.2589	–0.1469	0.0906	–0.1619	0.1104	–0.2877	–0.2135
24	0.2562	–0.1471	0.0895	–0.1628	0.1108	–0.2808	–0.2125
25	0.2793	–0.1651	0.0969	–0.1478	0.0837	–0.2854	–0.1874
26	0.2578	–0.1452	0.0911	–0.1600	0.1084	–0.2861	–0.2097
27	0.2601	–0.1468	0.0918	–0.1619	0.1113	–0.2886	–0.2141
28	0.2609	–0.1459	0.0924	–0.1611	0.1113	–0.2902	–0.2126
29	0.2612	–0.1460	0.0926	–0.1612	0.1111	–0.2899	–0.2126
30	0.2600	–0.1477	0.0910	–0.1629	0.1116	–0.2877	–0.2148
31	0.2483	–0.1462	0.0918	–0.1634	0.1174	–0.2886	–0.2084

Table IV *continued.*

No.	ATCH8	ATCH9	ATCH10	DIPV_X	DIPV_Y	DIPV_Z	DIPMOM	ESDL1
1	0.4177	–0.4132	–0.4033	2.1993	–1.7847	–0.0244	2.8324	–0.5466
2	0.4260	–0.3994	–0.3270	–1.2508	–1.4858	0.3883	1.9806	–7.2024
3	0.4274	–0.3942	–0.3242	0.1077	1.4632	0.4468	1.5337	–5.8408
4	0.4267	–0.3976	–0.3281	–3.1042	–2.0453	–1.0010	3.8498	–2.0706
5	0.4263	–0.3961	–0.3269	0.0267	–2.2836	3.6789	4.3301	–1.2090
6	0.4267	–0.3894	–0.3248	0.3801	–8.0741	0.3869	8.0923	–1.0867
7	0.4260	–0.3958	–0.3267	–1.5169	–0.5424	0.4442	1.6711	–1.0820
8	0.4275	–0.3983	–0.3238	–0.0356	3.7144	1.2436	3.9172	–8.4444
9	0.4257	–0.3981	–0.3279	–2.3652	–3.7578	–0.8217	4.5155	–5.2785
10	0.4272	–0.3976	–0.3243	–3.1640	1.5843	–2.7014	4.4518	–0.8667
11	0.4262	–0.3924	–0.3268	1.1464	0.0687	0.7662	1.3806	–18.5242
12	0.4268	–0.3877	–0.3272	–0.7323	0.7748	0.2510	1.0952	–3.6273
13	0.4283	–0.3970	–0.3249	–2.3472	5.1618	1.3489	5.8286	–3.6255
14	0.4264	–0.3954	–0.3229	2.8977	–0.2785	–0.0015	2.9111	–51.0177
15	0.4261	–0.3931	–0.3264	0.9609	–6.1255	–3.3030	7.0253	–1.3643
16	0.4268	–0.3906	–0.3264	0.2799	1.0840	–0.2142	1.1399	–5.5805
17	0.4188	–0.4125	–0.4043	1.0017	–2.6660	–0.0328	2.8481	–0.5455
18	0.4181	–0.4129	–0.4039	0.9936	–2.6456	–0.0186	2.8261	–0.5471
19	0.4176	–0.4135	–0.4033	0.9507	–2.3962	0.0587	2.5786	–0.5643
20	0.4181	–0.4063	–0.3998	3.4223	–0.3826	–0.0241	3.4437	–0.5128
21	0.4178	–0.4012	–0.4015	–1.1027	–8.9307	–0.0316	8.9986	–1.2692
22	0.4178	–0.3985	–0.3993	0.4214	–8.0769	–0.0305	8.0880	–0.9495
23	0.4257	–0.3971	–0.3243	3.4332	0.1762	0.2453	3.4464	–0.7424
24	0.4181	–0.4063	–0.3998	3.4223	–0.3826	–0.0241	3.4437	–0.5128
25	0.4173	–0.4015	–0.4013	–1.1208	–8.9224	–0.0355	8.9926	–1.2662
26	0.4245	–0.3937	–0.3248	2.2681	1.3472	0.6614	2.7197	–34.2046
27	0.4274	–0.3986	–0.3237	0.4356	2.6107	–1.5316	3.0580	–9.8302
28	0.4274	–0.3938	–0.3243	–0.2955	2.5579	0.4620	2.6161	–4.1539
29	0.4273	–0.3942	–0.3243	–1.0840	1.9937	–1.5566	2.7519	–4.1115
30	0.4257	–0.3958	–0.3246	3.1309	–1.3363	0.1914	3.4095	–0.6642
31	0.3493	–0.3284	–0.1500	3.5821	1.7919	0.6987	4.0658	–41.8125

Table IV *continued.*

No.	ESDL2	ESDL3	ESDL4	ESDL5	ESDL6	ESDL7	ESDL8	ESDL9
1	–0.8054	–1.7022	–0.2892	–1.2932	–1.4297	–0.4022	–0.4658	–0.6069
2	–3.7462	–1.6017	–5.5173	–4.0391	–1.3249	–1.3459	–0.5084	–0.6027
3	–3.5819	–0.8313	–2.7471	–1.2534	–1.0284	–1.1591	–0.7395	–0.6892
4	–1.3477	–1.3785	–1.8399	–2.3639	–0.9981	–0.6437	–0.4550	–0.5623
5	–0.8732	–1.1233	–1.0692	–1.5280	–0.8362	–0.5079	–0.4483	–0.5486
6	–0.4146	–0.7912	–0.6555	–0.5825	–0.5955	–0.4950	–0.3911	–0.5061
7	–0.8195	–1.0338	–0.9286	–1.3942	–0.7768	–0.4861	–0.4077	–0.5271
8	–4.9054	–0.8606	–3.9635	–1.4281	–1.3369	–1.5322	–0.8218	–0.7378
9	–2.0048	–1.6241	–3.5297	–2.5197	–1.1995	–1.1224	–0.4962	–0.5924
10	–0.7624	–0.9508	–1.0341	–1.2133	–0.7052	–0.4496	–0.5518	–0.5905
11	–10.2312	–0.9389	–8.6129	–2.1310	–2.3426	–2.9728	–1.2358	–0.9782
12	–2.4015	–0.7832	–1.7172	–1.0703	–0.7893	–0.8350	–0.5919	–0.6121

Table IV *continued.*

No.	ESDL2	ESDL3	ESDL4	ESDL5	ESDL6	ESDL7	ESDL8	ESDL9
13	–2.3137	–0.7935	–1.7701	–1.0545	–0.8627	–0.8342	–0.6293	–0.6234
14	–29.8732	–1.5157	–22.4600	–4.8865	–4.2065	–7.7191	–4.5312	–2.7104
15	–0.9999	–0.8610	–0.8580	–0.8798	–0.6758	–0.5146	–0.3688	–0.4993
16	–3.4701	–0.8240	–2.6015	–1.2303	–0.9914	–1.1163	–0.6970	–0.6730
17	–0.8034	–1.6953	–0.2885	–1.2909	–1.4224	–0.4025	–0.4663	–0.6060
18	–0.8030	–1.7001	–0.2896	–1.2900	–1.4288	–0.4022	–0.4646	–0.6062
19	–0.8395	–1.7833	–0.2900	–1.3583	–1.4943	–0.4054	–0.4852	–0.6206
20	–0.5490	–0.8149	–0.3995	–0.9536	–0.6531	–0.3904	–0.2376	–0.4387
21	–0.4326	–0.8419	–0.6660	–0.6042	–0.6639	–0.5282	–0.2530	–0.4480
22	–0.7380	–1.0875	–2.4801	–3.0831	–1.1205	–0.4709	–0.7914	–0.7384
23	–1.0745	–1.0148	–0.4334	–1.0363	–0.6888	–0.4243	–1.0779	–0.8538
24	–0.5490	–0.8149	–0.3995	–0.9536	–0.6531	–0.3904	–0.2376	–0.4387
25	–0.4321	–0.8415	–0.6647	–0.6029	–0.6637	–0.5273	–0.2528	–0.4482
26	–14.0039	–1.0883	–18.5399	–2.4267	–7.0824	–5.1572	–1.5257	–1.0239
27	–5.5302	–0.8699	–4.6660	–1.5162	–1.5364	–1.7318	–0.8566	–0.7574
28	–2.6584	–0.7949	–1.9782	–1.1159	–0.8589	–0.9134	–0.6187	–0.6247
29	–2.6324	–0.7949	–1.9609	–1.1127	–0.8565	–0.9078	–0.6215	–0.6257
30	–0.9460	–0.9866	–0.4211	–1.0243	–0.7004	–0.4143	–0.9300	–0.7845
31	–27.2335	–2.0914	–16.5747	–4.0222	–3.3515	–6.0766	–5.7077	–3.4349

Table IV *continued.*

No.	ESDL10	NSDL1	NSDL2	NSDL3	NSDL4	NSDL5	NSDL6	NSDL7
1	–0.3896	7.6719	4.2388	1.1007	6.0696	4.0671	1.0302	1.2752
2	–0.4706	0.7982	0.9939	0.9135	0.8790	0.8228	0.9414	0.3117
3	–0.4688	1.0031	1.0780	1.1817	1.7285	1.0289	1.1907	0.3703
4	–0.4129	0.9121	0.9646	0.9486	1.0544	0.8049	0.9603	0.3311
5	–0.3762	1.1601	1.1509	1.0352	2.1675	0.9393	1.0662	0.3733
6	–0.3280	1.1351	5.4842	3.7658	17.2603	18.2373	3.5904	0.3757
7	–0.3649	1.0388	1.0965	1.0528	0.9101	0.8766	1.0322	0.3549
8	–0.5404	0.9950	1.0630	1.1540	1.6896	1.0187	1.1572	0.3685
9	–0.4499	0.8166	1.0564	0.9536	0.8935	0.8389	0.9515	0.3178
10	–0.3473	1.1427	1.2053	1.1735	1.1778	0.9469	1.0980	0.3793
11	–0.7942	0.9909	1.0505	1.1364	1.6642	1.0084	1.1380	0.3681
12	–0.4057	1.0216	1.1065	1.2310	1.8079	1.0507	1.2366	0.3772
13	–0.4094	1.0239	1.1073	1.2209	1.8030	1.0518	1.2257	0.3766
14	–1.4855	3.5093	2.2010	1.1353	2.9021	1.1461	1.4756	0.7285
15	–0.3427	0.9949	1.1112	1.2251	1.1739	1.0567	1.1252	0.3657
16	–0.4597	1.0046	1.0785	1.1797	1.7299	1.0307	1.1852	0.3725
17	–0.3895	7.5333	4.1605	1.0990	5.9683	4.0048	1.0283	1.2598
18	–0.3893	7.7428	4.2790	1.1020	6.1255	4.1088	1.0315	1.2854
19	–0.3945	6.5450	3.6538	1.0581	5.1802	3.4714	1.0043	1.1158
20	–0.3277	5.8768	3.8047	1.1287	3.7149	1.3436	1.4154	1.0611
21	–0.3296	0.9932	1.9648	1.2609	2.5870	2.7579	1.2959	0.3571
22	–0.4324	1.1258	1.9410	1.2260	4.6949	1.0373	1.2497	0.3897
23	–0.3360	6.6018	3.5615	1.0783	4.3860	1.3004	1.8484	1.1676
24	–0.3277	5.8768	3.8047	1.1287	3.7149	1.3436	1.4154	1.0611

Table IV *continued.*

No.	ESDL10	NSDL1	NSDL2	NSDL3	NSDL4	NSDL5	NSDL6	NSDL7
25	–0.3296	0.9945	1.9733	1.2608	2.5997	2.7751	1.2974	0.3577
26	–1.5813	0.9953	1.0579	1.1390	1.6811	1.0183	1.1370	0.3722
27	–0.5798	0.9933	1.0596	1.1476	1.6826	1.0159	1.1507	0.3679
28	–0.4219	1.0152	1.0966	1.2085	1.7759	1.0430	1.2138	0.3740
29	–0.4211	1.0158	1.0973	1.2101	1.7782	1.0431	1.2176	0.3740
30	–0.3444	6.0402	3.4281	1.0804	4.0356	1.2915	1.6839	1.0902
31	–2.7701	0.9738	1.0278	1.1688	1.6686	0.9694	1.1517	0.3783

Table IV *continued.*

No.	NSDL8	NSDL9	NSDL10	VDWVOL	SURF_A	MOFI_X	MOFI_Y	MOFI_Z
1	1.4560	0.4936	0.6788	379.99982	488.07489	766.6057	37378.3047	36572.9805
2	1.6498	0.5105	0.8485	327.99991	386.58533	1686.4363	21047.6348	19971.3652
3	2.2705	0.7144	1.0592	320.59991	379.91208	2651.7319	19371.6113	17267.4023
4	1.7084	0.5357	0.8624	328.79990	395.36938	2776.7034	19569.5352	17354.9766
5	1.8967	0.5984	0.9381	342.29987	404.42609	3099.5256	21450.8535	18866.4336
6	5.7846	2.4673	2.1601	320.59991	378.27496	1726.7526	20579.1602	19403.0957
7	1.9231	0.6101	0.9397	313.79993	373.30637	2345.0713	17645.0371	15850.8125
8	2.1169	0.6596	1.0010	321.59988	386.11441	2267.2917	22328.7871	20831.5391
9	1.6883	0.5269	0.8613	328.79990	396.11472	1822.8052	20390.8359	19440.9727
10	2.0456	0.6542	0.9685	353.59985	413.67645	3400.4380	22357.9707	19442.1465
11	2.0422	0.6446	0.9576	304.79996	355.80188	1894.8627	15368.8545	13607.3057
12	2.4834	0.8010	1.1053	333.79993	385.53317	3188.1543	20791.5469	17782.8633
13	2.4008	0.7563	1.0960	331.99991	392.10461	2622.2456	21989.5332	20498.2715
14	2.0959	0.6674	1.0505	328.49991	394.02155	2529.6760	19250.0039	17192.3730
15	2.2353	0.7190	1.0460	353.59985	411.79279	1993.3540	23754.7051	22894.0762
16	2.2456	0.7114	1.0432	332.29990	391.05966	3615.8687	20164.7129	17081.3086
17	1.4602	0.4934	0.6803	250.40009	314.22668	574.2045	9189.6504	8711.1963
18	1.4571	0.4939	0.6791	282.80002	358.10831	623.5986	13796.0986	13431.9482
19	1.4370	0.4840	0.6736	396.19980	506.67865	793.2323	40830.8164	39243.7344
20	2.1902	0.8049	0.8552	340.19989	436.40833	1650.2062	25060.2734	23662.1348
21	2.1162	0.7561	0.8497	372.59979	477.42938	765.7248	37097.8516	36119.4805
22	2.0261	0.6627	0.8511	387.09976	493.83658	1628.6172	39861.3984	38461.7031
23	1.8164	0.5803	1.0563	280.40002	330.78622	1422.5823	10025.0400	8799.7803
24	2.1902	0.8049	0.8552	340.19989	436.40833	1650.2062	25060.2734	23662.1348
25	2.1135	0.7559	0.8496	404.99973	520.96588	808.4082	49091.1367	46927.3906
26	2.0056	0.6292	0.9671	328.29990	380.38226	3064.9253	19326.9492	16851.7227
27	2.0825	0.6477	0.9897	321.59988	385.27243	1846.9785	22716.9297	21426.6191
28	2.3728	0.7509	1.0921	353.99988	421.53546	3010.1609	28906.9922	26566.6621
29	2.3853	0.7541	1.0988	336.09991	401.82898	2447.4656	23459.0742	21598.6250
30	1.8805	0.5986	1.0391	307.79996	364.02640	1837.9072	15123.2236	13701.3262
31	2.2232	0.7991	1.0453	216.20004	259.55621	1631.9397	5819.5015	4289.1333

Table IV *continued.*

No.	PEAX_X	PEAX_Y	PEAX_Z	MOL_WT	S8_1DX	S8_1DY	S8_1DZ	S8_1CX
1	22.0590	0.5074	3.2329	376.54135	17.0896	13.6199	7.0784	11.1982
2	15.4115	1.9195	4.0845	417.24887	11.8036	9.9590	6.2589	8.5330
3	14.2916	1.8139	5.3462	419.22134	11.7737	10.0132	6.2309	8.5327
4	14.3081	1.8358	5.4702	420.31082	11.7604	10.1024	6.2214	8.5191
5	14.6589	1.7245	5.7286	436.31021	11.7727	10.6567	5.5288	8.5385
6	15.1625	1.8192	4.1767	419.22134	11.7962	9.9618	6.2654	8.5331
7	14.0235	1.8648	5.1119	399.23355	11.7825	9.9721	6.2570	8.5308
8	15.6953	2.1538	4.7621	418.32956	12.7140	9.8967	7.2800	8.8229
9	15.0954	2.2877	4.0771	420.31082	11.7836	9.9952	6.2577	8.5281
10	14.6221	1.6426	5.9303	452.30960	11.7258	10.6374	5.6009	8.5063
11	13.6757	0.9595	5.0251	362.34225	11.7909	9.5562	4.8335	8.5199
12	14.3769	1.0239	6.0164	431.23227	11.8248	11.6162	4.7136	8.5351
13	15.2597	2.5703	4.9018	429.77383	12.4982	9.2788	8.4438	8.6865
14	14.3277	1.6904	5.2696	414.80270	12.9533	10.0760	6.4081	9.1197
15	15.7683	2.5114	3.9863	452.30960	11.6899	9.6063	6.4384	8.4671
16	13.9506	1.7554	6.2265	435.28189	10.1820	11.9139	6.9458	7.3657
17	12.8093	0.9522	3.1573	264.32553	6.9012	9.7093	5.3613	5.2365
18	15.0923	1.4904	2.9081	292.37949	7.9625	11.8851	5.9513	5.7534
19	22.5436	2.2558	3.9062	390.56833	10.8214	18.7441	6.9792	7.1781
20	18.3366	1.3418	4.6663	350.45987	10.0295	16.4804	7.1094	6.7639
21	21.8901	1.1859	3.3963	378.51382	10.7617	18.8944	7.2670	30.9758
22	21.5989	1.1800	4.2919	412.95883	10.7824	18.9532	7.2213	31.3191
23	11.7708	1.2534	4.5915	314.34192	9.2551	8.3913	5.2754	6.3405
24	18.3366	1.3418	4.6663	350.45987	10.0295	16.4804	7.1094	6.7639
25	24.2130	2.8889	4.2780	406.56778	11.5158	21.1088	7.5424	33.7905
26	14.1238	1.8848	5.7771	418.23660	8.6220	11.7029	7.8834	5.9993
27	15.9624	1.8312	4.3473	418.32956	11.1225	11.9605	6.7951	7.1463
28	16.5302	1.8678	5.2789	484.83456	10.5676	13.4492	6.8999	6.8571
29	15.4201	1.8099	4.9030	452.77457	11.1316	11.6704	6.2340	7.1444
30	13.6143	1.6903	4.7317	364.35831	9.4340	10.5818	6.8240	6.4758
31	8.7625	0.9592	5.3520	277.66428	5.6061	7.7776	4.4484	3.8719

Table IV *continued.*

No.	S8_1CY	S8_1CZ	LOGP	M_PNT	SUM_F	SUM_R
1	–4.9646	–1.5653	7.239	81	0.25	–0.23
2	–1.2691	–1.2271	5.960	183	0.25	–0.23
3	–1.3552	–1.2850	6.994	207	0.67	0.16
4	–1.4426	–1.0338	7.372	143	0.20	–0.18
5	–1.6179	–1.1429	5.730	165	0.52	0.01
6	–1.2725	–1.2412	6.994	192	0.67	0.16
7	–1.3163	–1.2838	6.755	256	0.51	0.19
8	–1.9160	–1.5777	6.695	199	0.67	0.16
9	–1.3386	–1.2163	7.372	151	0.20	–0.18
10	–1.6703	–1.4219	5.670	195	0.54	0.22
11	–2.0179	1.2512	4.888	212	0.67	–0.23
12	–2.2636	1.2017	6.205	246	0.67	–0.23

Table IV *continued.*

No.	S8_1CY	S8_1CZ	LOGP	M_PNT	SUM_F	SUM_R
13	−1.5737	−1.1706	6.113	208	0.67	−0.23
14	−1.7659	−1.2340	6.180	159	0.67	−0.23
15	−1.2903	−1.7341	5.681	178	0.54	0.23
16	−2.4667	2.2822	6.838	222	0.67	−0.23
17	−5.0591	−0.5692	3.007	62	0.25	−0.23
18	−6.1533	−0.9282	4.065	78	0.25	−0.23
19	−9.5838	−1.5017	7.230	71	0.28	−0.26
20	−8.4629	−1.4630	8.466	90	0.67	0.16
21	16.4613	−1.6728	8.470	67	0.67	0.16
22	16.5005	−1.6306	9.300	81	1.08	0.01
23	−4.3674	1.1964	5.354	227	0.67	0.16
24	−8.4629	−1.4630	7.408	85	0.67	0.16
25	18.2964	0.0000	9.520	79	0.67	0.16
26	−5.7537	−1.6195	6.811	173	0.67	0.16
27	−5.9329	−1.2049	6.695	176	0.67	0.16
28	−6.6848	−1.2939	7.869	195	0.67	0.16
29	−5.7897	−1.2802	7.269	192	0.67	0.16
30	−5.4216	−0.9107	3.686	178	0.67	0.16
31	−3.7972	0.2244	4.654	170	0.67	0.16

REFERENCE

Selwood, D.L., Livingstone, D.J., Comley, J.C.W., O'Dowd, A.B., Hudson, A.T., Jackson, P., Jandu, K.S., Rose, V.S., and Stables, J.N. (1990). Structure–activity relationships of antifilarial analogues: A multivariate pattern recognition study. *J. Med. Chem.* **33**, 136–142.

Index